CORE-PLUS MATHEMATICS PROJECT

Contemporary Mathematics in Context

A Unified Approach

Arthur F. Coxford
James T. Fey
Christian R. Hirsch
Harold L. Schoen
Gail Burrill
Eric W. Hart
Ann E. Watkins
with
Mary Jo Messenger
Beth E. Ritsema
Rebecca K. Walker

New York, New York   Columbus, Ohio   Chicago, Illinois   Peoria, Illinois   Woodland Hills, California

Glencoe/McGraw-Hill

A Division of The **McGraw-Hill** *Companies*

This project was supported, in part, by the National Science Foundation.
The opinions expressed are those of the authors and not necessarily those of the Foundation.

Send all inquires to:
Glencoe/McGraw-Hill
8787 Orion Place
Columbus, OH 43240-4027

ISBN: 0-07-827537-7 (Part A)
ISBN: 0-07-827538-5 (Part B)

Contemporary Mathematics in Context
Course 1 Part A Student Edition

2 3 4 5 6 7 8 9 10 004/004 10 09 08 07 06 05 04 03 02

Core-Plus Mathematics Project Development Team

Project Directors
Christian R. Hirsch
Western Michigan University
Arthur F. Coxford
University of Michigan
James T. Fey
University of Maryland
Harold L. Schoen
University of Iowa

Senior Curriculum Developers
Gail Burrill
University of Wisconsin-Madison
Eric W. Hart
Western Michigan University
Ann E. Watkins
California State University, Northridge

Professional Development Coordinator
Beth E. Ritsema
Western Michigan University

Evaluation Coordinator
Steven W. Ziebarth
Western Michigan University

Advisory Board
Diane Briars
Pittsburgh Public Schools
Jeremy Kilpatrick
University of Georgia
Kenneth Ruthven
University of Cambridge
David A. Smith
Duke University
Edna Vasquez
Detroit Renaissance High School

Curriculum Development Consultants
Alverna Champion
Grand Valley State University
Cherie Cornick
Wayne County Alliance for Mathematics and Science
Edgar Edwards
(Formerly) Virginia State Department of Education
Richard Scheaffer
University of Florida
Martha Siegel
Towson University
Edward Silver
University of Michigan
Lee Stiff
North Carolina State University

Technical Coordinator
Wendy Weaver
Western Michigan University

Collaborating Teachers
Emma Ames
Oakland Mills High School, Maryland
Laurie Eyre
Maharishi School, Iowa
Annette Hagelberg
West Delaware High School, Iowa
Cheryl Bach Hedden
Sitka High School, Alaska
Michael J. Link
Central Academy, Iowa
Mary Jo Messenger
Howard County Public Schools, Maryland
Valerie Mills
Ann Arbor Public Schools, Michigan
Marcia Weinhold
Kalamazoo Area Mathematics and Science Center, Michigan

Graduate Assistants
Diane Bean
University of Iowa
Judy Flowers
University of Michigan
Gina Garza-Kling
Western Michigan University
Robin Marcus
University of Maryland
Chris Rasmussen
University of Maryland
Rebecca Walker
Western Michigan University

Production and Support Staff
Lori Bowden
James Laser
Michelle Magers
Cheryl Peters
Jennifer Rosenboom
Anna Seif
Kathryn Wright
Teresa Ziebarth
Western Michigan University

Software Developers
Jim Flanders
Colorado Springs, Colorado
Eric Kamischke
Interlochen, Michigan

Core-Plus Mathematics Project Field-Test Sites

Special thanks are extended to these teachers and their students who participated in the testing and evaluation of Course 1.

Ann Arbor Huron High School
Ann Arbor, Michigan
Kevin Behmer
Ginger Gajar

Ann Arbor Pioneer High School
Ann Arbor, Michigan
Jim Brink
Fay Longhofer
Brad Miller

Arthur Hill High School
Saginaw, Michigan
Virginia Abbott
Felix Bosco
David Kabobel
Dick Thomas

Battle Creek Central High School
Battle Creek, Michigan
Teresa Ballard
Rose Martin
Steven Ohs

Bedford High School
Temperance, Michigan
Ellen Bacon
David DeGrace
Linda Martin
Lynn Parachek

Bloomfield Hills Andover High School
Bloomfield Hills, Michigan
Jane Briskey
Homer Hassenzahl
Cathy King
Ed Okuniewski
Mike Shelly

Bloomfield Hills Middle School
Bloomfield Hills, Michigan
Connie Kelly
Tim Loula

Brookwood High School
Snellville, Georgia
Ginny Hanley
Marie Knox

Caledonia High School
Caledonia, Michigan
Daryl Bronkema
Jenny Diekevers
Thomas Oster
Gerard Wagner

Centaurus High School
Lafayette, Colorado
Sally Johnson
Gail Reichert

Clio High School
Clio, Michigan
Denny Carlson
Larry Castonia
Vern Kamp
Carol Narrin
David Sherry

Davison High School
Davison, Michigan
Evelyn Ailing
John Bale
Wayne Desjarlais
Darlene Tomczak
Scott Toyzan

Dexter High School
Dexter, Michigan
Kris Chatas
Widge Proctor
Tammy Schirmer

Ellet High School
Akron, Ohio
Marcia Csipke
Jim Fillmore
Scott Slusser

Firestone High School
Akron, Ohio
Barbara Crucs
Lori Zupke

Flint Northern High School
Flint, Michigan
John Moliassa
Al Wojtowicz

Goodrich High School
Goodrich, Michigan
Mike Coke
John Doerr

Grand Blanc High School
Grand Blanc, Michigan
Charles Carmody
Nancy Elledge
Tina Hughes
Steve Karr
Mike McLaren

Grass Lake Junior/Senior High School
Grass Lake, Michigan
Larry Poertner
Amy Potts

Gull Lake High School
Richland, Michigan
Virgil Archie
Darlene Kohrman
Dorothy Louden

Kalamazoo Central High School
Kalamazoo, Michigan
Sarah Baca
Gloria Foster
Bonnie Frye
Amy Schwentor

Kelloggsville Public Schools
Wyoming, Michigan
Jerry Czarnecki
Steve Ramsey
John Ritzler

Knott County Central High School
Hindman, Kentucky
Teresa Combs
P. Denise Gibson
Bennie Hall

Loy Norrix High School
Kalamazoo, Michigan
Mary Elliott
Mike Milka

Midland Valley High School
Langley, South Carolina
Ron Bell
Janice Lee

Murray-Wright High School
Detroit, Michigan
Anna Cannonier
Jack Sada

North Lamar High School
Paris, Texas
Tommy Eads
Barbara Eatherly

Okemos High School
Okemos, Michigan
Lisa Magee
Jacqueline Stewart

Portage Northern High School
Portage, Michigan
Pete Jarrad
Scott Moore

Prairie High School
Cedar Rapids, Iowa
Dave LaGrange
Judy Slezak

San Pasqual High School
Escondido, California
Damon Blackman
Ron Peet

Sitka High School
Sitka, Alaska
Cheryl Bach Hedden
Dan Langbauer
Tom Smircich

Sturgis High School
Sturgis, Michigan
Craig Evans
Kathy Parkhurst
Dale Rauh
JoAnn Roe
Kathy Roy

Sweetwater High School
National City, California
Bill Bokesch
Joe Pistone

Tecumseh High School
Tecumseh, Michigan
Jennifer Keffer
Kathy Kelso
Elizabeth Lentz
Carl Novak
Eric Roberts

Tecumseh Middle School
Tecumseh, Michigan
Jocelyn Menyhart

Traverse City East Junior High School
Traverse City, Michigan
Tamie Rosenburg

Traverse City West Junior High School
Traverse City, Michigan
Ann Post

Vallivue High School
Caldwell, Idaho
Scott Coulter
Kathy Harris

West Hills Middle School
Bloomfield Hills, Michigan
Eileen MacDonald

Ypsilanti High School
Ypsilanti, Michigan
Keith Kellman
Mark McClure
Valerie Mills
Don Peurach

Overview of Course 1

Part A

Unit 1 Patterns in Data

Patterns in Data develops student ability to make sense out of real-world data through use of graphical displays and summary statistics.

Topics include distributions of data and their shapes, as displayed in number line plots, histograms, box plots, and stem-and-leaf plots; scatterplots and association; plots over time and trends; measures of center including mean, median, mode, and their properties; measures of variation including percentiles, interquartile range, mean absolute deviation, and their properties; transformations of data.

Lesson 1 *Exploring Data*
Lesson 2 *Shapes and Centers*
Lesson 3 *Variability*
Lesson 4 *Relationships and Trends*
Lesson 5 *Looking Back*

Unit 2 Patterns of Change

Patterns of Change develops student ability to recognize important patterns of change among variables and to represent those patterns using tables of numerical data, coordinate graphs, verbal descriptions, and symbolic rules.

Topics include coordinate graphs, tables, algebraic formulas (rules), relationships between variables, linear functions, nonlinear functions, and *NOW-NEXT* recurrence relations.

Lesson 1 *Related Variables*
Lesson 2 *What's Next?*
Lesson 3 *Variables and Rules*
Lesson 4 *Linear and Nonlinear Patterns*
Lesson 5 *Looking Back*

Unit 3 Linear Models

Linear Models develops student confidence and skill in using linear functions to model and solve problems in situations that exhibit constant (or nearly constant) rate of change or slope.

Topics include linear functions, slope of a line, rate of change, intercepts, the distributive property, linear equations (including $y = a + bx$ and *NOW-NEXT* forms), solving linear equations and inequalities, using linear equations to model given data, and determining best-fit lines for scatterplot data.

Lesson 1 *Predicting from Data*
Lesson 2 *Linear Graphs, Tables, and Rules*
Lesson 3 *Linear Equations and Inequalities*
Lesson 4 *Looking Back*

Unit 4 Graph Models

Graph Models develops student ability to use vertex-edge graphs to represent and analyze real-world situations involving relationships among a finite number of elements, including scheduling, managing conflicts, and finding efficient routes.

Topics include vertex-edge graph models, optimization, algorithmic problem solving, Euler circuits and paths, matrix representation of graphs, graph coloring, chromatic number, digraphs, and critical path analysis.

Lesson 1 *Careful Planning*
Lesson 2 *Managing Conflicts*
Lesson 3 *Scheduling Large Projects*
Lesson 4 *Looking Back*

Overview of Course 1

Part B

Unit 5 Patterns in Space and Visualization

Patterns in Space and Visualization develops student visualization skills and an understanding of the properties of space-shapes including symmetry, area, and volume.

Topics include two- and three-dimensional shapes, spatial visualization, perimeter, area, surface area, volume, the Pythagorean Theorem, polygons and their properties, symmetry, isometric transformations (reflections, rotations, translations, glide reflections), one-dimensional strip patterns, tilings of the plane, and the regular (Platonic) solids.

Lesson 1 *The Shape of Things*
Lesson 2 *The Size of Things*
Lesson 3 *The Shapes of Plane Figures*
Lesson 4 *Looking Back*

Unit 6 Exponential Models

Exponential Models develops student ability to use exponential functions to model and solve problems in situations that exhibit exponential growth or decay.

Topics include exponential growth, exponential functions, fractals, exponential decay, recursion, half-life, compound growth, finding equations to fit exponential patterns in data, and properties of exponents.

Lesson 1 *Exponential Growth*
Lesson 2 *Exponential Decay*
Lesson 3 *Compound Growth*
Lesson 4 *Modeling Exponential Patterns in Data*
Lesson 5 *Looking Back*

Unit 7 Simulation Models

Simulation Models develops student confidence and skill in using simulation methods—particularly those involving the use of random numbers—to make sense of real-world situations involving chance.

Topics include simulation, frequency tables and their histograms, random-digit tables and random-number generators, independent events, the Law of Large Numbers, and expected number of successes in a series of binomial trials.

Lesson 1 *Simulating Chance Situations*
Lesson 2 *Estimating Expected Values and Probabilities*
Lesson 3 *Simulation and the Law of Large Numbers*
Lesson 4 *Looking Back*

Capstone Planning a Benefits Carnival

Planning a Benefits Carnival is a thematic, two-week, project-oriented activity that enables students to pull together and apply the important mathematical concepts and methods developed throughout the course.

Contents

Unit 3 Linear Models

Unit 4 Graph Models

Preface

The first three courses in the *Contemporary Mathematics in Context* series provide a common core of broadly useful mathematics for all students. They were developed to prepare students for success in college, in careers, and in daily life in contemporary society. Course 4 formalizes and extends the core program with a focus on the mathematics needed to be successful in college mathematics and statistics courses. The series builds upon the theme of *mathematics as sense-making*. Through investigations of real-life contexts, students develop a rich understanding of important mathematics that makes sense to them and which, in turn, enables them to make sense out of new situations and problems.

Each course in the *Contemporary Mathematics in Context* curriculum shares the following mathematical and instructional features.

- ***Unified Content*** Each year the curriculum advances students' understanding of mathematics along interwoven strands of algebra and functions, statistics and probability, geometry and trigonometry, and discrete mathematics. These strands are unified by fundamental themes, by common topics, and by mathematical habits of mind or ways of thinking. Developing mathematics each year along multiple strands helps students develop diverse mathematical insights and nurtures their differing strengths and talents.
- ***Mathematical Modeling*** The curriculum emphasizes mathematical modeling including the processes of data collection, representation, interpretation, prediction, and simulation. The modeling perspective permits students to experience mathematics as a means of making sense of data and problems that arise in diverse contexts within and across cultures.
- ***Access and Challenge*** The curriculum is designed to make more mathematics accessible to more students while at the same time challenging the most able students. Differences in student performance and interest can be accommodated by the depth and level of abstraction to which core topics are pursued, by the nature and degree of difficulty of applications, and by providing opportunities for student choice on homework tasks and projects.
- ***Technology*** Numerical, graphics, and programming/link capabilities such as those found on many graphing calculators are assumed and appropriately used throughout the curriculum. This use of technology permits the curriculum and instruction to emphasize multiple representations (verbal, numerical, graphical, and symbolic) and to focus on goals in which mathematical thinking and problem solving are central.
- ***Active Learning*** Instructional materials promote active learning and teaching centered around collaborative small-group investigations of problem situations followed by teacher-led whole-class summarizing activities that lead to analysis, abstraction, and further application of underlying mathematical ideas. Students are actively engaged in exploring, conjecturing, verifying, generalizing, applying, proving, evaluating, and communicating mathematical ideas.
- ***Multi-dimensional Assessment*** Comprehensive assessment of student understanding and progress through both curriculum-embedded assessment opportunities and supplementary assessment tasks supports instruction and enables monitoring and evaluation of each student's performance in terms of mathematical processes, content, and dispositions.

Unified Mathematics

Contemporary Mathematics in Context is a unified curriculum that replaces the traditional Algebra-Geometry-Advanced Algebra/Trigonometry-Precalculus sequence. Each course features important mathematics drawn from four strands.

The Algebra and Functions strand develops student ability to recognize, represent, and solve problems involving relations among quantitative variables. Central to the development is the use of functions as mathematical models. The key algebraic models in the curriculum are linear, exponential, power, polynomial, logarithmic, rational, and trigonometric functions. Modeling with systems of equations, both linear and nonlinear, is developed. Attention is also given to symbolic reasoning and manipulation.

The primary goal of the Geometry and Trigonometry strand is to develop visual thinking and ability to construct, reason with, interpret, and apply mathematical models of patterns in visual and physical contexts. The focus is on describing patterns with regard to shape, size, and location; representing patterns with drawings, coordinates, or vectors; predicting changes and invariants in shapes; and organizing geometric facts and relationships through deductive reasoning.

The primary role of the Statistics and Probability strand is to develop student ability to analyze data intelligently, to recognize and measure variation, and to understand the patterns that underlie probabilistic situations. The ultimate goal is for students to understand how inferences can be made about a population by looking at a sample from that population. Graphical methods of data analysis, simulations, sampling, and experience with the collection and interpretation of real data are featured.

The Discrete Mathematics strand develops student ability to model and solve problems involving enumeration, sequential change, decision-making in finite settings, and relationships among a finite number of elements. Topics incude matrices, vertex-edge graphs, recursion, voting methods, and systematic counting methods (combinatorics). Key themes are discrete mathematical modeling, existence (Is there a solution?), optimization (What is the best solution?), and algorithmic problem-solving (Can you efficiently construct a soluion?).

Each of these strands is developed within focused units connected by fundamental ideas such as symmetry, matrices, functions, and data analysis and curve-fitting. The strands also are connected across units by mathematical habits of mind such as visual thinking, recursive thinking, searching for and explaining patterns, making and checking conjectures, reasoning with multiple representations, inventing mathematics, and providing convincing arguments and proofs.

The strands are unified further by the fundamental themes of data, representation, shape, and change. Important mathematical ideas are frequently revisited through this attention to connections within and across strands, enabling students to develop a robust and connected understanding of mathematics.

Active Learning and Teaching

The manner in which students encounter mathematical ideas can contribute significantly to the quality of their learning and the depth of their understanding. *Contemporary Mathematics in Context* units are designed around multi-day lessons centered on big ideas. Lessons are organized around a four-phase cycle of classroom activities, described in the following paragraph—*Launch*, *Explore*, *Share and Summarize*, and *On Your Own*. This cycle is designed to engage students in investigating and making sense of problem situations, in constructing important mathematical concepts and methods, in generalizing and proving mathematical relationships, and in communicating both orally and in writing their thinking and the results of their efforts. Most classroom activities are designed to be completed by students working together collaboratively in groups of two to four students.

The launch phase promotes a teacher-led class discussion of a problem situation and of related questions to think about, setting the context for the student work to follow. In the second or explore phase, students investigate more focused problems and questions related to the launch situation. This investigative work is followed by a teacher-led class discussion in which students summarize mathematical ideas developed in their groups, providing an opportunity to construct a shared understanding of important concepts, methods, and approaches. Finally, students are given a task to complete on their own, assessing their initial understanding of the concepts and methods.

Each lesson also includes tasks to engage students in Modeling with, Organizing, Reflecting on, and Extending their mathematical understanding. These MORE tasks are central to the learning goals of each lesson and are intended primarily for individual work outside of class. Selection of tasks for use with a class should be based on student performance and the availability of time and technology. Students can exercise some choice of tasks to pursue, and at times they can be given the opportunity to pose their own problems and questions to investigate.

Multiple Approaches to Assessment

Assessing what students know and are able to do is an integral part of *Contemporary Mathematics in Context*, and there are opportunities for assessment in each phase of the instructional cycle. Initially, as students pursue the investigations that make up the curriculum, the teacher is able to informally assess student understanding of mathematical processes and content and their disposition toward mathematics. At the end of each investigation, the "Checkpoint" and accompanying class discussion provide an opportunity for the teacher to assess levels of understanding that various groups of students have reached as they share and summarize their findings. Finally, the "On Your Own" problems and the tasks in the MORE sets provide further opportunities to assess the level of understanding of each individual student. Quizzes, in-class exams, take-home assessment tasks, and extended projects are included in the teacher resource materials.

Acknowledgments

Development and evaluation of the student text materials, teacher materials, assessments, and calculator software for *Contemporary Mathematics in Context* was funded through a grant from the National Science Foundation to the Core-Plus Mathematics Project (CPMP). We are indebted to Midge Cozzens, Director of the NSF Division of Elementary, Secondary, and Informal Education, and our program officers James Sandefur, Eric Robinson, and John Bradley for their support, understanding, and input.

In addition to the NSF grant, a series of grants from the Dwight D. Eisenhower Higher Education Professional Development Program has helped to provide professional development support for Michigan teachers involved in the testing of each year of the curriculum.

Computing tools are fundamental to the use of *Contemporary Mathematics in Context*. Appreciation is expressed to Texas Instruments and, in particular, Dave Santucci for collaborating with us by providing classroom sets of graphing calculators to field-test schools.

As seen on page iii, CPMP has been a collaborative effort that has drawn on the talents and energies of teams of mathematics educators at several institutions. This diversity of experiences and ideas has been a particular strength of the project. Special thanks is owed to the exceptionally capable support staff at these institutions, particularly at Western Michigan University.

From the outset, our work has been guided by the advice of an international advisory board consisting of Diane Briars (Pittsburgh Public Schools), Jeremy Kilpatrick (University of Georgia), Kenneth Ruthven (University of Cambridge), David A. Smith (Duke University), and Edna Vasquez (Detroit Renaissance High School). Preliminary versions of the curriculum materials also benefited from careful reviews by the following mathematicians and mathematics educators: Alverna Champion (Grand Valley State University), Cherie Cornick (Wayne County Alliance for Mathematics and Science), Edgar Edwards (formerly of the Virginia State Department of Education), Richard Scheaffer (University of Florida), Martha Siegel (Towson University), Edward Silver (University of Michigan), and Lee Stiff (North Carolina State University).

Our gratitude is expressed to the teachers and students in our 41 evaluation sites listed on pages iv and v. Their experiences using pilot- and field-test versions of *Contemporary Mathematics in Context* provided constructive feedback and improvements. We learned a lot together about making mathematics meaningful and accessible to a wide range of students.

A very special thank you is extended to Barbara Janson for her interest and encouragement in publishing a core mathematical sciences curriculum that breaks new ground in terms of content, instructional practices, and student assessment. Finally, we want to acknowledge Eric Karnowski for his thoughtful and careful editorial work and express our appreciation to the editorial staff of Glencoe/McGraw-Hill who contributed to the publication of this program.

To the Student

Contemporary Mathematics in Context may be quite different from other math textbooks you have used. With this text, you will learn mathematics by doing mathematics, not by memorizing "worked out" examples. You will investigate important mathematical ideas and ways of thinking as you try to understand and make sense of realistic situations. Because real-world situations and problems often involve data, shape, change, or chance, you will learn fundamental concepts and methods from several strands of mathematics. In particular, you will develop an understanding of broadly useful ideas from algebra and functions, from statistics and probability, from geometry and trigonometry, and from discrete mathematics. You also will see connections among these strands—how they weave together to form the fabric of mathematics.

Because real-world situations and problems are often open-ended, you will find that there may be more than one correct approach and more than one correct solution. Therefore, you will frequently be asked to explain your ideas. This text will provide you help and practice in reasoning and communicating clearly about mathematics.

Because solving real-world problems often involves teamwork, you often will work collaboratively with a partner or in small groups as you investigate realistic and interesting situations. You will find that two to four students working collaboratively on a problem can often accomplish more than any one of you would working individually. Because technology is commonly used in solving real-world problems, you will use a graphing calculator or computer as a tool to help you understand and make sense of situations and problems you encounter.

You're going to learn a lot of useful mathematics in this course—and it's going to make sense to you. You're going to learn a lot about working cooperatively and communicating with others as well. You're also going to learn how to use technological tools intelligently and effectively. Finally, you'll have plenty of opportunities to be creative and inventive. Enjoy.

Patterns in Data

Unit 1

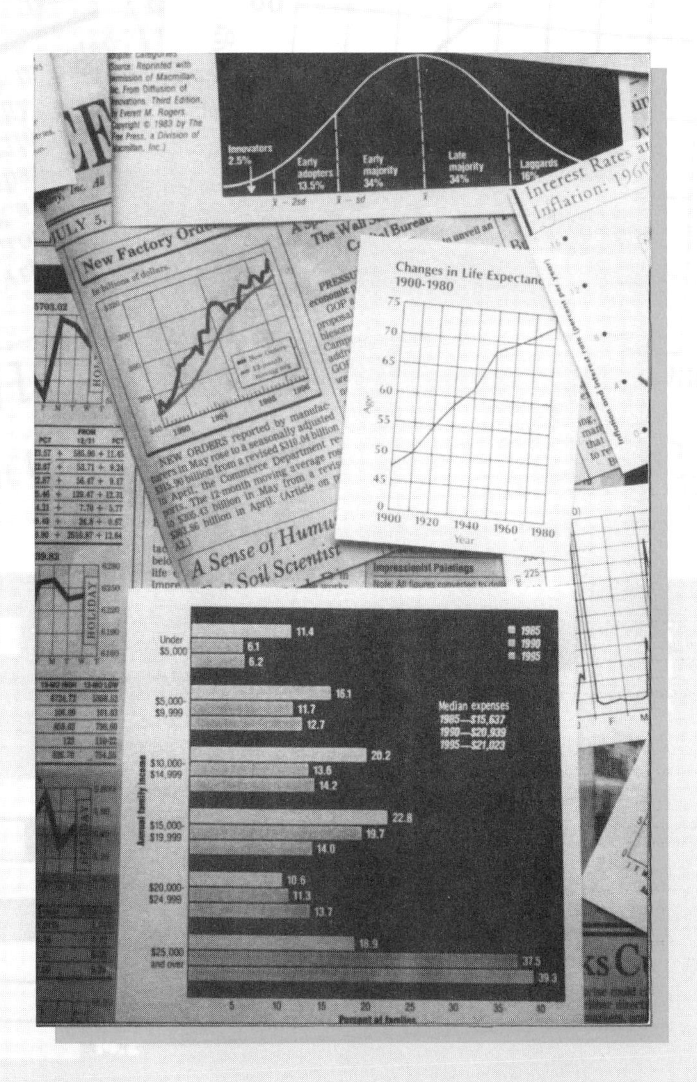

Lesson 1

Exploring Data

The main theme of this text is that mathematics provides a powerful way of making sense of the world in which you live. Contemporary mathematics is rooted in the study of patterns involving data, shape, change, and chance—aspects of situations you experience every day. It involves investigating and expressing relationships among these patterns in ways that make sense to you. Once understood, the patterns and relationships can be applied in a variety of ways to other common and complex situations.

This first unit is about statistics—collecting, organizing, summarizing, and making inferences from data.

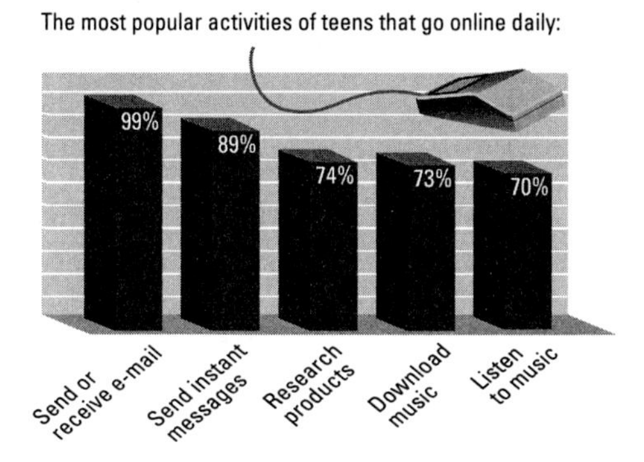

Source: Pew Internet & American Life Project

Think About This Situation

As a class, examine the data above from a study of teens who use the Internet. The findings are based on a phone survey of youth ages 12 to 17.

a If a similar study was conducted with online students in your school, do you think the results would be similar? Why or why not?

b Who might find this information important to know? How might they use it?

c The teen interviews were based on a "callback survey" of Internet households identified in a tracking poll as ones in which both the parents or guardians and children had Internet access. Why might the results give an inaccurate picture of the online teen population?

INVESTIGATION 1 Collecting and Analyzing Data

As you use this text, besides working as a whole class and individually, you often will work in pairs and in small groups. It is important that you have confidence in one another, share ideas, and help each other when asked. When working in groups, assigning specific jobs to group members helps the work run smoothly. Here is one way to do that.

Role	Responsibility
Coordinator	Obtains necessary resources. Recommends data gathering methods and units of measure appropriate for the situation. Communicates with other coordinators or the teacher on behalf of the group.
Measurement Specialist	Performs the actual measurements as needed.
Recorder	Records measurements taken and shares information with other groups.
Quality Controller	Carefully observes the measuring and recording processes and suggests when measurements should be double-checked for accuracy.

Work with your teacher to do the following.

- Form into groups of four people each if possible.
- Assign each group member one of the above roles.

Complete the data-gathering activities that follow by working together in your small groups. Each group will need a measuring tape or a measuring stick and a string.

Project 1

1. Measure, in centimeters, the height and armspan of each person in your group. Record your data in a table like the one shown here.

Student Name	Height	Armspan

   a. What one height would be most typical or representative of your group?

   b. Do you think the typical height in your group would be the same as the typical height of the entire class? Explain your reasoning.

   c. How much variation is there in the armspans of your group?

   d. Do you think the amount of variation in armspans of your class would be less than or greater than the variation in your group? Explain your reasoning.

2. Now, combine your data for heights and armspans with the data from other groups and record them in another table. Call it the Class Data Table.

   a. Use the Class Data Table to find a typical height for the entire class. Do you still agree with your answer to Activity 1 Part b?

   b. After examining the Class Data Table, do you still agree with your response to Activity 1 Part d?

3. On the basis of the data from your class, does there seem to be a relationship between height and armspan?

Project 2

Before starting this next project, everyone in your group should change roles. One way to arrange this is to have each group member pass his or her job to the student on the left. Project coordinators should decide together on a common method and a common unit of measure to be used by every group.

1. Measure the circumference (distance around) of the thumb and wrist of each person in your group. Record your data in a table like the one shown here.

Student Name	Thumb Circumference	Wrist Circumference

2. Combine your data of thumb and wrist circumferences with the data from other groups and record them in the Class Data Table. Again, record all measurements in the same unit.

3. Based on the class data, what relationship, if any, do you see between the wrist and thumb size of a person?

4. In the novel *Gulliver's Travels*, the Lilliputians made clothes for Gulliver. They estimated that twice around the thumb is once around the wrist and twice around the wrist is once around the neck.

   a. Does the first part of this statement seem true for your class data? Why or why not?

   b. Describe how you could check the second part of the Lilliputians' estimate.

Project 3

Rotate group jobs before beginning the following project. Common units of measure for every group should be decided upon by the project coordinators. Agreement should also be reached on a method for measuring stride length.

1. Determine the shoe length and stride length for each person in your group. Record the data in a table like the one shown here.

Student Name	Shoe Length	Stride Length

Why do you think you were asked to determine shoe length rather than shoe size?

2. Combine your data for shoe lengths and stride lengths with the data from other groups and record them in the Class Data Table.

   a. What is the smallest shoe length in the class? What is the largest shoe length?

   b. Does there appear to be a relationship between shoe length and stride length?

Keep your class data from Projects 1, 2, and 3. You will need them later in this unit.

Checkpoint

In the three projects, you collected and analyzed measurement data about your group and about your classmates.

a What kinds of relationships did you find in the data from the three projects?

b What pitfalls are possible when you generalize results from a small sample of people, such as your group, to a larger population, such as your class?

c How well did your group cooperate in completing these projects? Write one comment about each group member in the following form.

We appreciated it when . . .

Each person from your group should be prepared to report to the whole class on the group's responses to these questions.

On Your Own

Determine the thumb dominance of a sample of your family or friends by having people fold their hands on top of a desk with their fingers interlocked. The dominant thumb (right or left) is on top of the other.

a. Record your finding for each person in a table along with information on his or her hand dominance (left-handed or right-handed).

b. Does there appear to be a relationship between hand dominance and thumb dominance? If so, write a sentence describing the relationship.

c. How confident are you in your conclusion?

d. What might you do to have more confidence in your conclusion?

INVESTIGATION 2 Describing Patterns in Data

To learn from one another in investigating situations, it is important that your group work together as a team. Following the guidelines below will help your group work well together. It will also help you build skills that have become very important for people who work in fields such as health care, business, and industry.

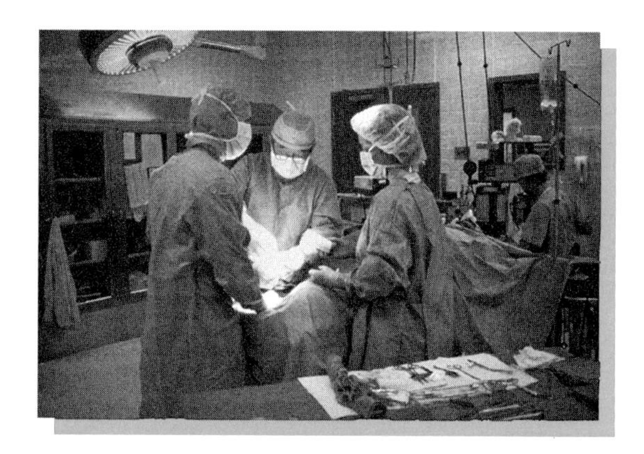

Group Guidelines

- Each group member contributes to the group's work.
- Each member of the group is responsible for listening carefully when another group member is talking.
- Each member of the group has the responsibility and the right to ask questions.
- Each group member should help others in the group when asked.
- Each member of the group should be considerate and encouraging.
- Work together until everyone in the group understands and can explain the group's results.

1. How well did your group work together in completing the projects in the last investigation? In your group, discuss responses to the following questions.

a. How could your group encourage everyone to contribute ideas?

b. How could your group ensure that everyone else listens while someone is talking?

c. How could your group ensure that each person understands and agrees with the group's decisions?

In this investigation, you will continue to collaborate with one another in small groups. The activities will not require taking and recording measurements. Nevertheless, it is helpful if each group member has an assigned role at the beginning. Roles often used by groups of four people are described on the next page. When working in groups of three, the roles of quality controller and coordinator can be combined. Decide on a role for each member of your group.

Role	Responsibility
Reader	Reads and explains the questions or problems on which the group will be working.
Recorder	Writes a summary of the group's decisions and ideas, and reads them back to the group to ensure agreement and accuracy. Shares the group's summary with other groups or the entire class.
Quality Controller	Monitors the group's results and makes sure that the group produces high quality work of which all the members will be proud.
Coordinator	Keeps the group on track and makes sure everyone is participating. Communicates with the teacher on behalf of the group.

Making sense of information and data is a common feature of many careers. It is also important for a thorough understanding of radio, television, and newspaper reports, which often include information gathered from opinion polls or surveys. As an example, consider the following report of a survey of the pop music industry.

The *Los Angeles Times* once asked 22 important people in the pop music industry to determine the pop world's "hottest properties." Among the 22 people were the presidents of Warner Bros. Records and Interscope Records. The artists and groups were ranked by giving 10 points for each first-place vote, 9 points for a second-place vote, and so on. The following is the list of the twenty artists and groups who received the most points.

Industry Leaders' Hot Properties

Rank	Artist	Points	Rank	Artist	Points
1	Eminem	90	11	Creed	37
2	Dr. Dre	73	12	U2	34
3	The Beatles	67		Lauryn Hill/Fugees	34
4	Dave Matthews Band	59	14	Marc Anthony	28
5	Madonna	51	15	Red Hot Chili Peppers	27
6	Destiny's Child	50	16	Lenny Kravitz	22
7	Shania Twain	49	17	Kid Rock	21
8	Limp Bizkit	46		Sade	21
	Faith Hill	46	19	DMX	20
10	Celine Dion	45	20	Britney Spears	18

Source: Hilburn, Robert. "For now, yesterday is here to stay." *Los Angeles Times*, April 22, 2001.

2. In your group, discuss each of the following questions. Record the answer that your group decides is the best response.

a. Eminem could have received his 90 points by getting 7 first-place votes, 2 third-place votes, and a seventh-place vote. Find two other ways that Eminem could have received his 90 points.

b. What is the largest possible number of points an artist or group could have received?

c. Does it appear that the rankings of the 22 executives were a lot alike, or do you think there was substantial disagreement in their rankings?

3. Students in one class at City High were asked to *describe* the distribution of points for the top 20 artists and groups.

Joshua's response: "Eminem was highest with 90 points and Britney Spears was 20th with 18 points. My favorite group is U2 at 12th place. I don't think that is an accurate assessment of how 'hot' the group is. They are definitely going to be around much longer than some of those rated above them."

Sarah's response: "With 90 points, Eminem was much higher than the other artists and groups. Nevertheless, this is an average of only about 4 points per executive, or an average ranking of 7th. The rest of the artists and groups were clustered between 73 and 18 points."

a. Whose response does your group think was the most complete and helpful? Why?

b. What would you do to improve the response that you chose as the best?

4. When asked to *describe* the distribution of points for the top 20 artists and groups, Paul and María first made a **stem-and-leaf plot** like the one shown below.

Industry Leaders' Hot Properties

Stem	Leaf
1	8
2	0 1 1 2 7 8
3	4 4 7
4	5 6 6 9
5	0 9
6	7
7	3
8	
9	0

1 | 8 represents 18.

a. In your group, discuss the organization of this stem-and-leaf plot. Is the plot an accurate display of the data in the chart? If not, explain how to correct it.

Paul's description: "Based on the stem-and-leaf plot, most artists and groups were ranked pretty low. Only a few artists and groups had high ratings."

María's description: "You can see in the stem-and-leaf plot that there were a whole bunch in the 20s and this amount decreased a lot by the time you got up to the 60s. The 90 is all by itself."

b. What are the strengths and weaknesses of Paul's response?

c. What are the strengths and weaknesses of María's response?

d. As a group, write what you think would be a better response than either Paul's or María's. Compare your response with those of other groups.

5. Twenty-two students from the Student Council at City High ranked their top ten artists and groups in a similar manner to the people asked by the *Los Angeles Times.* Their results are shown in the chart below. Make a stem-and-leaf plot for these data.

Student Council's Hot Properties

Rank	Artist	Points	Rank	Artist	Points
1	Destiny's Child	71	11	Faith Hill	36
2	Madonna	70	12	Limp Bizkit	34
3	Lenny Kravitz	65	13	Dr. Dre	29
4	Creed	64	14	Eminem	28
5	U2	57		Shania Twain	28
6	Dave Matthews Band	56		Kid Rock	28
7	The Beatles	54	17	Sade	26
8	Britney Spears	47	18	DMX	21
9	Lauryn Hill/Fugees	43	19	Marc Anthony	18
10	Red Hot Chili Peppers	42	20	Celine Dion	11

6. Students were asked to make a stem-and-leaf plot for the Student Council rankings and to *compare* the plot to the one in Activity 4.

Desmond's response: "The new stem-and-leaf plot has the same large number of artists and groups in the 20s. The rest only go as high as 70s instead of 90s."

Regina's response: "The new plot doesn't have a gap and is more evenly spread out, except the 20s."

a. What are the strengths and weaknesses of Desmond's response?

b. What are the strengths and weaknesses of Regina's response?

c. As a group, write what you think would be a better response than either Regina's or Desmond's. Compare your response to those of other groups.

7. For homework, students at City High were asked to *explain why* stem-and-leaf plots are helpful in understanding data such as the rankings and points from the *Los Angeles Times*.

   Cecilia wrote this explanation: "Stem-and-leaf plots help you understand the information because you can see quickly how the data are spread out or clumped together. You can see any gaps where there are no data. By using the numbers instead of tally marks or Xs, you can still read the number of points that the judges assigned. Since these data sets are small, the stem-and-leaf displays were concise, easy to make, and easy to read."

   Ms. Thomas, the mathematics teacher, decided that Cecilia's explanation was excellent, so she gave her 5 points, the maximum points possible.

   Here is Jesse's explanation: "The stem-and-leaf plot has the stems on the left and the leaves on the right. For the numbers that have two digits, the tens digit is the stem and the ones digit is the leaf. For numbers that have three digits, the hundreds and tens digits together are the stem and the ones digit is the leaf. The numbers are in order from smallest at the top to largest at the bottom."

   As a group, use Cecilia's response as a guide to evaluate Jesse's explanation on a scale of 1 to 5, with 5 being the top score. Explain why you assigned the score you did.

Checkpoint

In this investigation, you explored characteristics of good written responses.

a What are the important features of a good response when you are asked to *describe* something?

b What are the important features of a good response when you are asked to *explain* your reasoning?

c What are the important features of a good response when you are asked to *compare* two or more things?

Each person from your group should be prepared to share your group's characteristics of good descriptions, explanations, and comparisons.

On Your Own

Make a stem-and-leaf plot of the heights reported in the Class Data Table. Then write a description of the information you can see from looking at your plot.

By now you may be wondering: "What's all the fuss about writing good explanations? This is a math course!" Verbal explanations connect the abstraction of mathematics with its meaning in the real world. So being able to describe your conclusions and explain your reasoning is important for at least two reasons: (1) it helps you better understand your own thinking about the mathematics you are studying; and (2) it is a skill, like teamwork, that businesses and industries look for in new employees.

The final activity in this lesson will help you and your group become more familiar with these new aspects of doing mathematics.

Reassign your group roles. Study the chart below giving average monthly earnings in 1999 for adults in the United States.

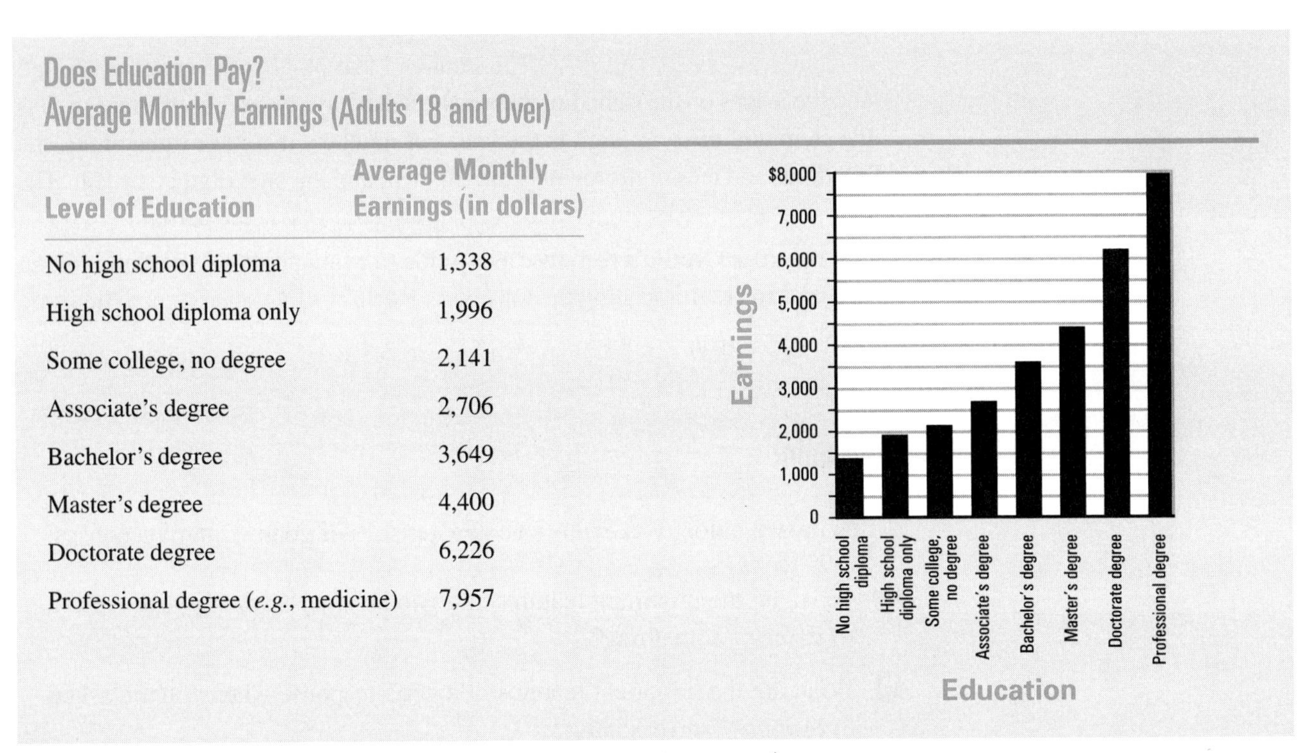

Does Education Pay?
Average Monthly Earnings (Adults 18 and Over)

Level of Education	Average Monthly Earnings (in dollars)
No high school diploma	1,338
High school diploma only	1,996
Some college, no degree	2,141
Associate's degree	2,706
Bachelor's degree	3,649
Master's degree	4,400
Doctorate degree	6,226
Professional degree (*e.g.*, medicine)	7,957

Source: U.S. Census Bureau, *Statistical Abstract of the United States: 2001* (120th edition). Washington, DC, 2000.

8. Discuss each question and then come to a single response that each member of your group agrees with and understands.

a. On the average, what would a U.S. worker make in a year if the person was not a high school graduate?

b. On the average, what is the difference in yearly earnings between a high school graduate with no further education and a person without a high school diploma?

9. Look at the pattern of change in average monthly earnings for the categories "no high school" through "doctorate degree."

a. From which level of education to the next is the largest jump in average monthly earnings?

b. From which level to the next is the second largest jump in average monthly earnings?

c. Did you use the table or *bar graph* in answering Parts a and b? Explain why you chose the display(s) you did.

10. Now it's your turn. Write a question that interests your group and can be answered using these data. Write the answer to your question, too.

11. Write a summary of the conclusions you can draw using the information in the table and the graph. Be careful: some professional degrees may not require as many years of education as some master's and doctorate degrees.

Checkpoint

Review your work on "Does Education Pay?"

a Compare the information shown in the table and the bar graph. In what ways is the graph better than the table for displaying the data on average monthly earnings? In what ways is it worse?

b How could your analysis of these data be used to plan for your own future?

c How has your group work improved since the first day of class? How could you further improve your group work next time?

Be prepared to share your group's responses with the whole class.

In this lesson, you have reviewed how to make stem-and-leaf plots and how to measure lengths and record your results efficiently. You also have begun to learn how to analyze and write descriptions of distributions of data. In the next lesson, you will learn more about plotting and describing distributions.

On Your Own

Think about the significance of reasoning from data and working collaboratively as you complete the following tasks.

a. Over a lifetime, how much more money could a person expect to earn with a bachelor's degree than with only a high school diploma? Assume the retirement age is the traditional 65 years.

b. Find one example of how working in groups (teamwork) is used in business or industry. Write a brief summary of your findings.

Lesson 2

Shapes and Centers

Everyday, people are bombarded by data on television, on radio, in newspapers, and in magazines. For example, states release report cards for schools and statistics on crime and unemployment; cities are rated to determine the "best places to live"; sports writers report batting averages and select the top NFL quarterback; and consumers shop for "best buys." Making sense of data is important in everyday life and in most professions today. Read carefully the following news story.

AAA Revs Up Car List
GM, Chrysler Each Hold Three Spots

The American Automobile Association picked three General Motors cars for its list of the best 2001 models. GM's winners were the Buick LeSabre in the $20,000-$25,000 category, the Cadillac DeVille in the $40,000-$50,000 range, and the GMC Yukon Denali in the SUV over $25,000 category. All three of these vehicles were also on last year's list although the Cadillac Deville was in the $35,000-$40,000 category then.

Chrysler also had three models on the list but one of them was in the Cool Car category. The Chrysler PT Cruiser was named as the cool car of the year. Last year's cool car was the Audi TT Quattro.

CHRYSLER PT CRUISER

The 42-million member AAA used a 10-point scale to evaluate over 200 cars, minivans, SUVs, and trucks in 20 categories. The top score of 174 went to the Mercedes Benz S500 at over $50,000, a repeat winner. In addition to the three GM cars, there were four other repeat winners: Ford Focus, $12,500-$15,000; Acura 3.2TL, $25,000-$30,000; Dodge Grand Caravan, Minivan; and Nissan Xterra, SUV under $25,000.

New to this year's list are the Nissan Sentra in the under $12,500 range, the Chrysler Sebring Sedan in the $15,000-$20,000 range, the Saab 9-5 in the $30,000-$35,000 range, and the Volvo XC in the $35,000-$40,000 range.

To make their ratings, the staff of the American Automobile Association identified the 20 characteristics that they felt were most important to consider when buying a car. Groups of three to six staff people tested the cars for a two-week period. They rated them on each of the categories using a scale of 1 to 10. Ratings of 1 to 5 are negative; 6 to 10 are positive. The results for some of the cars in the $20,000 to $25,000 price range are summarized in the following table.

American Automobile Association Ratings of 2001 Cars ($20,000 to $25,000 Price Range)

Car	Acceleration	Transmission	Braking	Steering	Ride	Handling	Driveability	Fuel Economy	Comfort	Interior Space	Driving Position	Instruments	Controls	Visibility	Entry/Exit	Quietness	Cargo Space	Exterior	Interior	Value
Buick LeSabre	9	9	7	8	9	7	8	7	8	9	8	8	8	8	9	9	9	7	8	9
Chevrolet Camaro	10	7	7	9	5	8	7	4	7	5	8	8	8	4	5	6	5	8	7	7
Chrysler Concorde	8	9	6	7	9	8	8	6	9	9	8	7	8	7	9	9	7	7	8	9
Chrysler PT Cruiser	7	8	5	8	8	7	8	4	8	8	8	7	7	7	9	8	9	8	8	9
Dodge Intrepid	8	9	6	8	9	8	9	6	9	9	8	7	8	7	9	9	5	7	8	9
Ford Mustang GT	9	8	6	9	7	9	8	4	8	5	9	9	9	6	4	6	5	8	8	8
Mazda Miata	8	9	9	9	6	9	8	6	8	6	8	8	8	6	6	6	3	8	8	9
Toyota MR2	9	8	7	8	6	9	8	7	7	6	8	8	7	7	7	6	1	6	8	8
Toyota Prius	5	8	5	6	8	6	7	10	7	7	8	8	8	8	8	9	5	8	8	6
Volvo V40 Wagon	8	8	7	9	7	8	8	6	8	7	8	8	8	7	8	8	10	9	9	9
Mazda 626	9	9	7	9	7	9	9	6	8	7	8	8	7	8	8	8	7	8	8	9

A rating of 1 is the lowest. A rating of 10 is the highest.

Source: *2001 New Car & Truck Buying Guide.* Heathrow, FL: AAA Publishing, 2000.

Think About This Situation

Suppose your family wants to buy a used 2001 car that originally sold for between $20,000 and $25,000.

a According to the article, which car did the American Automobile Association (AAA) think was best? Do you think there are any important characteristics that they have omitted?

b There is so much information in the table above that it is difficult to see which car might be best. How could you organize or summarize this information to make it more useful in deciding which car to buy?

c Based on the rating table, do you agree with the conclusions drawn by AAA and reported in the article? Why or why not?

INVESTIGATION 1 Shapes of Distributions

1. With your small group, select one of the cars to study more carefully.

a. If you were a dealer selling the car you selected, what feature(s) would you stress? Why?

b. If you were the manufacturer of the car, what feature(s) would you try to improve? Why?

c. Using the scale of 1 to 10, what overall rating would you assign the car? Explain your reasoning.

An examination of any of the columns of the table reveals that the ratings vary from car to car. The ratings are all 8 or 9 in *driving position*, but the ratings vary from 1 to 10 in *cargo space*. The pattern of variation or **distribution** of the ratings is best seen in a plot.

2. Examine this **number line plot** of the AAA ratings of *exterior workmanship* (*exterior* in table). Describe the overall shape of the distribution by answering the following questions:

a. What is the lowest rating? What is the highest rating? What cars are associated with these ratings?

b. Are there any **outliers**—unusually high or low ratings that lie away from the other ratings?

c. Does the plot have one or more peaks?

d. Does the shape of the plot appear to be **symmetric**—right and left sides look almost like mirror images of each other? If not, does the plot appear to be stretched more to the higher or more to the lower ratings?

e. Do you see any gaps in the overall pattern of the plot?

f. Where is the center of the distribution?

g. How spread out are the ratings?

Once a distribution is displayed in a plot, you can describe its important features by discussing questions similar to those in Activity 2.

3. Explore how number line plots can help you compare distributions.

a. Make a number line plot of the *entry/exit* ratings.

b. In what ways is the distribution of *entry/exit* ratings like that of the *exterior workmanship* ratings? In what ways is it different?

c. Part of your group should make a number line plot of the *quietness* ratings. The rest of your group should prepare a plot of the *fuel economy* ratings. Compare the two distributions in terms of the key questions for describing distributions.

Number line plots can be used to get quick visual displays of data. They enable you to see any patterns or unusual features in the data. They are most useful when working with small sets of data. **Histograms** can be used with data sets of any size. In a histogram, the **horizontal axis** is marked with a numerical scale. The **vertical axis** represents counts or frequencies. (Note that neither of these last two characteristics is true of the bar graph displayed on page 12.)

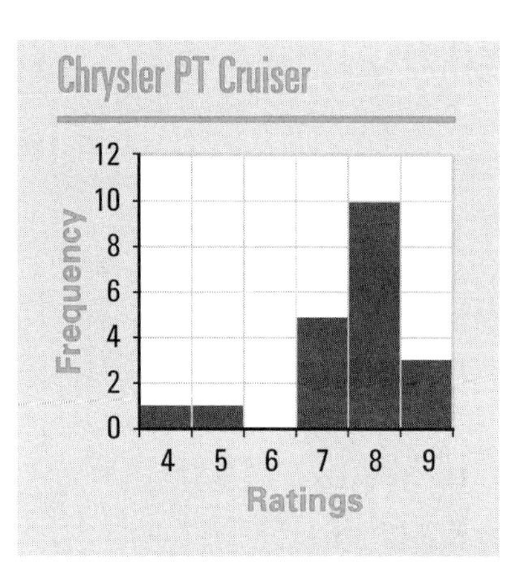

4. Examine this histogram of the ratings for the Chrysler PT Cruiser.

a. What does the bar at 9 represent?

b. Why is there no bar at 6?

c. Describe the distribution of the ratings.

5. Make a number line plot and a histogram of the ratings for the Volvo V40 Wagon.

a. What does the height of a bar on the histogram tell you? Give an example.

b. Describe the distribution of these ratings.

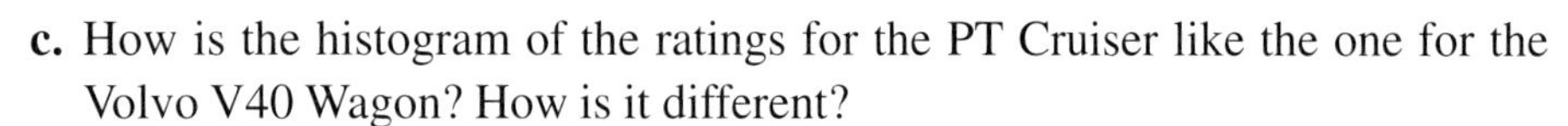

c. How is the histogram of the ratings for the PT Cruiser like the one for the Volvo V40 Wagon? How is it different?

6. The percentages of persons who live below the poverty level vary from state to state. Shown below is a histogram of these percentages for the fifty states for 1999. The frequency, or number of states, is on the vertical axis.

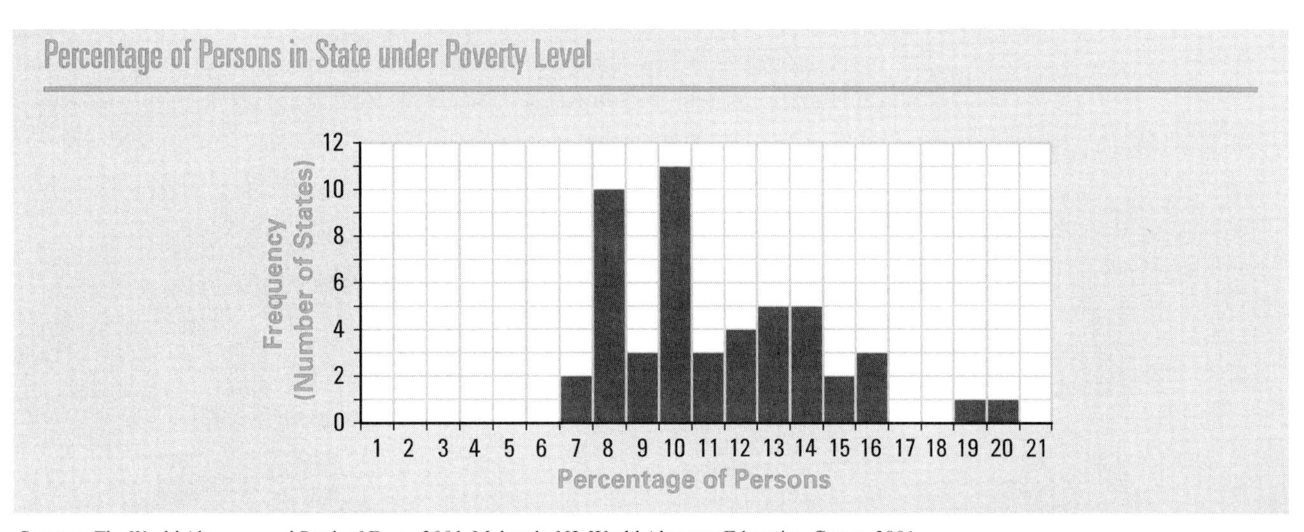

Source: *The World Almanac and Book of Facts, 2001*. Mahwah, NJ: World Almanac Education Group, 2001.

a. What does the bar above 16 represent?

b. What is the percentage of persons who live in poverty in the state with the largest percentage?

c. What is the percentage in the states with the smallest percentage?

d. How many states have 15% or more of persons living in poverty?

e. Describe the distribution of these percentages.

Checkpoint

You can see important characteristics of a distribution from a histogram or number line plot.

a When would you prefer to make a histogram rather than a number line plot?

b Make a sketch of what you think the histogram of the ages of all women married for the first time last year in the United States might look like. Describe this distribution.

c How might the histogram of the ages of all men married for the first time last year in the United States compare to that of the women?

Be prepared to share your group's thinking and results with the whole class.

On Your Own

The two histograms below show the performance of two social studies classes at Central High School on the same quiz. The quiz scores were grouped into intervals of length 5. *A score on the edge of a bar is counted in the bar on the right.*

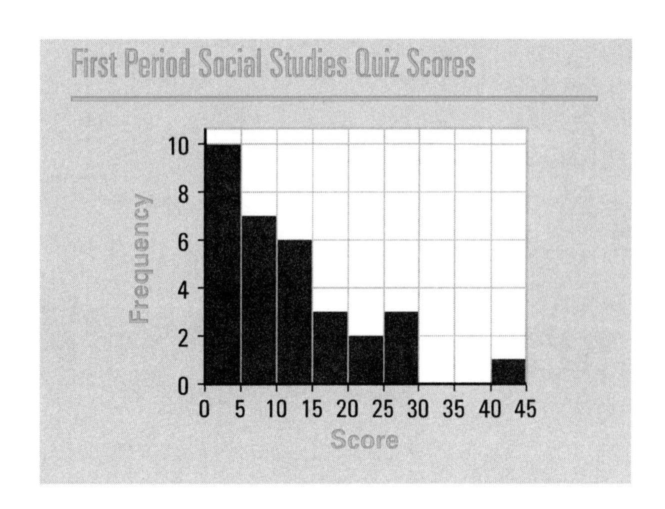

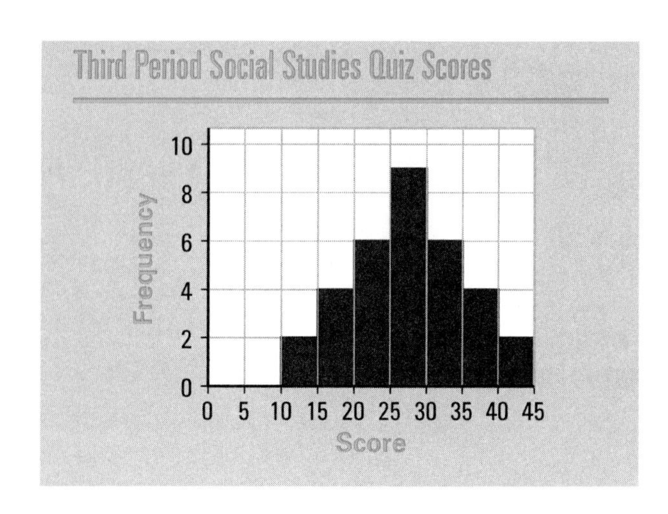

a. What percentage of the students in each class scored 30 or more?

b. Where is each distribution centered? What conclusion can you draw by comparing the two histograms?

c. Does either of the plots show gaps in the distribution of scores?

d. Which of the distributions is approximately symmetric? Would you expect the data to have this shape? Why or why not?

e. Which of the distributions is stretched to the right or left? What might account for the distribution having this shape?

INVESTIGATION 2 Producing Plots with Technology

In the previous investigation, you learned that a histogram of a distribution can reveal certain features of the data: where the data are centered, how much the data vary, whether there are gaps or clusters in the data, and the existence of outliers or other unusual values. Some distributions are mound-shaped or **approximately normal**, where the distribution has one peak and tapers off on both sides. Normal distributions are symmetric—the two halves look like mirror images of each other. Some distributions have the data stretched towards the right. These distributions are said to be **skewed to the right**. Some distributions have the data stretched towards the left. These distributions are said to be **skewed to the left**. (See the diagram below.)

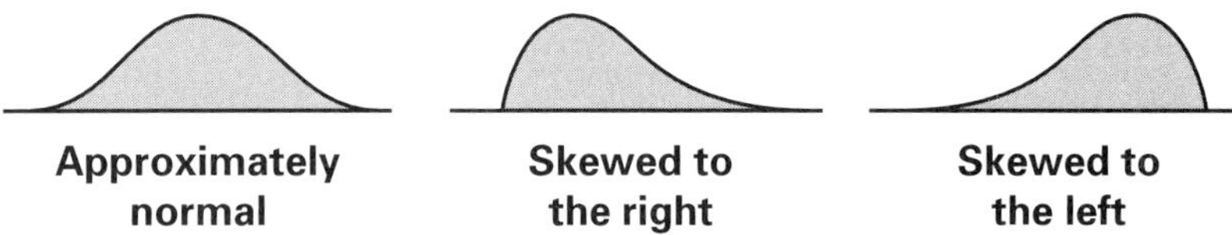

Producing a graphical display is the first step toward understanding data. You can use computer software or a graphing calculator to produce various plots of data. This generally requires the following three steps.

- Enter the data into an array or list(s). Be sure to clear any previous unwanted data.
- Set a *viewing window* for the plot. This is usually done by specifying the minimum and maximum values and scale on the horizontal (x) axis. Depending on the nature of the data, you may also need to specify the minimum and maximum values and scale on the vertical (y) axis. Some technological tools have automatic scaling for data plots.
- Select the type of plot desired.

Examples of the screens involved are shown here. Your calculator or software may look different.

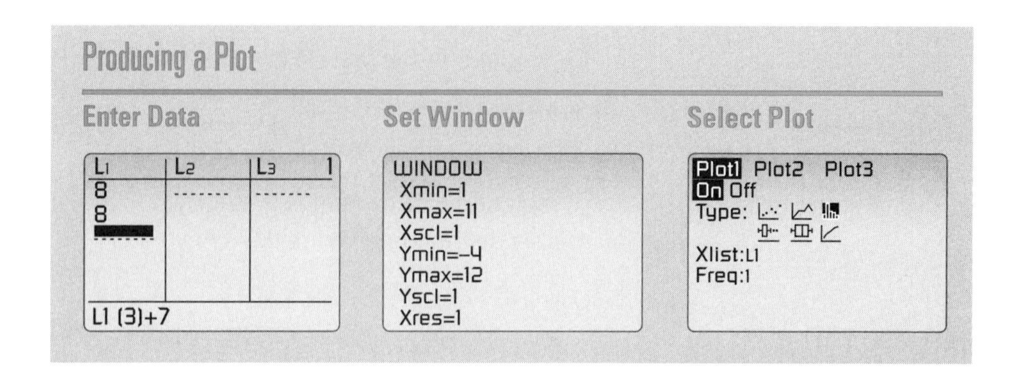

In this investigation, you will learn how to produce and analyze a histogram on your graphing calculator or computer.

1. Produce a histogram of the AAA ratings for the Volvo V40 Wagon.

a. Most calculators and computer software can display values along a graph, using a feature such as "Trace." Use the trace feature to display values as you move the cursor along the graph.

- What information is given for each bar?
- If a value occurs on the edge of a bar, in what bar is it counted?

b. Use the trace feature to find how many ratings of 9 there are.

c. Describe the overall shape of the distribution of ratings.

d. What is the most common rating? How can you tell from the histogram?

e. Change the "Xscl" value to 2. What does this do to the histogram? Why? Does it change the basic shape of the histogram? (Software and calculators vary. Your teacher may give you different instructions.)

f. Experiment with other widths for each bar. Which value seems to be the best for displaying these data?

2. Divide the remaining ten cars listed on page 15 among your group.

a. Individually, make a histogram of the ratings for each of your cars using your calculator or computer. Make a sketch of each histogram.

b. With your small group, examine the overall shape of the distributions of car ratings.

- For which cars is the distribution of ratings almost symmetric?
- Which cars have a distribution that is skewed? What does this tell you about the ratings for that car?
- Which distribution is spread out the most? The least? What does it mean if the ratings for a car have a large spread?

c. For which car(s) is the distribution of ratings centered at the lowest value?

d. Based on the histograms, select the top four cars you would consider buying. How did the plots help you make your choice?

Histograms can help you see the shape of a distribution. Choosing the width of the bars (Xscl) determines the number of bars. If the data set is small, histograms with different numbers of bars can give different visual impressions. When you explore the following data set you will examine several possible histograms and decide which you think is best. You also may come to better understand the nutritional differences among what may be some of your favorite foods.

The table on the following page gives nutritional information about some fast foods. Total calories, amount of fat in grams, amount of cholesterol in milligrams, and amount of sodium (salt) in milligrams are provided. Foods are organized according to type of meat: ground beef, chicken, and roast beef.

3. As a group, examine the data in the table. What information do you find surprising or interesting?

4. Many people order chicken in fast food restaurants because they believe it will have less fat than a hamburger. Does it appear from the table that they are right or wrong? Explain your response.

5. Consider the data on the total number of calories for the fast foods listed.

a. Make a histogram of the values. Use the values Xmin = 0, Xmax = 1100, Ymin = –2, Ymax = 10, and Yscl = 1. Experiment with different choices of Xscl. Which value of Xscl gives a good picture of the distribution?

b. Why isn't 137 the best value to use for Xmin?

c. Describe the pattern of the data that you can see from the histogram. Consider the usual features such as the spread, approximate location of the center, symmetry or skewness, gaps, and any unusual values.

How Fast Foods Compare

Company	Fast Food	Total Calories	Fat (grams)	Cholesterol (mg)	Sodium (mg)
McDonald's	Cheeseburger	350	14	45	830
Wendy's	Single	420	20	70	930
McDonald's	Quarter Pounder	430	21	70	840
McDonald's	Big Mac	590	34	85	1,090
Burger King	Hamburger	340	16	50	540
Wendy's	Big Bacon Classic	580	31	95	1,500
Burger King	Whopper	680	39	80	940
Hardee's	All Star	660	43	100	1,260
Burger King	Double Whopper w/Cheese	1,020	65	170	1,460
Hardee's	Grilled Chicken	350	16	65	860
Hardee's	Chicken Fillet	480	23	55	1,190
Wendy's	Grilled Chicken	480	21	65	1,000
Wendy's	Chicken (regular)	410	15	65	1,280
Burger King	BK Broiler Chicken	550	25	105	1,110
McDonald's	Chicken McGrill	450	18	60	970
McDonald's	Chicken McNuggets (6)	290	17	55	540
Burger King	Chicken Sandwich	660	39	70	1,330
Subway	Roasted Chicken Breast Salad	137	3	36	730
Subway	6" Asiago Caesar Chicken Sub	311	6	48	880
Subway	6" Roasted Chicken Breast Sub	391	15	46	1,000
Arby's	French Dip	410	16	45	1,200
Arby's	Regular Roast Beef	330	14	45	890
Arby's	Super Roast Beef	450	21	45	1,060

Source: *McDonald's Nutrition Facts*, McDonald's Corporation, 2001; www.wendys.com; *Nutritional Information*, Burger King Corp., 1999; *Hardee's Nutritional Information*, 2000; www.subway.com; *Comprehensive Guide of Quality Ingredients*, Arby's, Inc., 2001.

6. Now consider the data on the amount of cholesterol in the listed fast foods.

   a. Make a histogram of the values. Use Ymin = –4, Ymax = 10, and Yscl = 2. Experiment with your own values of Xmin, Xmax, and Xscl. Which values give a good picture of the distribution?

   b. Describe the histogram of the amount of cholesterol in fast foods.

   c. Change the maximum y value to 5. Is this a good choice? Why or why not?

Checkpoint

In this investigation, you learned how to produce histograms using technology. You also gained more experience in analyzing histograms.

a What does the shape of a distribution tell you about the data?

b If you are producing a histogram with your calculator or computer, how will you decide on the best choice of width for the bars?

c How will you decide on a reasonable value for the maximum y value?

Be prepared to share your group's thinking on ways of producing and interpreting histograms.

It takes a lot of practice to efficiently choose an appropriate viewing window for a technology-produced plot. You will get more practice in later investigations.

On Your Own

Consider the data on the amount of fat in the fast foods.

a. Which fast food item has the least amount of fat? Which has the greatest amount?

b. Use your calculator or computer to produce a histogram of these data. Set Ymin = –4 and Yscl = 1.

   - How can you use the numbers from Part a to determine Xmin, Xmax, and Xscl?
   - What procedure can you use to find a good value for Ymax?

c. Write a short description of the histogram so that a person who had not seen the histogram could draw an approximately correct sketch of it.

MORE

Modeling • Organizing • Reflecting • Extending

Modeling

These tasks provide opportunities for you to apply the ideas you have learned in the investigations. Each task asks you to model and solve problems in other situations.

1. Ratings of automobile characteristics such as comfort or visibility are subjective. They are based on how the raters feel about the characteristic. Other ratings such as acceleration or fuel economy can be based on more objective measures. The table below gives data on the mileage (miles per gallon) for city and highway driving of the rated cars.

City and Highway Mileage

	LeSabre	Camaro	Concorde	PT Cruiser	Intrepid	Mustang GT	Miata	MR2	Prius	V40 Wagon	Mazda 626
City (mpg)	19	16	19	20	19	17	23	25	45	22	21
Highway (mpg)	30	27	29	26	29	24	28	30	52	32	27

a. Make a number line plot of the city mileages of the rated cars.

b. Make a similar display of the highway mileages of the rated cars.

c. Compare your plots for Parts a and b. How are they similar? How are they different?

d. What two kinds of infomation are included in the table, but not in your plots for Parts a and b?

2. Refer to the armspan data in the Class Data Table prepared in Lesson 1.

a. Use your calculator to produce a histogram of the armspans of all the students in your class.

b. Describe the distribution of armspans.

c. Enter the armspans of all the males in your class in one list and the armspans of all the females in your class in a different list. Produce and make sketches of separate histograms for the armspans of females and of males. Use the same viewing window for each plot.

d. How is the histogram of the armspans for females like the one for males? How is it different?

e. Suppose armspan is one of the characteristics you will use in sizing rain ponchos for sale to students in your school. Describe how you would use the histograms to help make decisions about the range of measurements that correspond to the sizes small, medium, large, and extra large.

3. Worker compensation is a major factor in the economic competitiveness of companies in world markets. The following table gives hourly compensation costs (in U.S. dollars) for production workers from selected countries. Hourly compensation costs include not just hourly salary, but also vacation, holidays, benefits such as insurance, and other costs to the employer.

Hourly Compensation Costs for Production Workers (in U.S. dollars)
(Selected Countries, 1999)

Country	1999	Country	1999
Australia	15.89	Mexico	2.12
Austria	21.83	Netherlands	20.94
Belgium	22.82	New Zealand	9.14
Canada	15.60	Norway	23.91
Denmark	22.96	Singapore	7.18
Finland	21.10	South Korea	6.71
France	17.98	Spain	12.11
Germany	26.18	Sweden	21.58
Hong Kong	5.44	Switzerland	23.56
Ireland	13.57	Taiwan	5.62
Italy	16.60	United Kingdom	16.56
Japan	20.89	United States	19.20

Source: *The New York Times Almanac, 2001*. New York: Penguin Putnam, Inc., 2000.

a. How do you think the U.S. Bureau of Labor Statistics might have determined the estimate of hourly compensation costs for production workers in the United States?

b. What was the average cost for a 40-hour week to the employer of a production worker in Switzerland in 1999? What was the yearly cost if the worker was paid for 52 weeks, 40 hours each week? Did the worker actually receive this amount? Explain.

c. Make a histogram of the hourly compensation costs shown in the previous chart. Make each bar represent an interval of $2. Write a summary of the information conveyed by the histogram.

d. How did hourly compensation in the United States compare to that of the other countries?

e. Experiment with changing the widths of the bars and redrawing the histogram. Use 5, 10, and 1 for the widths. Compare your results. Which width reveals the most information about the distribution of compensation costs?

4. In the early part of this century, the United States was viewed as a "melting pot" for people emmigrating from countries with differing cultures. More recently, cultural diversity has come to be viewed by many as a strength of this country. One measure of this diversity is the variety of languages spoken in this country.

The following table gives the 25 most commonly spoken languages in the United States after English, the number of Americans five years or older who speak each language, the percent change from 1980 to 1990, and the state with the highest percent of speakers.

Languages in the United States

Language	Number of Speakers in 1990	Percent Increase since 1980	State with Highest % of Speakers	Language	Number of Speakers in 1990	Percent Increase since 1980	State with Highest % of Speakers
Spanish	12,339,172	50.1	New Mexico	Hindi	331,484	155.1	New Jersey
French	1,702,176	8.3	Maine	Russian	241,798	38.5	New York
German	1,547,099	–3.7	North Dakota	Yiddish	213,064	–33.5	New York
Italian	1,308,648	–19.9	New York	Thai/Lao	206,266	131.6	California
Chinese	1,249,213	97.7	Hawaii	Persian	201,865	84.7	California
Tagalog	843,251	86.6	Hawaii	French Creole	187,658	654.1	Florida
Polish	723,483	–12.4	Illinois	Armenian	149,694	46.3	California
Korean	626,478	127.2	Hawaii	Navajo	148,530	20.6	New Mexico
Vietnamese	507,069	149.5	California	Hungarian	147,902	–17.9	New Jersey
Portuguese	429,860	19.0	Rhode Island	Hebrew	144,292	45.5	New York
Japanese	427,657	25.0	Hawaii	Dutch	142,684	–2.6	Utah
Greek	388,260	–5.4	Massachusetts	Mon-Khmer	127,441	676.3	Rhode Island
Arabic	355,150	57.4	Michigan				

Source: *USA Today,* April 28, 1993. Copyright 1993, USA Today. Reprinted with permission.

a. What is the most common language, other than English, spoken in the United States? What does its percent increase tell you? Which state has the highest percentage of speakers of this language? What might explain this fact?

b. Which language had the largest *percent* increase in number of speakers? Which language had the largest increase in the *number* of speakers of the language?

c. What does the –19.9% for Italian mean? What might account for this and other negative percent increases?

d. How many people in the United States spoke Japanese in 1980? How many people spoke Spanish in 1980?

e. Make a histogram that displays the number of speakers of the different languages. What does the height of a bar tell you? Write three sentences explaining to some adult in your home what the plot tells you.

f. Your histogram includes only the top 25 languages. Draw a sketch of what you think a histogram would look like if it included the top 50 languages. Explain why you think your sketch is reasonable.

Organizing

These tasks will help you organize the mathematics you have learned in the investigations and connect it with other mathematics.

1. Copy this number line plot of the acceleration ratings. Transform it into a histogram by drawing bars around the Xs. You will also need to add a vertical axis with a frequency scale.

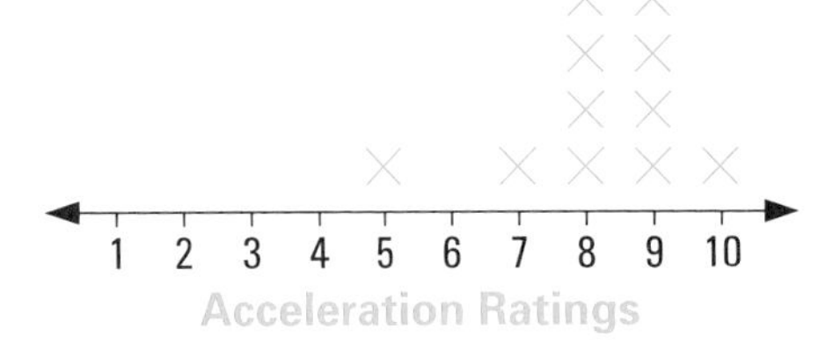

2. The two histograms below both represent the heights of the same tenth-grade soccer team at Greendale High School.

a. How do the histograms differ?

b. How many students were between 64 and 65 inches tall?

c. Compare how well the two plots convey information about the data.

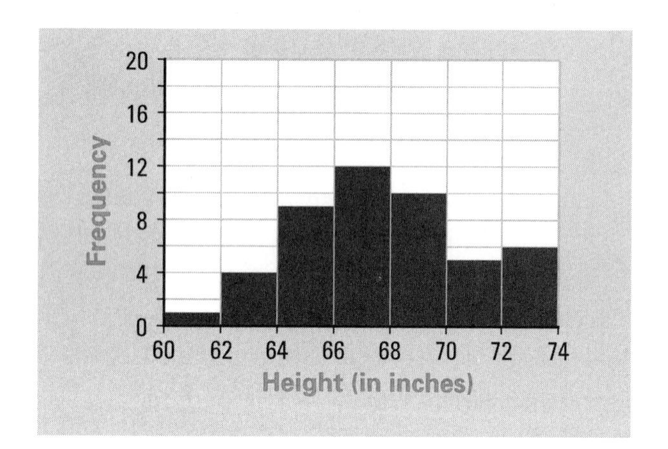

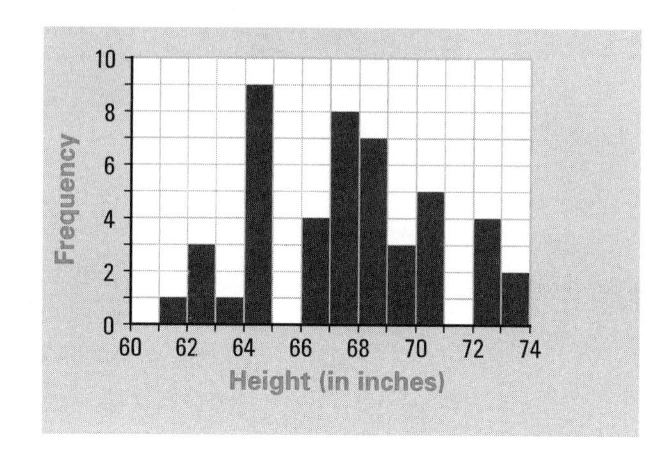

3. Make a sketch of what you think the histograms of the following distributions would look like. Classify each of the distributions as skewed to the right, skewed to the left, approximately normal, or rectangular-shaped.

a. the last digits of the phone numbers of students in your school

b. the heights of adult women

c. the weights of adult men

d. the ages of all people who died in the United States last week

4. Describe a situation different from those in this investigation which would yield data whose distribution is

a. skewed to the right;

b. skewed to the left;

c. approximately normal;

d. rectangular;

e. **bimodal** (two peaks).

5. Sketch a distribution of scores that might be described as follows.

"The distribution ranges from 10 to 100 and is skewed to the left with most of the data clustered between 60 and 90. No one received a score in the 30s. About $\frac{2}{3}$ of the scores are greater than 50."

Reflecting

These tasks will help you think about what the mathematics you have learned means to you. These tasks also will help you think about what you do and do not understand.

1. What kinds of graphical representations of data are most common in your social studies textbook? In your science textbook? What have you learned in these two investigations that might be of help to you in these other courses?

2. In the book *On the Shoulders of Giants*, David Moore, a statistician at Purdue University, describes *data* as "numbers with a context." What do you think he means by this statement?

3. Ask two adults what factors they would take into account when buying a car. Would they make use of comparison tests and ratings published in automotive or consumer magazines? Summarize what they say.

4. As you saw in Investigation 2, fast foods vary considerably in the amounts of calories, fat, cholesterol, and sodium they contain. People concerned about their health may prefer foods lower in fat, cholesterol, and sodium. Taking the number of calories into account, which of the foods listed in that investigation seem to best satisfy these conditions?

5. Think back on your work in Investigation 2.

a. What did you find most interesting?

b. What, if anything, seemed to cause you difficulty? How did you overcome the difficulty?

Extending

Tasks in this section provide opportunities for you to explore further or more deeply the mathematics you are learning.

1. With a partner, choose a topic from the list below and survey at least 30 people for their response.

- amount of money people carry with them
- time it takes to get to school each day
- number of hours per week spent on the phone
- number of hours per week spent watching television
- number of miles driven per week
- amount of money spent per week
- number of hours spent participating in athletics per week

Make a number line plot or histogram of your data. Write a paragraph describing the topic you investigated and the results. Include an assessment of how accurate you think the answers are.

2. Making sense of data is a skill that is becoming more and more important for a thorough understanding of news reports. Read the news story below.

Cleaning Up the Coast: It's 12 Years of Garbage

By Robert Braile,
Globe Correspondent

PORTSMOUTH—Sometimes, saving nature can be as easy as reaching down, picking up a soda can or cigarette butt, and putting it where it belongs. That's what 1,000 people, mostly students, will do next weekend on New Hampshire's coastal beaches in the state's 12th annual coastal cleanup, which last year amassed 13,000 pounds of trash.

People from as far away as Littleton and Claremont have already called the New Hampshire Coastal Program, the event's sponsor, seeking to participate in the Friday and Saturday cleanups. "It's a real opportunity for people to come together, give something back to their communities, and practice environmental stewardship for the entire state," said Cynthia Lay, the program's outreach coordinator.

The event is part of a global cleanup the same day involving more than 500,000 volunteers, who will descend upon 4,500 beaches, lake shores, riverbanks and other coastal areas in 55 states and territories in this country, and in more than 90 other countries, to pick up trash. Last year in America alone, 160,000 volunteers picked up more than 3 million pounds of trash at more than 3,000 sites on nearly 7,000 miles of coastline.

In New Hampshire, there will also be a special Friday morning and afternoon cleanup involving school children and an underwater cleanup for divers on Saturday off the docks of Prescott Park. The former event drew about 300 students from kindergarten through the 11th grade last year. The latter event drew 15 drivers last year who pulled 3,000 pounds of trash from the Piscataqua River, placing all of it in the park to allow others to see just how much debris ends up in the water, "an amazing success and a lot of fun," Lay said.

In New Hampshire, past cleanups have turned up young lobsters trapped in bottles, seals and other marine mammals wrapped in fishing lines, and other unfortunate encounters, Lay said. So why do we still litter? "Because education is a continual process, people change with the times, and those kinds of messages, like that television commercial, are only remembered for certain periods of time, by only certain sectors of the public," Lay said.

Littering on land and sea does indeed persist. In last year's New Hampshire cleanup, 38,556 cigarette butts were picked up, leading the list of most common items. Some 6,446 pieces of rope, 6,383 pieces of glass, 4,781 pieces of plastic, 3,876 food bags or wrappers, 3,611 pieces of paper, 3,036 cans, 2,918 pieces of foamed plastic, 2,085 caps or lids, 1,858 bottles, 1,313 straws, and 1,253 other plastic items, followed. They all comprise the so-called "dirty dozen" of New Hampshire's coastal trash.

"Human hands and a human face lie behind every piece of garbage that enters the marine environment," said Roger McManus, President of the Center for Marine Conservation, a national group that has sponsored the International Coastal Cleanup since 1986. That year, some 2,800 volunteers picked up 124 tons of trash on the Texas coasts.

"The responsibility for this world-wide problem does not belong solely to shipping companies, fishing fleets and governments," McManus continued. "Ultimately, some individual had to throw that trash overboard, into the street, down the toilet, or into the storm drain. People are the problem, but through the International Coastal Cleanup, people are also the solution."

"We don't see ourselves as connected to our rivers and oceans," Lay said. "That's why the cleanup is such a significant event. It's an educational experience. People learn about the impacts our activities can have on the environment, which then helps them to change their behaviors for the future."

Even students, who make up about 70 percent of the volunteers, need to hear that message, Lay said. "It's interesting, listening to the students who show up on Friday," she said. "They'll be walking around, picking up trash, and you'll hear them say things like, 'Who would do this? This is disgusting.' Then we'll all go have lunch afterwards at the cookout, and after lunch there'll be litter they've left all over the place. That's when we do some teaching, and answer their question. 'Who would do this?' we say. 'You would. That's why it's important to change.'"

Source: *The Boston Globe*, September 12, 1999.

Using information from the article and data from the New Hampshire Coastal Cleanup Program, the following data displays can be made.

Cleaning up the Beaches

Plastics continue to top the list of the beach debris collected in the annual New Hampshire coastal cleanup. Some findings:

The kinds of debris

How the 95,906 bits (13,000 pounds) of debris break down:

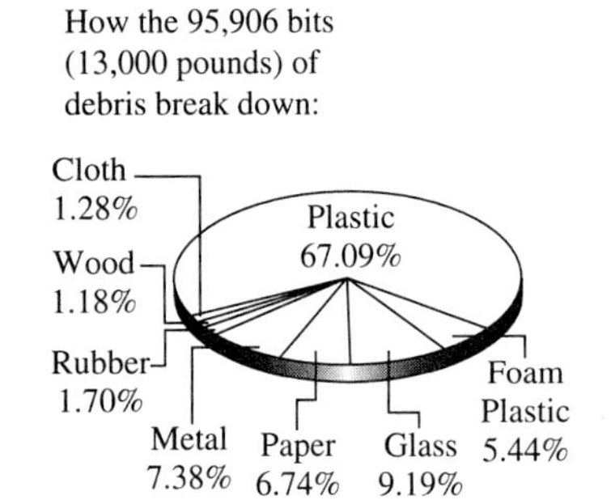

Some progress reported

Collection shows a decrease in the number of pieces of foam plastic, paper, and metal collected.

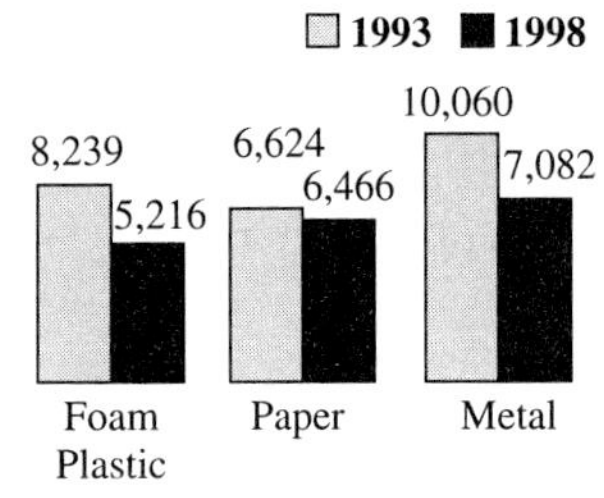

Number of individual items (top 12)

Item	Number	Item	Number
Cigarette butts	38,556	Cans	3,036
Rope pieces	6,446	Foamed plastic pieces	2,918
Glass pieces	6,383	Caps/lids	2,085
Plastic pieces	4,781	Bottles	1,858
Food bags/wrappers	3,876	Straws	1,313
Paper pieces	3,611	Other plastic items	1,253

Source: webster.state.nh.us/coastal/trash; Boston Globe.

a. Who collected the information and how?

b. The first and third displays appear to tell different stories about the types of debris. How might this be explained in terms of the original data source?

c. The article indicates that 6,383 pieces of glass were collected in the cleanup. How might you reconcile this fact with the *circle graph*?

d. Do the decreases reported in the second display necessarily mean that progress has been made? Explain your thinking.

e. Describe what the article tells you about who participates in the annual beach cleanup.

3. Some graphing calculators and computer software programs have an option that sorts data stored in a list. Explore the sorting capabilities of your calculator or software.

a. How might you use these capabilities to help find the middle value for a list of data? The most common value?

b. Share your findings with the class.

INVESTIGATION 3 Measures of Center

In Lesson 1, you collected and analyzed data about some of the physical characteristics of your class, such as height and stride length. Human services agencies often contract with research groups to conduct large survey studies of behavior patterns of various segments of the population. Data from such surveys are used to help shape policies and programs. The results of a survey of children's eating habits presented in *USA Today* in April 2001 are given on the following page.

Kids Gobbling Empty Calories

By Nanci Hellmich

Teens are eating 150 more calories a day in snacks than they did two decades ago. And kids of all ages are munching on more of the richer goodies between meals than children did in the past.

These latest findings have national nutritionists, weight-control experts, and concerned parents wondering whether snacking has run amok in the USA, contributing to the rising obesity rates of children.

Researchers at the University of North Carolina-Chapel Hill analyzed government data on the eating habits of more than 21,000 children, ages 2 to 18, from 1977 to the mid-1990s and found:

- Kids are consuming 25% of their daily calories between meals, compared with 18% in 1977. That means kids are eating about a meal's worth of calories from snacks.
- Teens have increased their munching from 1 snack a day in 1977 to almost 2.
- Teens are getting about 610 calories a day from snacks, compared with about 460 calories a day in 1977.
- The snacks that kids are eating are more energy-dense, that is, richer foods packed with more sugar, fat, and calories per ounce than snacks that kids ate 20 years ago. Kids today might be more likely to have French fries and a Coke for a snack, for instance, instead of a cookie and a glass of milk.

Source: *USA Today*, April 30, 2001.

Think About This Situation

As a class, consider the findings reported above on the snacking habits of children and teens.

a The article says that, "Teens have increased their munching from 1 snack a day in 1977 to almost 2."
- Does this mean all teens or a typical teen?
- How could this number have been determined?

b What exactly could the article mean by the statement that "Teens are getting about 610 calories a day from snacks?"

c How could information like this be collected? Which information seems almost impossible to get?

Hectic after-school schedules and unpredictable dinner times may contribute to the increase in snacking among teens. These facts likely also contribute to the popularity of fast food restaurants and their menus of hamburgers, French fries, and soft drinks. If you eat a hamburger, you consume more than just bun, beef, and toppings. Reproduced below is information from Investigation 2 (page 22) on the cholesterol in burgers and chicken from fast food restaurants.

Cholesterol (mg)

Burgers	45	70	70	85	50	95	80	100	170		
Chicken	65	55	65	65	105	60	55	70	36	48	46

A quick scan of the data in the table suggests that the distributions of amount of cholesterol are somewhat different. A *back-to-back stem-and-leaf plot* is often a good way to compare two distributions.

1. Prepare a back-to-back stem-and-leaf plot of the cholesterol in the burger and chicken entrees.

a. Copy the stems shown at the right. Put the leaves for the chicken entrees to the right of the stems and put the leaves for the burgers to the left of the stems.

b. In general, is there more cholesterol in a burger or in a chicken entree? Explain your reasoning.

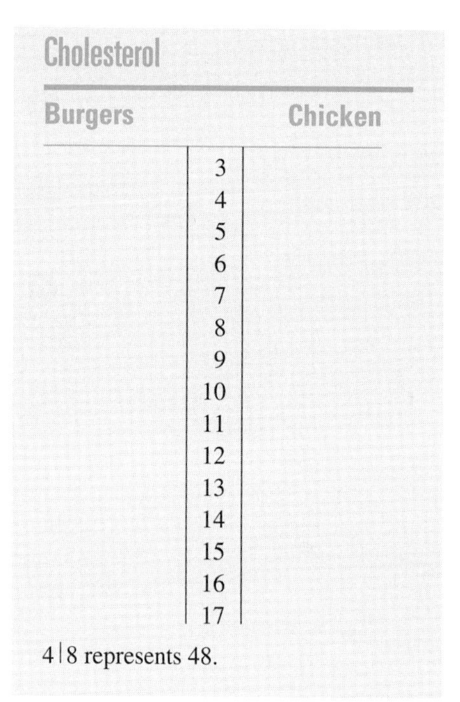

Cholesterol

Burgers		Chicken
	3	
	4	
	5	
	6	
	7	
	8	
	9	
	10	
	11	
	12	
	13	
	14	
	15	
	16	
	17	

4 | 8 represents 48.

2. When asked to find a "typical" amount of cholesterol in a burger, Meliva, Kyle, and Yü responded with different answers.

a. If Meliva, Kyle, and Yü continued their same line of thinking, what would each say about the amount of cholesterol in a "typical" chicken entree?

b. Did any one of the three give a better answer than the other two? Why or why not?

c. Whose answer could help you determine if most brands of burgers have 85 mg of cholesterol or more? Explain your reasoning.

d. Why is Kyle's answer greater than Meliva's?

Checkpoint

When displaying and summarizing data, you will need to choose between several possibilities.

a Compare a stem-and-leaf plot to a histogram. How are they alike and how are they different?

b How does a back-to-back stem-and-leaf plot help you compare two distributions?

c Describe three methods of estimating a "typical" value for a distribution.

Compare your thinking and descriptions with those of other groups.

Summaries that describe a typical or representative value for a distribution are called **measures of center**. You may already be familiar with three measures of center.

- The **median** is the midpoint of an *ordered* list of data—half the values are at or below it and half are at or above it.
- The **mode** is the value that occurs most frequently.
- The **mean**, or arithmetic average, is the sum of the values divided by the number of values.

On Your Own

Think about the significance of different measures of center as you complete these tasks.

a. By comparing means, would you conclude there is more cholesterol in the burgers or in the chicken selections? Does the same conclusion hold if you compare medians? Modes?

b. For a popular CD, what does the median price in the stores in your area tell you? The mode price? The mean price? What piece of information about the distribution would you rather know? Why?

The mean lies at the "balance point" of the number line plot, histogram, or stem-and-leaf plot. That is, if the histogram were made of bricks stacked on a light tray, the mean is where you would place one hand to hold the tray so the tray would balance.

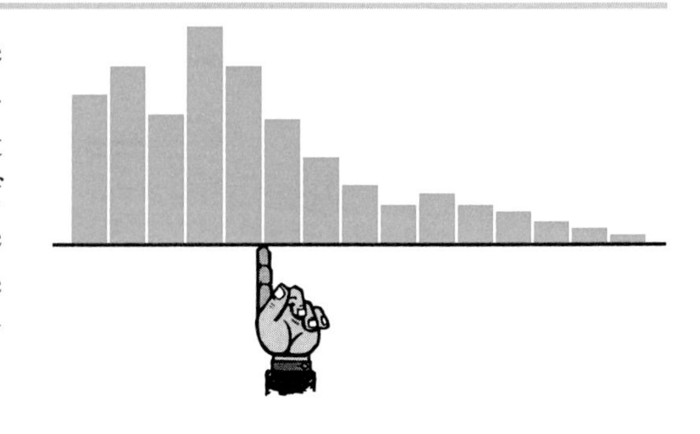

© A.M.P.A.S®

3. The histogram at the right shows the ages of the actresses whose performances won in the Best Leading Role category at the annual Academy Awards (Oscars). Estimate the mean age of the winners.

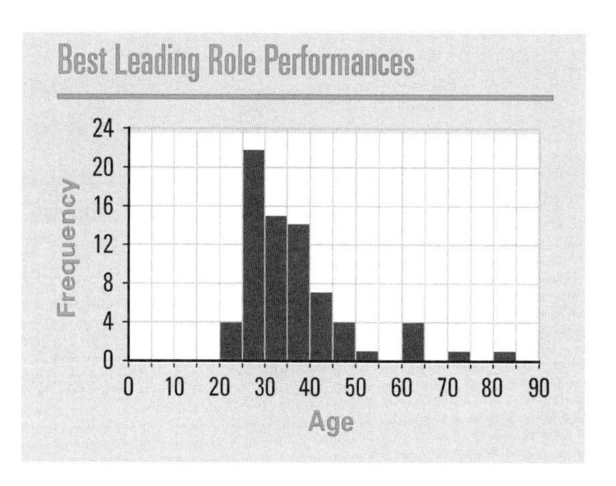

4. The histogram at the right shows a set of 40 values.

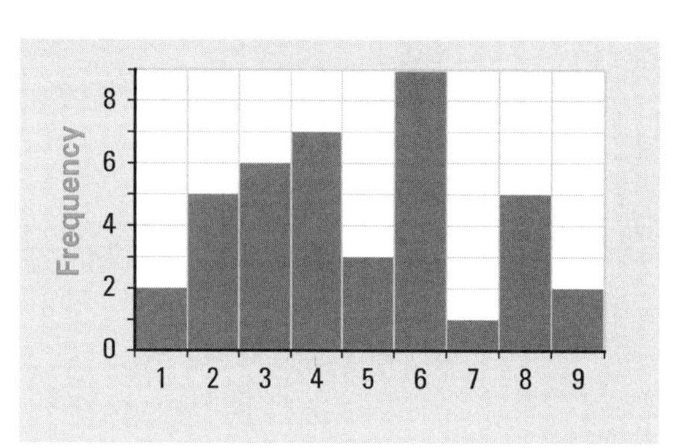

a. How many 5s are there in this set of values?

b. What is the mode of this set of values?

c. Find the median of this set of values. Locate the median on the horizontal axis of the histogram.

d. Find the area of the bars to the left of the median.

e. Find the area of the bars to the right of the median.

f. What do you conclude about the location of the median on a histogram?

g. Estimate the mean of the distribution by estimating the balance point of the histogram.

h. Find the mean using your calculator or computer. How close was your estimate in Part g?

5. Once data have been entered into your calculator or computer, you can calculate measures of center quickly.

a. Use your calculator or computer to find the mean number of milligrams of cholesterol in the chicken items. Mark the mean on the stem-and-leaf plot you prepared in Activity 1.

b. Use your calculator or computer to find the median number of milligrams of cholesterol. Mark the median on the plot.

c. With a partner, explore other ways of calculating the median number of milligrams of cholesterol on your calculator or computer using sorted lists.

d. Most technological tools have no operation for calculating the mode of a distribution. Explain how sorting data would make the process of determining a mode easier.

e. Using data lists on your calculator or computer, find the minimum and the maximum amounts of cholesterol in the chicken items.

6. Find the mean and median of the following set of values.

1, 2, 3, 4, 5, 6, 70

a. Change the outlier 70 to 7. Then find the mean and median of the new set of values. Which changed more, the mean or the median?

b. Is the mean or the median more **resistant to outliers**? That is, which tends to change less if an outlier is added to a set of values? Explain your reasoning.

c. Is the mode resistant to outliers? Why or why not?

7. Describe a situation where it would be better if the teacher uses the median test score when computing your grade. Describe a situation where the mean is better.

8. Refer to your histograms of the total calories and amounts of cholesterol and fat for the fast foods in Investigation 2 of this lesson.

a. Find the mean and median of each distribution. Divide up the work within your group. Using different colored pencils or pens, locate and label the mean and median on the horizontal axis of each histogram.

b. Write down any observations you can make about connections between the shape of the distribution and the locations of the mean and the median. Test your observations using the data on the amount of sodium in the fast foods listed and with at least two sets of data that you make up yourselves.

Checkpoint

Whether you use the mean, median, or mode depends on why you are computing a measure of center.

a What are the advantages and disadvantages of each measure of center for summarizing a set of data?

b Describe how to find or estimate the mean, median, and mode from a histogram.

Be prepared to share your group's thinking with the whole class.

On Your Own

The table below gives the percentage of households that own their own home in some countries in the western hemisphere.

a. Find the mean and the median of this set of data.

b. Are there any outliers in this data set that might affect the mean? If so, how do they affect the mean?

c. Write a sentence or two explaining what the mean tells you about home ownership.

d. Write a sentence or two explaining what the median tells you about home ownership.

Homeownership Data

Country	Percentage Who Own Home	Country	Percentage Who Own Home
Barbados	70.7	Jamaica	46.7
Belize	58.8	Mexico	61.1
Canada	62.6	Panama	75.5
Dominica	66.4	Saint Kitts-Nevis	56.7
Grenada	74.5	St. Vincent/Grenadines	72.1
Guatemala	64.7	Trinidad/Tobago	63.6
Haiti	73.2	United States	64.2

Source: *United Nations Statistical Chart on World Families*. New York: United Nations, 1993.

MORE

Modeling ○ Organizing ○ Reflecting ○ Extending

Modeling

1. Listed below are the 25 fastest-growing franchises in the U.S. based on the number of new franchise units added from 1998 to 1999.

Fastest-Growing Franchises

Company	Type of Business	Minimum Start-up Costs
McDonald's	hamburgers, chicken, salads	$477,800
Coverall North America Inc.	commercial cleaning	6,300
Taco Bell Corp.	Mexican quick-service restaurant	236,400
Subway	submarine sandwiches & salads	63,400
Jani-King	commercial cleaning	8,200
Mail Boxes Etc.	postal/business/communications services	113,000
The Quizno's Corp.	submarine sandwiches, soups, salads	170,000
ByeByeNow.com Travel Inc.	travel agency	1,500
Jiffy Lube Int'l. Inc.	fast oil change	174,000
Curves for Women	women's fitness & weight-loss centers	20,600
Worldsites	Internet services	34,000
Century 21 Real Estate Corp.	real estate	35,900
RE/MAX Int'l. Inc.	real estate	20,000
Auntie Anne's Inc.	hand-rolled soft pretzels	156,000
Sonic Drive In Restaurants	drive-in restaurant	621,300
GNC Franchising Inc.	vitamin & nutrition stores	130,200
Quik Internet	Internet services	63,700
Popeyes Chicken & Biscuits	Cajun-style fried chicken & biscuits	700,000
Great Clips Inc.	family hair salons	87,200
Church's Chicken	southern fried chicken & biscuits	194,800
Liberty Tax Service	income-tax preparation services	25,600
Adventures in Advertising Inc.	promotional products/advertising specialties	13,500
Papa Murphy's	take-&-bake pizza	148,800
PostNet Postal & Business Services	postal, business & communications center	96,700
Jan-Pro Franchising Int'l. Inc.	commercial cleaning	1,000

Source: www.entrepreneur.com

a. What kinds of businesses occur most often in this list? What are some possible reasons for their popularity?

b. Make a graph that would be appropriate for displaying the distribution of minimum start-up costs.

c. Describe the shape, center, and spread of the distribution.

d. Why might a measure of center of minimum start-up costs be somewhat misleading to a person who wanted to start a franchise? (Hint: What franchises are these?)

2. Juanita, a recent high school graduate seeking a job at United Tool and Die, was told that "the average salary is over $21,000." Upon further inquiry, she obtained the following information about the number of employees at various salary levels.

Type of Job	Number Employed	Individual Salary
President/Owner	1	$200,000
Business Manager	1	60,000
Supervisor	2	45,000
Foreman	5	26,000
Lathe & Drill Operator	50	16,000
Secretary	2	14,000
Custodian	1	9,000

a. What percentage of employees earn over $21,000?

b. What is the total payroll?

c. Is the reported average salary correct?

d. Suppose the president's annual salary is increased by $5,000. How will this change affect the mean? The median? The mode?

e. In a different company of 54 employees, the median salary is $24,000, the mode is $18,000, and the mean is $26,000. What is the total payroll?

3. The following table gives the total and per capita energy consumption by nation for 1999. The information was obtained from the Energy Information Administration's International Energy Database.

Nation	Total Energy Consumption (Quadrillion (10^{15}) Btu)	Per Capita Consumption (million Btu)
Australia	4.74	249.87
Austria	1.39	169.93
Bulgaria	0.84	102.19
Canada	12.52	409.95
Costa Rica	0.14	39.00
Cuba	0.39	35.01
Cyprus	0.10	128.21
Denmark	0.89	167.29
Ecuador	0.36	29.01
France	10.26	173.60
Germany	13.98	170.30
Greece	1.28	120.41
Hungary	1.07	106.26
Japan	21.71	171.61
Netherlands	3.85	243.52
Nigeria	0.90	8.26
Norway	1.89	424.72
Poland	3.84	99.35
Portugal	1.02	102.20
Saudi Arabia	4.34	207.66
Switzerland	1.23	172.51
United Kingdom	9.92	167.85
United States	97.05	355.90

a. Study the numbers in the table and write down at least two observations about the position of the U.S.

b. Which country consumed the least amount of energy during 1999? What might explain this?

c. What was the per capita consumption in the Netherlands?

d. Using the data provided, estimate the populations of the United States and Cuba in 1999.

e. Use the histogram shown here to help describe the distribution of the energy consumption.

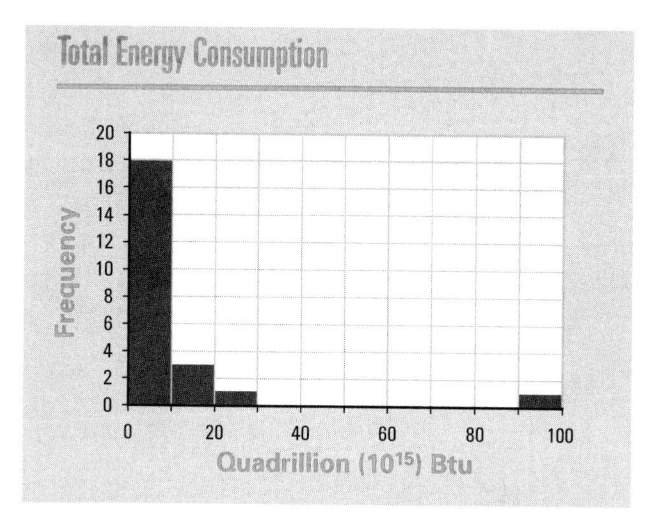

f. Does it make sense to find an average energy consumption?

g. Locate the position of the United States on the histogram. How does it compare to the other countries? What might explain the very high total energy consumption in the United States?

4. A histogram of the per capita consumption from Modeling Task 3 is shown below.

a. What observations can you make from it?

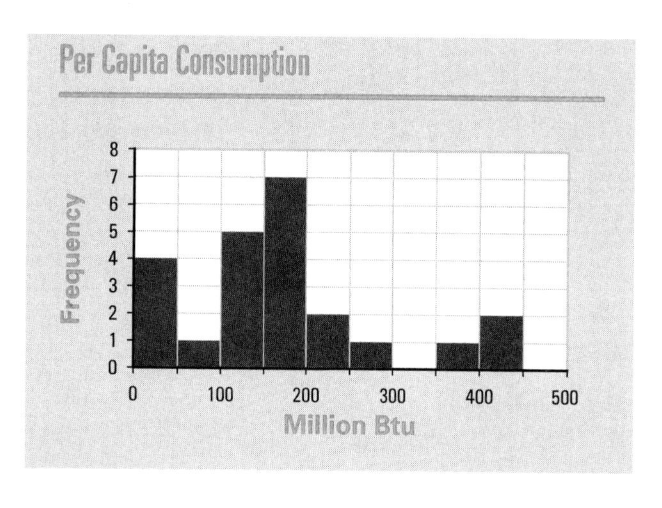

b. How does the histogram of the per capita consumption compare to the total energy consumption histogram in Part e of Modeling Task 3? Explain any difference.

c. Find the median per capita consumption for 1999. Write a sentence describing what the median tells you.

d. List the five nations that had the greatest total energy consumption. List the five nations that had the greatest per capita consumption. Compare the two lists. What might explain the differences?

5. For each of the following four distributions:

- Estimate the mean and median. Which is larger?
- Write a sentence describing what the median tells about the data.
- Write a sentence describing what the mean tells about the data.

a. The distribution represents the percentage of the population that is between the ages of 5 and 14 for a selected set of countries.

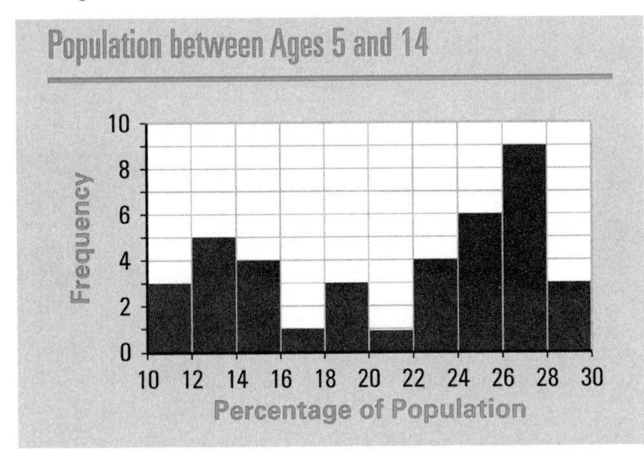

b. The distribution represents the 1998 reported sales, in millions of dollars, from clothing and accessories stores in each state and the District of Columbia in one year.

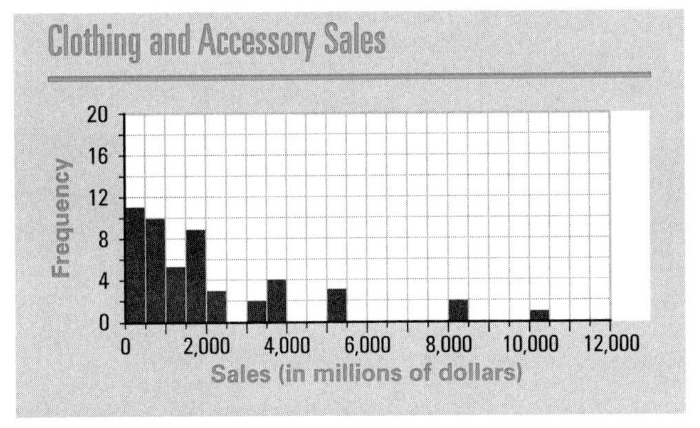

Source: U.S. Bureau of the Census, *Statistical Abstract of the United States: 2000* (120th edition). Washington, DC, 2000.

c. The distribution is the amount paid, in thousands of dollars per year, by various companies for local television advertising.

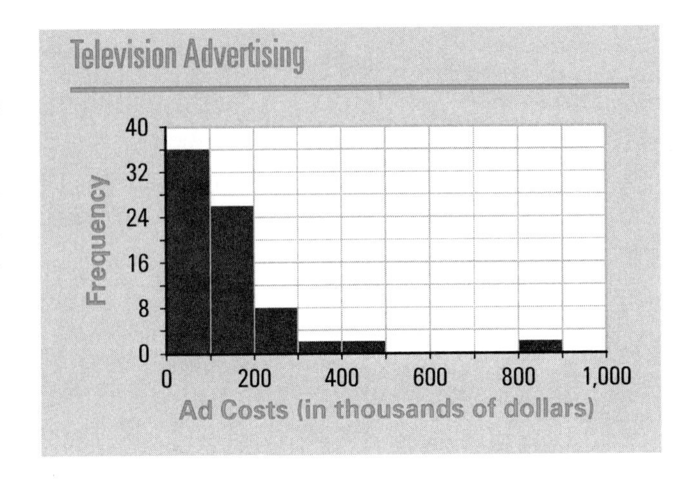

d. The distribution represents the vertical jump, in inches, of basketball players in an NBA draft.

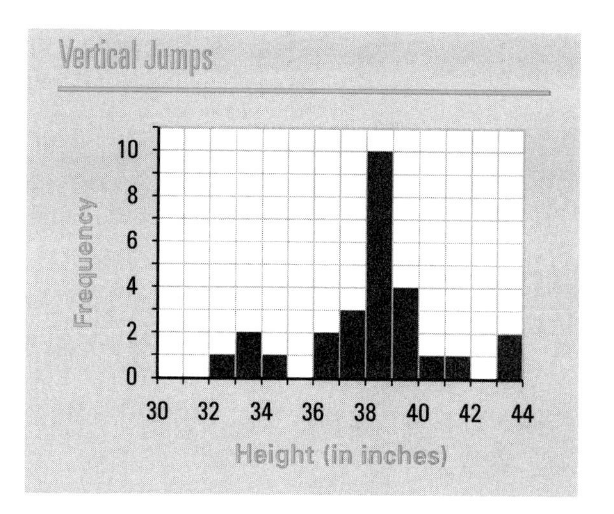

Organizing

1. Use the fifteen numbers below for Parts a through e.

13, 23, 20, 14, 27, 21, 29, 31, 12, 10, 11, 21, 5, 19, 36

Create a new set of numbers from the one above by adding or deleting just one number so that the new set will have

a. a mean larger than the median;

b. a median of 20.5;

c. no unique mode;

d. a median of 20;

e. a mean of 25.

2. For each situation below, find five whole numbers between 1 and 10 inclusive that illustrate the situation. You may repeat numbers in each set.

a. The mean and median of the set are the same.

b. The mean is less than the median of the set.

c. The set has the largest possible mean. What is the median of this set? What is the mode of this set?

d. A unique mode exists, and the mean, median, and mode are all the same, but the shape of the distribution is different from that in Part c.

e. Exchange papers with a classmate. If you have different sets, do both of your sets illustrate the same situation?

3. Matt received an 80 and an 85 on his first two English tests this quarter.

a. If a grade of B requires a mean of at least 85, what must he get on his next test to have a grade of B?

b. Suppose, on the other hand, that a B requires a median of at least 85. What would Matt need on his next test to have a grade of B?

4. Think about the connections between the shape of a distribution and the locations of measures of center.

a. Describe the characteristics of a histogram whose mean and median are nearly the same.

b. Describe the characteristics of a histogram whose mean is greater than its median.

c. Describe the characteristics of a histogram whose mean, median, and mode are nearly the same.

d. Which of your descriptions fit the following graph?

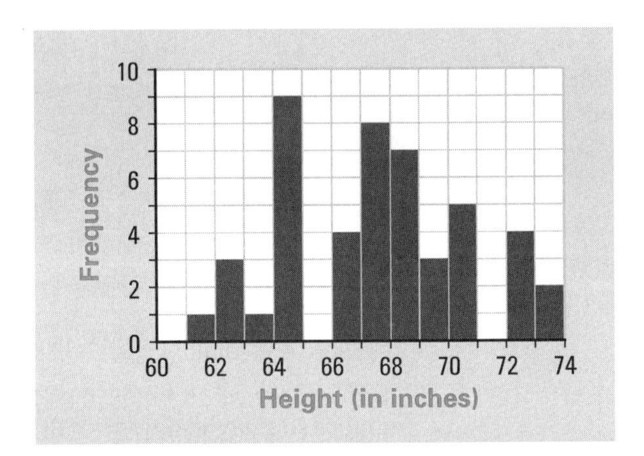

Reflecting

1. The National Longitudinal Survey of Youth asked Americans in their thirties about jobs they held between their 18th and 30th birthdays. The mean number of jobs was 7.5 and the median was 7.0. The mean number of years they worked between these two birthdays was 8.6 and the median was 9.4.

a. Why is the mean number of jobs higher than the median number of jobs?

b. Why is the mean number of years worked less than the median number of years worked?

2. You have seen in this investigation that each measure of center has certain advantages and disadvantages.

a. Which measure of center will always be the same as one of the values in the data? Why is this not true of the others?

b. Which measure of center will be most affected by extreme values in the data? Explain your reasoning.

3. Describe how you could find the median height of students in your mathematics class by making at most two measurements.

4. The median is typically reported when giving a measure of center for house prices in a region and for family incomes. Why do you think this is the case?

Extending

1. The planning committee for a class party decided to show a "classic" movie. They asked sixteen students to rank the following movies from 1 to 5 (1 was their least favorite movie and 5 was their favorite). The rankings for the males and the females were recorded separately.

Movie Rankings

Movie	Girls' Rankings					Boys' Rankings				
	1	2	3	4	5	1	2	3	4	5
Little Mermaid	//	//		///	/	//	/		/	////
Forrest Gump	///		///	/	/	///	///	/	/	
Raiders of the Lost Ark	/	/	///	///		/	///	///	/	
Titanic	/	///			////	/	/	//	//	//
Star Wars	/	//	//	/	//	/		/	////	//

a. Using the total number of points, which movie would be the overall favorite?

b. Using the mean for each movie, find the group's favorite using

- just the girls;
- just the boys;
- all the students.

c. Is there any difference in the outcome using the means instead of total points?

d. Do you think your observation in Part c will always be the case? Explain your reasoning.

2. Suppose that a survey of families counted how many children were in each family. The results are recorded in this table.

Family Size

Number of Children	Number of Families
0	15
1	22
2	36
3	21
4	12
5	6
7	1
10	1

a. Make a histogram of the distribution. Estimate the average number of children per family from the histogram.

b. Calculate the average number of children per family. You can do this on some calculators and spreadsheet software by entering the number of children in one list and the number of families in another list. The following instructions work with some calculators. Your calculator or computer may work differently.

- Enter the data in lists L1 and L2.
- While still in the lists, position the cursor on top of L3 and type L1 × L2 and then press ENTER. What appears in list L3?
- Using list L3, find the average number of children per family.

c. Find the average number of children for the families of students in your class. How is the distribution from your class sure to be different from the one above?

3. Get a set of weights and a meter or yard stick. Experiment by placing several weights at various positions on the meter stick and finding the balance point. Explain how this experiment illustrates the statement: "The mean of a set of data is the balance point of the distribution."

4. Many people who have dropped out of the traditional school setting can earn an equivalent to a high school diploma. A GED (General Educational Development Credential) is given to a person who passes a test for a course to complete high school credits. There were 830,738 people in the United States and its territories who received GEDs in 1999. The break-down by age of those receiving GEDs was as follows:

Earning a GED

Age	19 yrs and under	20–29 yrs	30–39 yrs	40–49 yrs	50 yrs and over
% completing GED	45%	35%	12%	5%	3%

Source: GED Testing Service, American Council on Education, 2001.

a. Estimate the mean age of someone who receives a GED and explain how you arrived at your estimate.

b. Estimate the median age of someone who receives a GED and explain how you arrived at your estimate.

Lesson 3 Variability

Whenever two people observe an event, they are likely to see different things. If two people measure something to the nearest millimeter, they may well get two different measurements. If two people conduct the same experiment, they will get slightly different results. In fact, there is **variability** in nearly everything; no two leaves or snowflakes are exactly alike. Because variability is so common, it is important that you begin to understand what causes variability and how it can be measured and interpreted.

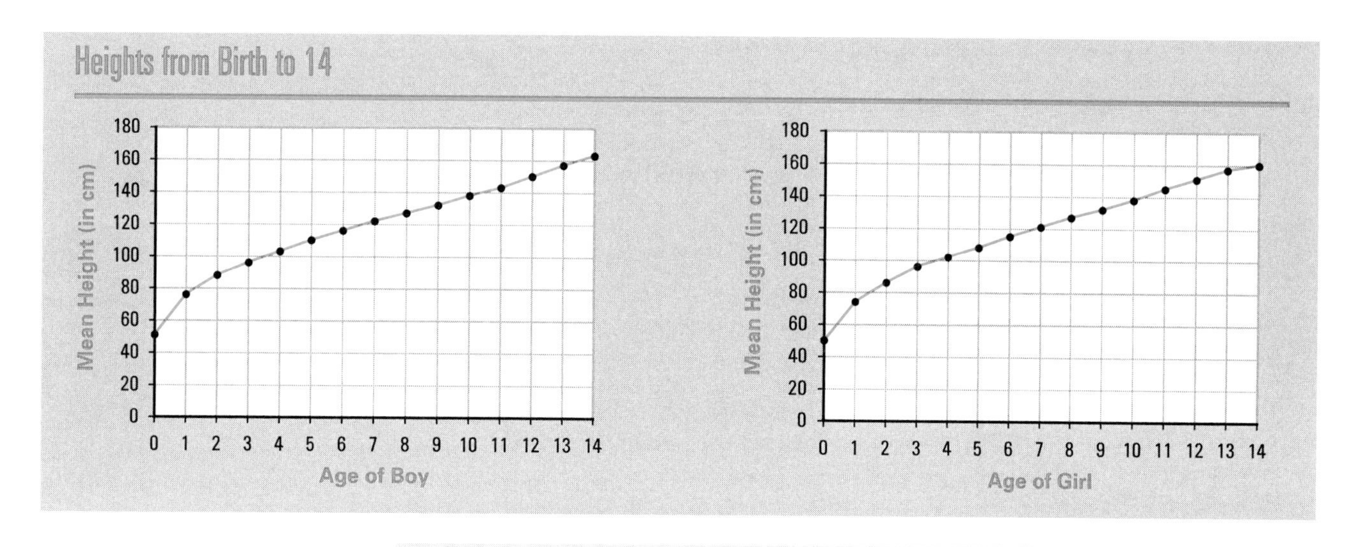

Think About This Situation

The data in the growth charts above come from a physician's handbook. Use the plots to answer the following questions.

a Is it reasonable to call a 14-year-old boy "taller than average" if his height is 165 cm? Is it reasonable to call a 14-year-old boy "tall" if his height is 165 cm? What additional information about 14-year-old boys would you need to know to be able to say that he is "tall"?

b At what height would you be willing to call a 14-year-old girl "tall"? Do you have enough information to make this judgment?

c During which year do children grow most rapidly?

INVESTIGATION 1 Measuring Variability: The Five-Number Summary

If you are in the 40th **percentile** of height for your age, that means that 40% of people your age are your height or shorter than you are and 60% are taller. Shown below are physical growth charts for boys and girls, 2 to 18 years in age. The charts were developed by the National Center for Health Statistics.

The curved lines for the height (top) and weight (bottom) tell a physician what percentile a boy or girl is in. The percentiles are the small numbers 5, 10, 25, 50, 75, 90, and 95 towards the right ends of the curved lines. For example, suppose John is a 17-year-old boy who weighs 60 kg or 132 pounds. John is in the 25th percentile of weight for his age. Twenty-five percent of 17-year-old boys weigh the same or less than John and 75% weigh more than John. If John's height is 180 cm or almost 5'11", he is in the 75th percentile of height for his age.

1. Based on the information given about John, how would you describe John's general appearance?

Physical Growth Percentiles, Boys 2 to 20 Years

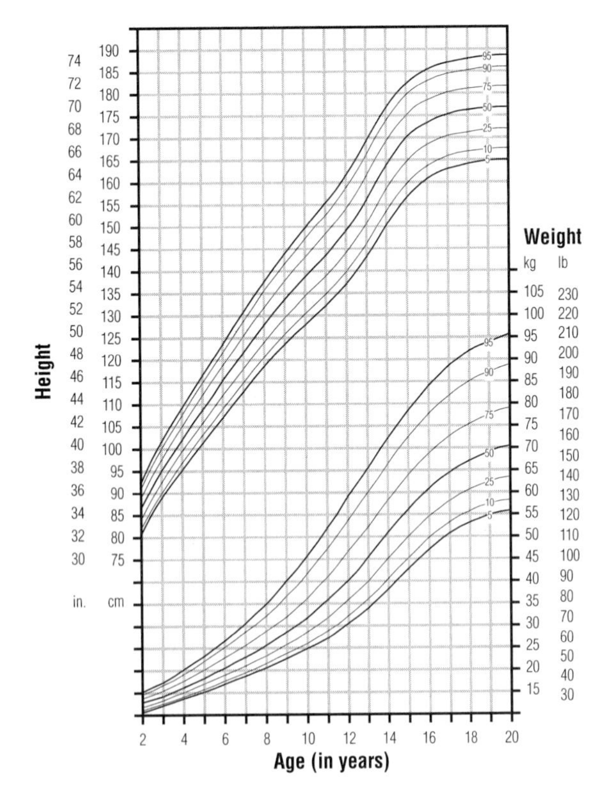

Source: www.cdc.gov/growthcharts/

Physical Growth Percentiles, Girls 2 to 20 Years

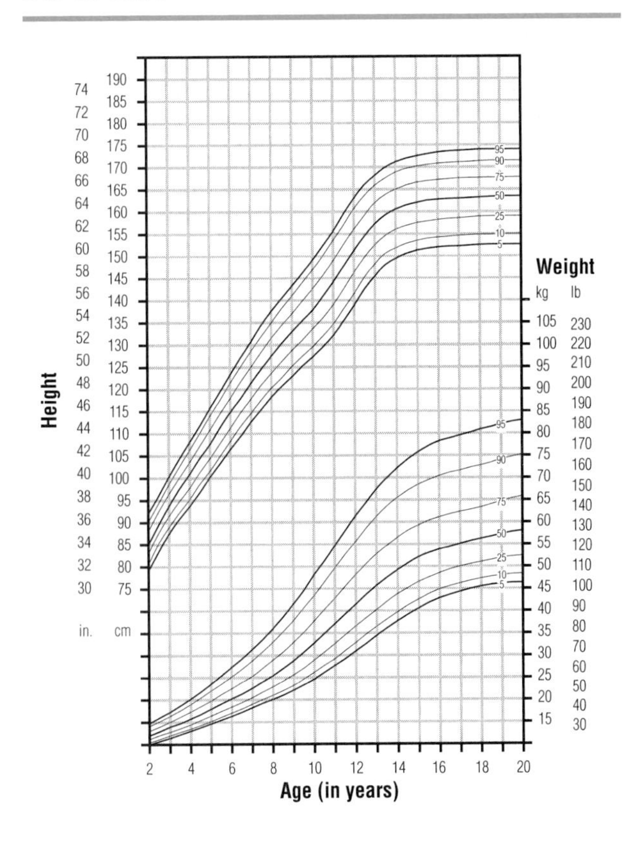

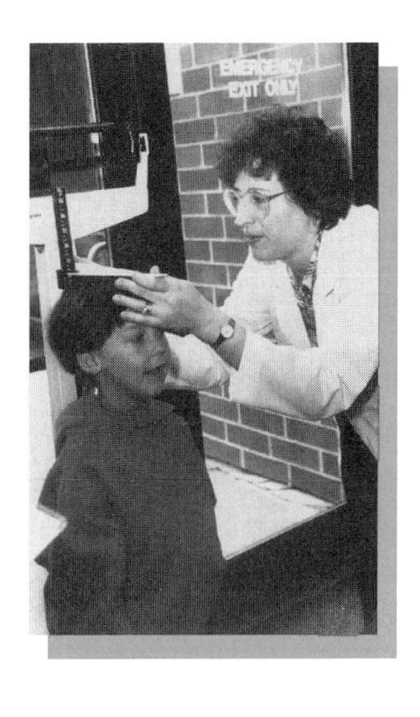

2. With your group, spend some time learning to read the CDC growth charts. They contain an amazing amount of information!

   a. What is the approximate percentile for a 9-year-old girl who is 128 cm tall?

   b. About how tall does a 12-year-old girl have to be so that she is as tall or taller than 75% of the girls her age?

   c. How tall would a 14-year-old boy have to be so that you would consider him "tall" for his age? How did you make this decision?

   d. How tall would a 14-year-old girl have to be so that you would consider her "tall" for her age? How did you make this decision?

   e. What is the 25th percentile of height for 4-year-old boys? The 50th percentile? The 75th percentile?

   f. How can you tell from the height and weight chart when children are growing the fastest? When is the increase in weight the greatest for girls? For boys?

3. Some percentiles have special names.

   a. What is another name for the 50th percentile?

   b. The 25th percentile is sometimes called the **lower quartile**. Estimate the lower quartile of height for 6-year-old girls.

   c. The 75th percentile is sometimes called the **upper quartile**. Estimate the upper quartile of height for 6-year-old girls.

The quartiles together with the median give some indication of the center and spread of a set of data. A more complete picture of the distribution of a set of data is given by the **five-number summary**: the **minimum value**, the **lower quartile** (Q_1), the **median** (Q_2), the **upper quartile** (Q_3), and the **maximum value**.

4. From the charts, estimate the five-number summary for 13-year-old girls' heights and for 13-year-old boys' heights. Some estimates will be more difficult than others.

The distance between the first and third quartiles is called the **interquartile range (IQR)**. The IQR is a measure of how spread out or variable the data are. The distance between the minimum value and the maximum value is called the **range**. The range is another, typically less useful, measure of how variable the data are.

5. Refer back to your estimates in Activity 4 and the CDC growth charts.

   a. What is the interquartile range of the heights of 13-year-old girls? Of 13-year-old boys?

   b. What happens to the interquartile range of heights as children get older?

   c. In general, do boys' heights or girls' heights have the larger interquartile range or are they about the same?

   d. What happens to the interquartile range of weights as children get older?

6. Can you estimate the range of the heights of 18-year-old boys? Why is the interquartile range more informative than the range?

For the children's heights, you were able to estimate quartiles from the chart. Next you will learn how to compute quartiles from sets of data.

When you explored the AAA ratings for cars in Lesson 2, Investigation 1, you saw variability in the ratings of each car. Below are the ratings for the Volvo V40 Wagon and the Chevy Camaro and their corresponding histograms.

Volvo V40 Wagon Ratings	Chevy Camaro Ratings
8, 8, 7, 9, 7, 8, 8, 6, 8, 7, 8, 8, 8, 7, 8, 8, 10, 9, 9, 9	10, 7, 7, 9, 5, 8, 7, 4, 7, 5, 8, 8, 8, 4, 5, 6, 5, 8, 7, 7

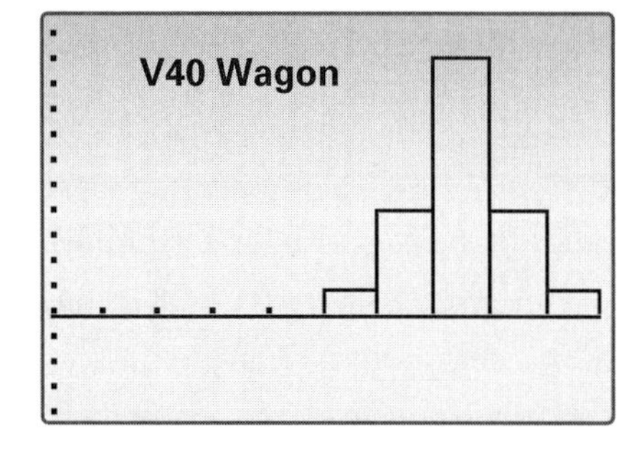

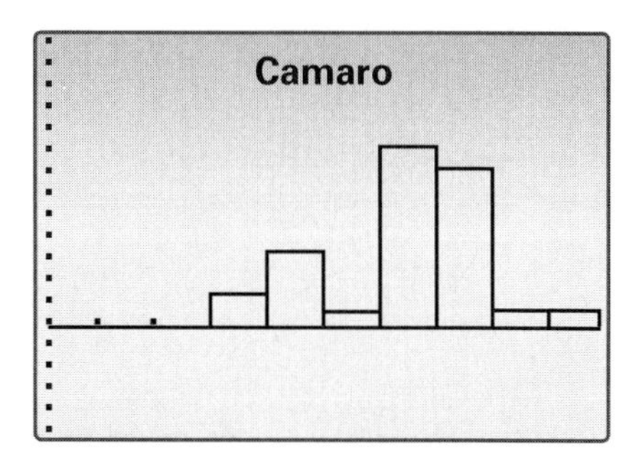

7. Which of the cars has greater variability in its ratings? Explain your reasoning.

8. Put the ratings for the Volvo V40 Wagon in an ordered list and find the median. Mark the position of the median on your ordered list.

   a. How many ratings are on each side of the median?

   b. Once the data are ordered, you can divide them into quarters to find the quartiles. Find the midpoint of each half of the ratings for the V40 Wagon. Mark the positions of the quartiles on your ordered list of the ratings.

   c. What fraction or percentage of the ratings is less than or equal to the lower quartile?

   d. What fraction or percentage of the ratings is greater than or equal to the upper quartile?

9. You can find the five-number summary quickly using technology.

   a. Find the five-number summary for the Camaro's ratings using your calculator or computer software.

   b. What is the 25th percentile of the ratings for the Camaro? The 75th percentile?

Checkpoint

A percentile gives the location of a value in a set of data while the range and IQR are measures of how spread out the data are.

a Will the range be changed if an outlier is added to a data set? Will the interquartile range be changed?

b Why does the interquartile range tend to be a more useful measure of a data set's variability than the range?

c If you get 75 points out of 100 on your next math test, can you tell what your percentile is? Explain.

d Give an example of when you would want to be in the 10th percentile rather than in the 90th.

e Give an example of when you would want to be in the 90th percentile rather than in the 10th.

Be prepared to share your group's thinking and examples with the rest of the class.

On Your Own

The ratings for the Toyota Prius are reproduced below.

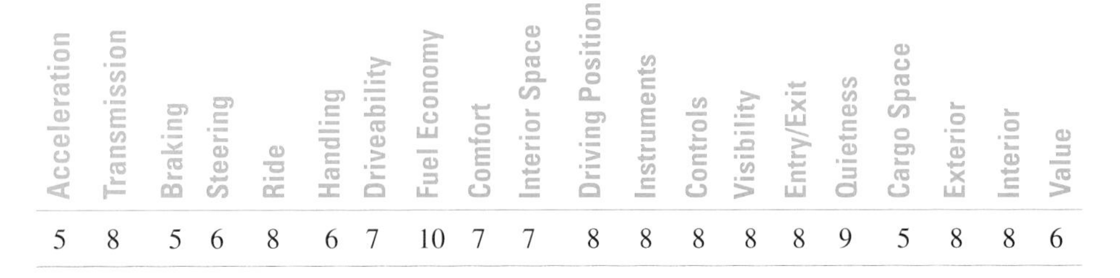

Acceleration	Transmission	Braking	Steering	Ride	Handling	Driveability	Fuel Economy	Comfort	Interior Space	Driving Position	Instruments	Controls	Visibility	Entry/Exit	Quietness	Cargo Space	Exterior	Interior	Value
5	8	5	6	8	6	7	10	7	7	8	8	8	8	8	9	5	8	8	6

a. Find the five-number summary for the Prius. What is the interquartile range?

b. Refer to your results in Activity 8 and Activity 9 (page 50). Of the ratings for the Prius, the Camaro, and the V40 Wagon, which has the greatest interquartile range? What would this tell you as a possible buyer?

INVESTIGATION 2 Picturing Variability

Your heart continually pumps blood through your body. This pumping action can be felt on the side of your neck or your wrist where an artery is close to the skin. The small swelling of the artery as the heart pushes the blood is called your *pulse*.

1. Take your pulse for 60 seconds. Record the pulse rate (number of beats per minute) for each member of your group.
   - **a.** Record your data and data from other groups on the Class Data Table you started in Lesson 1.
   - **b.** Find the five-number summary for the pulse rates for your class.

The five-number summary can be displayed in a **box plot**. To make a box plot of your pulse rates, first make a number line. Below this line draw a box from the lower quartile to the upper quartile; then draw line segments connecting the box to each extreme (the maximum and minimum values). Draw a vertical line in the box to indicate the median. The segments at either end are often called **whiskers**, and the plot is sometimes called a *box-and-whiskers plot*. Here are the results for one class.

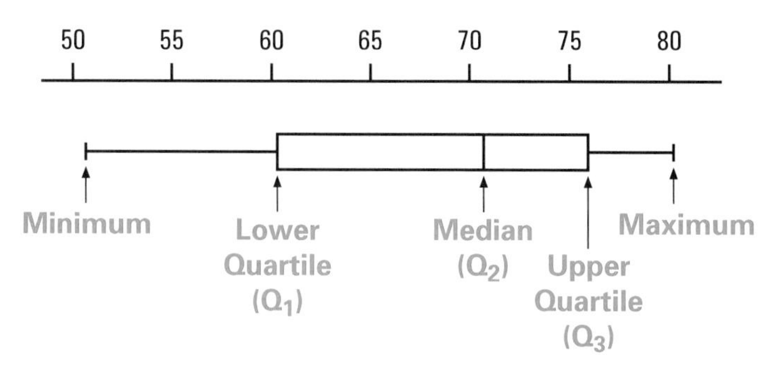

2. Study the sample box plot above.
   - **a.** Is the distribution represented by the box plot skewed to the left or to the right, or is it symmetric?
   - **b.** Draw a box plot of the pulse rates for your class.
   - **c.** Is the distribution of pulse rates for your class skewed to the left or to the right, or is it symmetric?
   - **d.** What is the length of the box for your class data? What is the mathematical term for this length?

3. Box plots are most useful when comparing two or more distributions.
   - **a.** How do you think box plots of class pulse rates before and after exercising would compare?
   - **b.** Have some members of your group do some sort of mild exercise for a short time. Take their pulse readings immediately afterwards. Record the data on the Class Data Table.

c. Combine your new data with data from other groups.

d. Construct a box plot of the new pulse rates. Place it below your first box plot.

e. How are the box plots different? Write a summary of your conclusions about pulse rates of the class before and after exercise.

f. What other plot could be used to compare the two distributions? Make this plot. Can you see anything interesting that you could not see from the box plots?

You can produce box plots with technology by following a procedure similar to that of making histograms. After entering the data and specifying the viewing window, select Box Plot as the type of graph desired.

4. Refer to the ratings for the V40 Wagon and Camaro given in Investigation 1 on page 50.

a. Using your calculator or computer software, make a box plot of the ratings.

b. Use the trace feature to find the five-number summary for the V40 Wagon. Compare the results with your computations in Activity 8 for Investigation 1 of this lesson.

c. Draw the box plot below a histogram of the V40 Wagon's ratings. Use the same scale for both.

d. How does the box plot compare to the histogram?

e. Produce a box plot of the ratings for the Camaro. Draw the box plot below a histogram of the Camaro's ratings. Use the same scale for both.

f. Compare the box plots for the V40 Wagon and the Camaro. What do the different lengths of the boxes tell you about the variability in their AAA ratings? Is either distribution skewed? Symmetric?

Reproduced below are the ratings for the Concorde, Mustang, and Miata.

Concorde Ratings	Mustang Ratings	Miata Ratings
8, 9, 6, 7, 9, 8, 8, 6, 9, 9, 8, 7, 8, 7, 9, 9, 7, 7, 8, 9	9, 8, 6, 9, 7, 9, 8, 4, 8, 5, 9, 9, 9, 6, 4, 6, 5, 8, 8, 8	8, 9, 9, 9, 6, 9, 8, 6, 8, 6, 8, 8, 8, 6, 6, 6, 3, 8, 8, 9

5. Produce box plots for the Concorde, Mustang, and Miata ratings. Draw each box plot below a corresponding histogram. Divide up the work among your group. Including the V40 Wagon and the Camaro, your group now has five different box plots to compare.

a. Use the box plots to determine which of the five cars has the largest interquartile range.

b. Which cars have almost the same interquartile range? Does this mean their distributions are the same? Explain your thinking.

c. Why do the Concorde and Mustang have no whisker at the upper end?

d. Why is the lower whisker for the Miata so long?

e. Based on the box plots, which of the five cars seems to have the best ratings? How did the plots help you make your decision?

Checkpoint

In this investigation, you have learned how to display the five-number summary on a box plot.

a How does a box plot convey how close together data are in a distribution?

b What does a box plot tell you that a histogram does not?

c What does a histogram tell you that a box plot does not?

Be prepared to share your group's thinking about the usefulness of box plots.

Box plots can be used to compare several distributions. Some computer software and graphing calculators allow you to display several box plots on the same screen. When making box plots, remember to include labels and a scale.

On Your Own

Refer to the data on the amount of fat in the fast foods listed in Investigation 2 on page 22. Produce a box plot of these data.

a. What is the five-number summary of these data?

b. What is the interquartile range for these data, and what information does it tell you?

c. Where is the Burger King Double Whopper with Cheese located on the plot?

d. Choose a single item (such as your favorite, if you have one) and describe its relation to the other foods in the list in terms of fat content.

MORE

Modeling • Organizing • Reflecting • Extending

Modeling

1. The table below gives the percentiles of recent SAT mathematics scores for national college-bound seniors. The highest possible score is 800 and the lowest possible score is 200. Only scores that are multiples of 50 are shown in the table, but all multiples of 10 from 200 to 800 are possible.

College-Bound Seniors

SAT Math Score	Percentile	SAT Math Score	Percentile	SAT Math Score	Percentile
750	98	550	61	350	7
700	94	500	45	300	2
650	87	450	28	250	1
600	76	400	15	200	0

Source: College Board Online. www.collegeboard.org/sat/cbsenior/cbs/cbs00/totalm00.html

a. What percentage of students get a score of 650 or lower on the mathematics part of the SAT?

b. What is the lowest score you could get on the mathematics part of the SAT and still be in the top 39% of those who take the test?

c. Estimate what score you would have to get to be in the top half of the students who take this test.

d. Estimate the 25th and 75th percentiles and the interquartile range. In a sentence or two, explain what this interquartile range means.

2. Refer to the height data in the Class Data Table prepared in Lesson 1.

a. Produce a box plot of these data.

- Make a sketch of the plot. Write the five-number summary at the appropriate places.
- Mark where your height would be on the plot.
- If your teacher's height was added to the data, how would that change the box plot?

b. Produce separate box plots for the heights of females and of males in your class. Put these on the same screen as the first plot, if possible.

c. What is the median height for the class? For females? For males?

d. Which of the three height data sets—total class, females, or males—has the largest interquartile range? What does this tell you about the distribution of heights?

3. The following table gives the price and size of chocolate bars as reported in *Consumer Reports*.

a. The cost per ounce is missing for Ghirardelli dark chocolate and for Hershey's milk chocolate. Compute those values.

b. Organize the data in the "Cost per Oz" column by making a stem-and-leaf plot.

c. At about what percentile is Hershey's Pot of Gold Almond Chocolate Bar? Your favorite chocolate bar?

d. Are there any outliers in the cost-per-ounce data?

e. Examine the stem-and-leaf plot and make a sketch of what you think the box plot of the same data will look like. Then, make the box plot, either by hand or using technology, and check your sketch.

f. What information about chocolate bars can you learn from the stem-and-leaf plot that you cannot from the box plot? What information about chocolate bars can you learn from the box plot that you cannot from the stem-and-leaf plot?

g. Which chocolate bars would you label as "good buys"? Explain your reasoning.

h. Why is it more reasonable to plot the cost-per-ounce data than the price data?

Choosing a Chocolate Bar

Dark Chocolate	Price	Size (oz)	Cost per oz
Dove	$ 0.50	1.3	$ 0.38
Chocolate Lindt Excellence Swiss Bittersweet	1.95	3.0	0.65
Tobler Tradition Swiss Bittersweet	1.81	3.5	0.52
Ghirardelli	2.00	3.0	
Perugina	2.00	3.0	0.67
Sarotti Extra Semisweet	1.99	3.5	0.57
Hershey's Special Dark	0.88	2.6	0.34
Milk Chocolate			
Dove	0.49	1.3	0.38
Guylian No Sugar Added	2.90	3.0	0.97
Sport Alpenmilch	2.00	3.5	0.57
Ghirardelli	1.92	3.0	0.64
Master Choice	0.99	3.5	0.28
Hershey's Symphony	0.87	2.4	0.36
Lindt Swiss	1.85	3.0	0.62
Tobler Swiss	1.76	3.5	0.50
Fannie May	1.00	2.0	0.50
Perugina	2.00	3.0	0.67
Cadbury's Dairy Milk	1.22	5.0	0.24
Nestlé	0.71	2.5	0.28
Newman's Own Organics	1.97	3.0	0.66
Sarotti	1.99	3.5	0.57
Hershey's	0.86	2.6	
Milk Chocolate with Extras			
Ghirardelli with Almonds	2.15	3.0	0.72
Toblerone with Honey and Almond Nougat	1.85	3.5	0.53
Hershey's Symphony with Almonds & Toffee Chips	0.87	2.4	0.36
Lindt Swiss with Chopped Hazelnuts	1.85	3.0	0.62
Cadbury's Roast Almond	1.21	5.0	0.24
Hershey's Pot of Gold Almond	1.29	2.8	0.46
Perugina with Hazelnuts	2.00	3.0	0.67
Hershey's with Almonds	0.84	2.6	0.32
Hershey's Sweet Escapes Chocolate Toffee Crisp	0.45	1.4	0.32

Source: *Consumer Reports*, March 1999.

4. The following table gives the percentage of the population in each state and the District of Columbia who are Hispanic, Black, American Indian, Native Alaskan, Asian, or Pacific Islander. It also indicates which presidential candidate got the majority of votes cast in the 2000 presidential election in each state.

Population by Race: 1999

State	% Hispanic, Black, American Indian, Native Alaskan, Asian, or Pacific Islander	Bush/ Gore	State	% Hispanic, Black, American Indian, Native Alaskan, Asian, or Pacific Islander	Bush/ Gore
Alabama	27.9	B	Montana	9.0	B
Alaska	27.9	B	Nebraska	10.6	B
Arizona	32.4	B	Nevada	29.7	B
Arkansas	19.3	B	New Hampshire	3.8	B
California	50.1	G	New Jersey	31.6	G
Colorado	21.7	B	New Mexico	52.5	G
Connecticut	19.7	G	New York	34.9	G
DC	71.1	G	North Carolina	26.7	B
Delaware	25.3	G	North Dakota	7.3	B
Florida	32.0	B	Ohio	14.4	B
Georgia	33.7	B	Oklahoma	20.4	B
Hawaii	71.4	G	Oregon	12.5	G
Idaho	10.0	B	Pennsylvania	13.9	G
Illinois	28.8	G	Rhode Island	13.4	G
Indiana	12.0	B	South Carolina	32.1	B
Iowa	5.6	G	South Dakota	10.5	B
Kansas	13.7	B	Tennessee	18.9	B
Kentucky	8.9	B	Texas	44.7	B
Louisiana	36.4	B	Utah	11.4	B
Maine	2.3	G	Vermont	2.4	G
Maryland	35.7	G	Virginia	27.6	B
Massachusetts	15.6	G	Washington	17.2	G
Michigan	19.1	G	West Virginia	4.2	B
Minnesota	8.8	G	Wisconsin	10.5	G
Mississippi	38.3	B	Wyoming	9.6	B
Missouri	14.3	B			

Source: U.S. Bureau of the Census, *Statistical Abstract of the United States: 2000* (120th edition). Washington, DC, 2000.

a. Which state has the largest percentage of people who are Hispanic, Black, American Indian, Native Alaskan, Asian, or Pacific Islander? The largest number?

b. If you find the mean of these 51 percentages, will that necessarily give you the percentage for the United States as a whole? Give a small example to illustrate your answer.

c. Make a box plot of the percentages for states that favored Bush in the 2000 election and a parallel box plot for the states (and DC) that favored Gore.

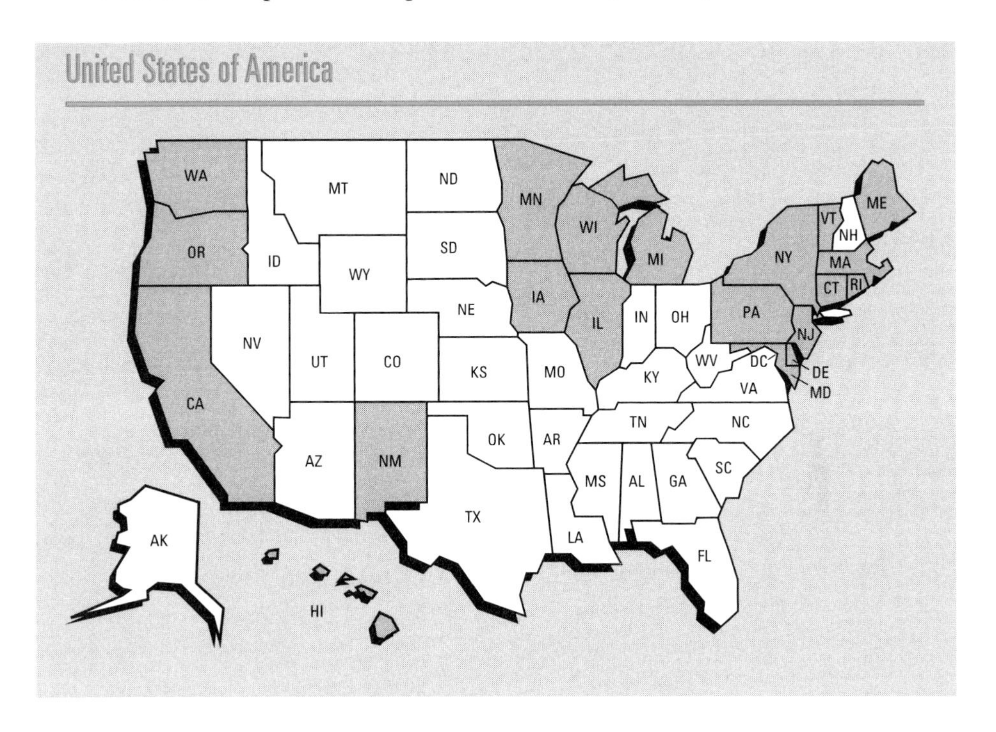

d. Describe the differences between the two box plots. Why do you think there are these differences?

e. What other plot could be used to compare the two distributions? Make this plot that shows the two distributions. Can you see anything interesting that you could not see from the box plots?

Organizing

1. Below is a box plot and a histogram for the Chrysler PT Cruiser, created from the data in Lesson 2, Investigation 1. What is missing from the box plot of the Chrysler PT Cruiser? Why? Does the histogram give you a clue that this would happen?

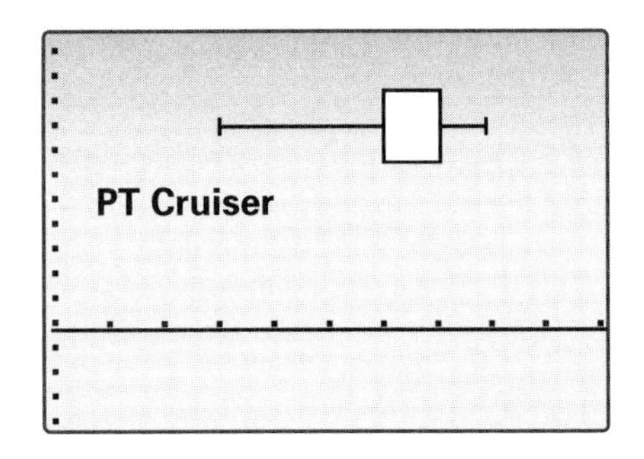

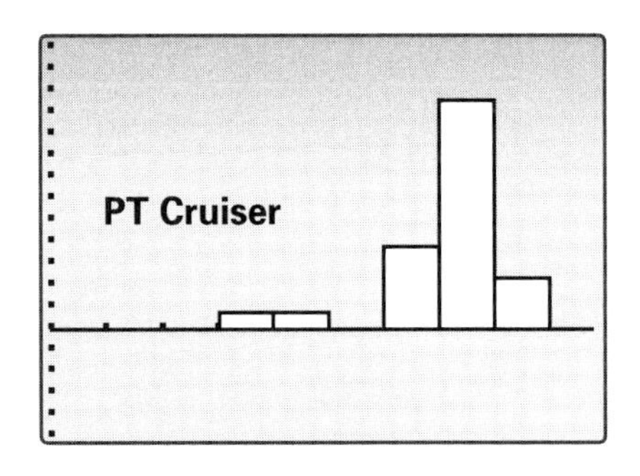

2. Consider the box plot at the right.

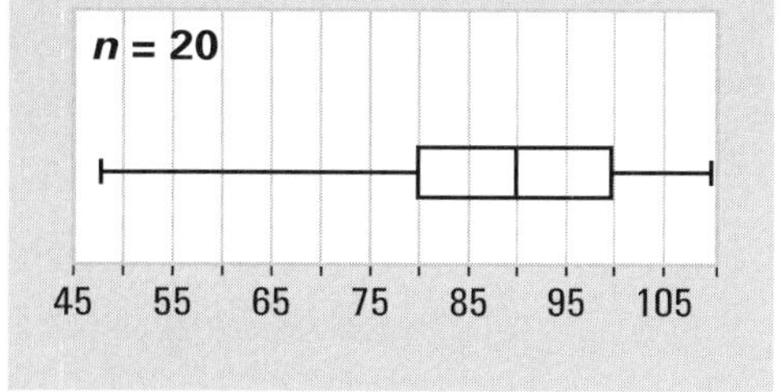

   a. What do you suppose the "$n = 20$" on the plot means?

   b. How many values are between 50 and 80? Between 80 and 100? Greater than 80?

   c. Is it possible for the box plot to represent the values below? Explain your reasoning.

   50, 60, 60, 75, 80, 80, 82, 83, 85, 90, 90, 91, 91, 94, 95, 95, 98, 100, 106, 110

   d. Create a set of values that could be represented by this box plot.

3. The box plots below represent the amounts of money (in dollars) carried by all of the people surveyed in four different places at a mall.

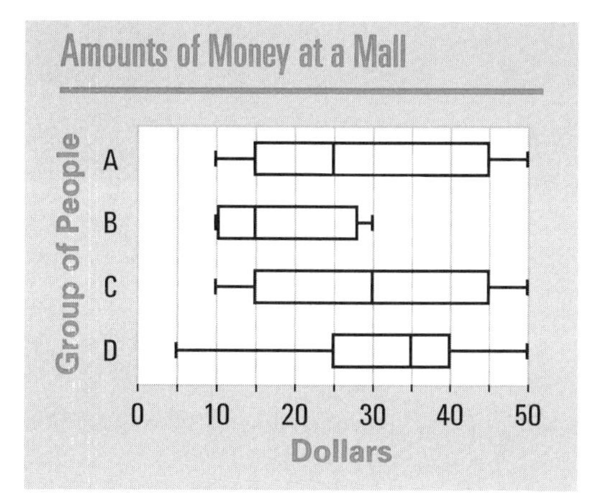

   a. Which group of people has the smallest range in the amounts of money? The largest?

   b. Which group of people has the smallest interquartile range in the amounts of money? The largest?

   c. Which group of people has the largest median amount of money?

   d. Which group of people has the most symmetric distribution of amounts of money?

   e. Which group of people do you think might be high school students standing in line for tickets at a movie theater on Saturday night? Explain your reasoning.

   f. Match each "Amount of Money" box plot with the histogram that seems most appropriate.

   i.

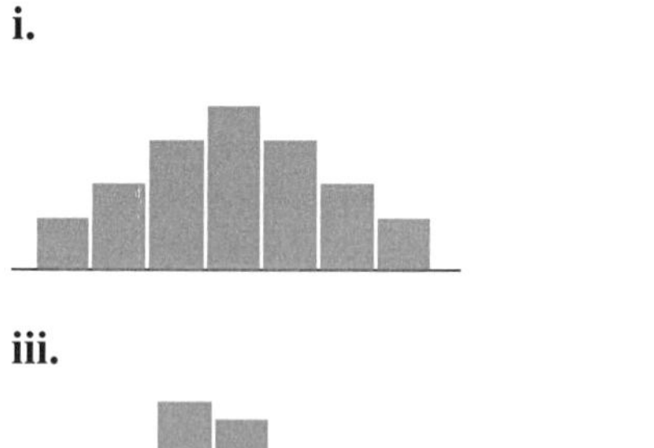

   ii.

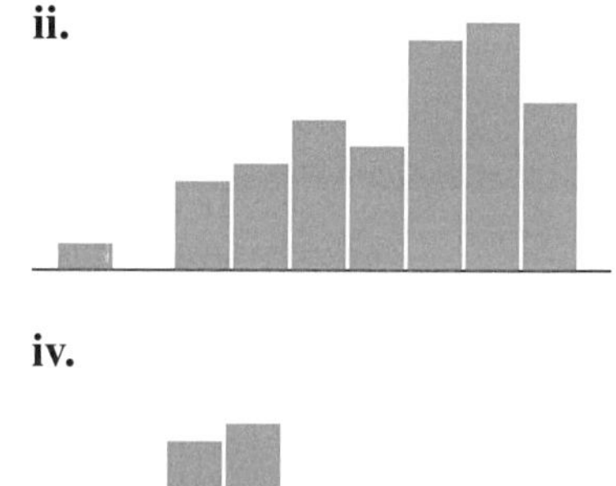

   iii.

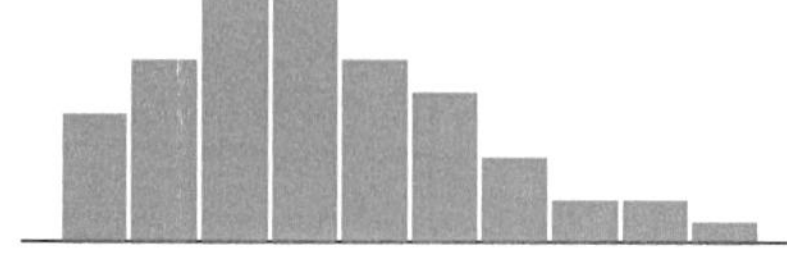

   iv.

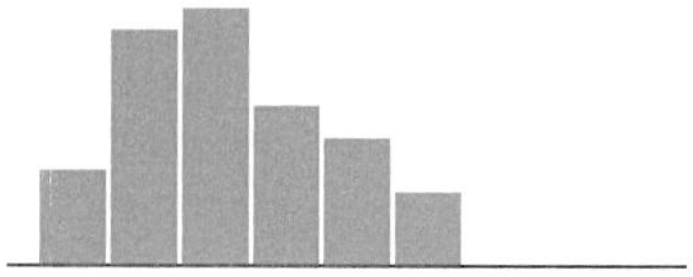

4. Draw histograms that might describe the same sets of values as those represented by these box plots.

a. $n = 20$

0 10 20 30 40 50

b.

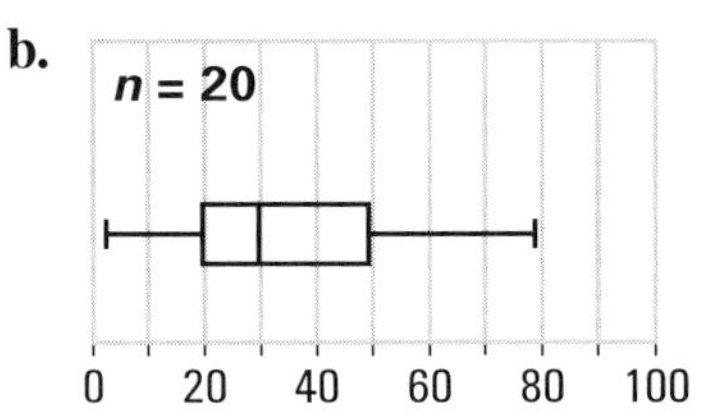

Reflecting

1. Describe at least two ways to use your calculator or computer to produce the five-number summary for a data set. Which do you prefer? Why?

2. Suppose when you finished Activity 1 on page 52, you found that your pulse rate was below the lower quartile. What does this mean?

3. In what situations would you use a histogram to display data? A box plot?

4. Why is the word "quartile" a good name for the 25th, 50th, and 75th percentiles? What do you suppose a "decile" is?

5. Is a maximum always an outlier? Is an outlier always a maximum or minimum? Explain your answers.

Extending

1. The histogram below displays the results of a survey filled out by varsity athletes in football and women's and men's basketball from schools around Detroit, Michigan. These results were reported in a school newspaper.

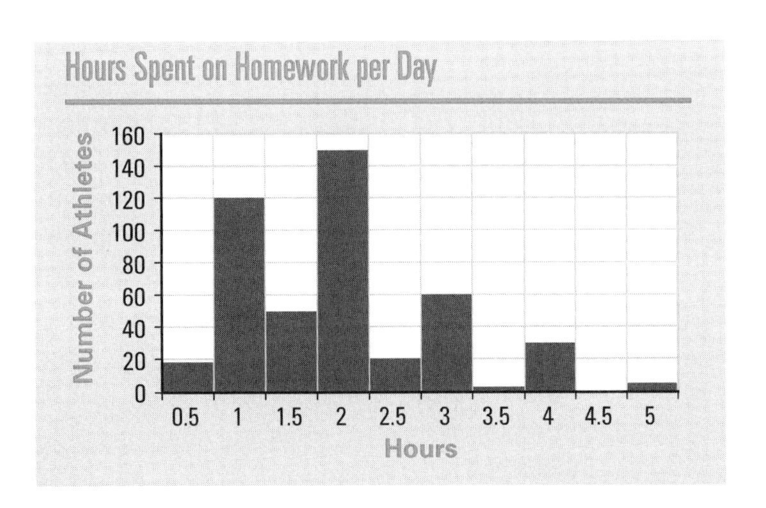

a. Estimate the median and the quartiles. Draw a box plot that also might represent the data.

b. What is an unusual feature of this distribution? What do you think is the reason for this?

c. Write a paragraph describing the distribution of the number of hours these athletes spend doing homework.

2. If your family has records of your growth, plot your own growth in height on a copy of the appropriate National Center for Health Statistics growth chart. What percentile are you in now for your age? How much has the percentile varied over your lifetime?

3. These box plots represent the scores of 80 seniors and 80 juniors on a performance test. List the characteristics you know will be true about a box plot for the combined scores of seniors and juniors. For example, what will the minimum be?

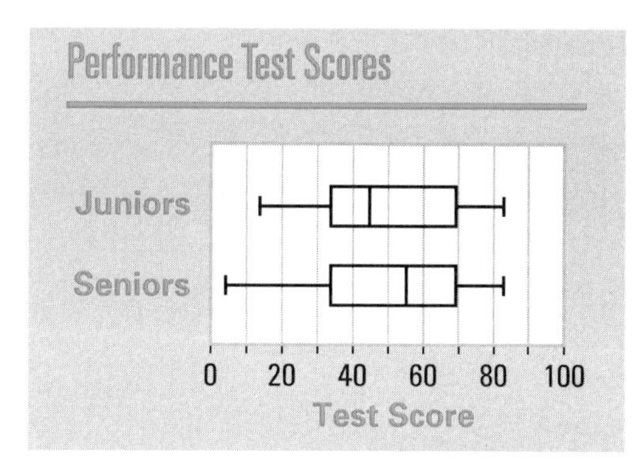

4. An **outlier** is a value that lies far away from most other values in a distribution. One rule for identifying possible outliers is to find values in a data set larger than

$$Q_3 + 1.5 \times (Q_3 - Q_1)$$

or smaller than

$$Q_1 - 1.5 \times (Q_3 - Q_1)$$

a. Refer to the data on the number of calories in fast foods listed in Lesson 2, Investigation 2 on page 22.

- Calculate the mean and the median for this data set. Which is larger? Why?
- Determine whether this set of data has any possible outliers. If so, remove the outlier(s) from the data set. Calculate the mean and median for the new data set. Compare the two means and the two medians.

b. Jolaina found outliers by using a box plot. She measured the length of the box and marked off 1 box length to the right of the original box and 1 box length to the left of the original box. If any of the values extended beyond these new boxes, these points were considered outliers. Jolaina had a good idea but made one mistake. What was it?

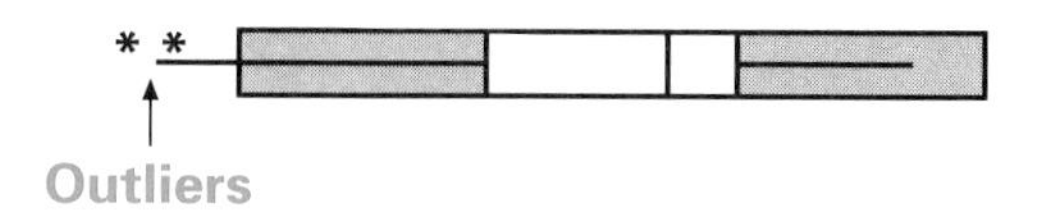

INVESTIGATION 3 MAD About the Mean

In the last investigation, you examined the variability within a set of data by looking at graphical displays such as stem-and-leaf plots, histograms, and box plots and by computing the interquartile range. You may have found box plots particularly useful in displaying the difference in variability between two or more distributions. In this investigation, you will learn about the MAD, another measure of variation.

1. As a class activity, your teacher will look at a watch with a second hand and say "start." After a certain length of time he or she will say "stop."

a. Write down on paper your (individual) estimate of how many seconds passed between the "start" and the "stop." It's not fair to look at your watch!

b. Your teacher will tell you how many seconds it actually was. How far off was your estimate? Make a stem-and-leaf plot of the *errors* made by your class. Don't use any negative numbers. An error of 6 seconds too long and an error of 6 seconds too short will both go on the stem-and-leaf plot as 6.

c. Find the average error for students in your class.

d. Try the experiment again with a different length of time. Is the new average error larger or smaller than before? Why is this the case?

2. Next, as a class choose one pair below for which you do not know the answers, but have a way to find them.

- length of your book in inches and height of the doorway in inches
- your teacher's height in centimeters and your principal's height in centimeters
- the number of students who were absent from your school last Tuesday and the number of students who were absent last Friday
- the cost of running the athletic program at your school and the profit from tickets and concessions

a. Each member of your class should make his or her own estimate of the two values.

b. How far off were you from the correct answers? Make a back-to-back stem-and-leaf plot of the *absolute value* of the errors (record both –5 and 5 as 5) you and your classmates made in estimating the two values. Which of the two estimates seems to have the larger errors?

c. Find the average error for each estimated value. What does the average error tell you about the goodness of the estimates of each value?

d. For which value was the average error smaller? Give a possible reason for this.

3. Refer to your Class Data Table prepared for Lesson 1.

a. Calculate the mean height for your group.

b. How much does each member of your group vary from the mean height?

c. On average, how much did the members of your group differ from the mean? How did you find this average?

d. Compare your results with each of the other groups. Which group had the greatest variability among its members? Which one had the least?

The average of the distances from the mean that you found in Activity 3 is called the **mean absolute deviation** or **MAD**.

4. Each set of values below has a mean of 10 and a range of 16.

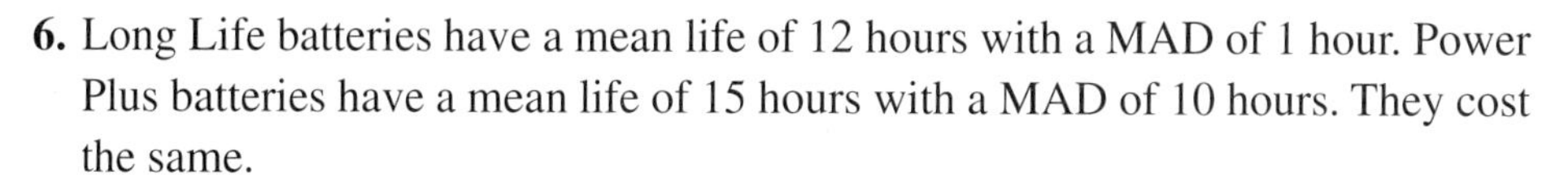

Set 1	Set 2
2, 4, 6, 8, 10, 12, 14, 16, 18	2, 2, 2, 2, 10, 18, 18, 18, 18

a. Which of the two sets has values that vary more from the mean? Explain.

b. Check your answer to Part a by computing the MAD for each set.

5. What is the MAD for each set of values below? Explain how to get the answer without doing any calculations.

a. 4, 4, 4, 4, 4, 4, 4, 4

b. 5, 5, 5, 5, 5, 5, 15, 15, 15, 15, 15, 15

6. Long Life batteries have a mean life of 12 hours with a MAD of 1 hour. Power Plus batteries have a mean life of 15 hours with a MAD of 10 hours. They cost the same.

a. Give an example of a situation where you would prefer to use Long Life batteries.

b. Give an example of a situation where you would prefer to use Power Plus batteries.

7. Emperor and Squealer are different breeds of pig. For Emperor's breed, the mean weight is 120 pounds with a MAD of 15 pounds. For Squealer's breed, the mean weight is 133 pounds with a MAD of 30 pounds. Emperor weighs in at 150 pounds and Squealer weighs in at 163 pounds. Is it Emperor or Squealer who is large for his breed? Explain your reasoning.

Checkpoint

In this investigation, you learned about the MAD—a measure of variation about the mean of a distribution.

a Why is it important to measure variability?

b Describe two ways of measuring and reporting variability in a data set.

c Which would you expect to have a smaller mean absolute deviation, the distribution of current prices of blue jeans or the distribution of current prices of new compact discs? Explain your reasoning.

Be prepared to share your group's descriptions and thinking with the entire class.

Measures of center, such as the mean and median, and measures of variability, such as the mean absolute deviation, the interquartile range, and others, are called **summary statistics**. They provide a quick summary of a set of data. However, they are most informative when accompanied by an appropriate graphical display.

On Your Own

At exactly the same instant, Tony and Jenny checked the time on the clocks and watches at their houses. The times on Jenny's ten clocks were 8:16, 8:10, 8:14, 8:16, 8:12, 8:15, 8:13, 8:17, 8:15, and 8:22.

a. Calculate the mean time on Jenny's clocks. Calculate the mean absolute deviation.

b. Tony had the same mean on his ten clocks as Jenny did, but his mean absolute deviation was 10 minutes. Find an example of ten times that could be the times on Tony's clocks.

c. What do the mean and the MAD tell you about how useful it is to look at a clock at Tony's house and at Jenny's house?

INVESTIGATION 4 Transforming Measurements

Calculating the mean absolute deviation (MAD) for a set of data can be tedious and time consuming. In this investigation you will first learn how to compute the MAD on your calculator or computer. You will then explore how transformations such as changing from inches to centimeters affect the mean and the MAD.

1. Each member of your group should measure the length of the same desk or table to the nearest tenth of a centimeter. Do your measurements independently, and don't look at the measurements recorded by the other group members. Get results from classmates so that your group has a total of at least ten measurements for the same object.

a. Make a number line plot of the measurements.

b. Enter the data into a list on your calculator or spreadsheet software and calculate the mean. Mark the mean on the number line plot.

c. Compute the mean absolute deviation of the measurements.

d. What do the mean and MAD tell you about the accuracy of your measurements?

2. Suppose that a group of 10 students collected the same measurements that your group did in Activity 1, except the end of their ruler was damaged. Consequently, each of their measurements is exactly 0.2 cm longer than those collected by your group.

a. What do you think they got for their mean and mean absolute deviation?

b. *Transform* the data in the calculator or computer list by adding 0.2 cm to each of your measurements.

c. Describe how a number line plot of the transformed data would compare to the plot you made in Part a of Activity 1.

d. Compute the mean and mean absolute deviation of the transformed measurements.

e. In what way is the mean of the transformed measurements related to the original mean?

f. In what way is the mean absolute deviation of the transformed measurements related to the original mean absolute deviation?

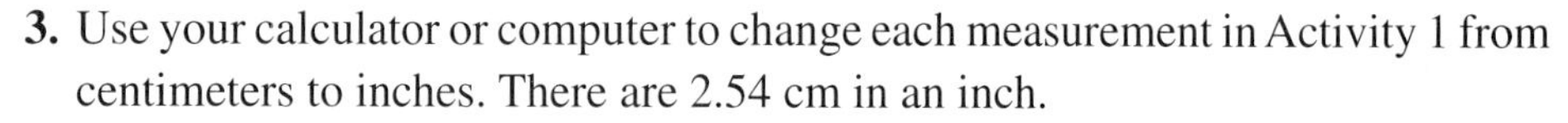

3. Use your calculator or computer to change each measurement in Activity 1 from centimeters to inches. There are 2.54 cm in an inch.

a. Predict what the new mean and the mean absolute deviation will be for the measurements in inches.

b. Compute the mean and the mean absolute deviation of these new transformed measurements.

c. In what way is the mean of the transformed measurements related to the original mean?

d. In what way is the mean absolute deviation of the transformed measurements related to the original mean absolute deviation?

e. Suppose that one student mistakenly multiplied by 2.54 when transforming the measurements. What do you think this student got for the mean and mean absolute deviation of the transformed measurements? Check your prediction.

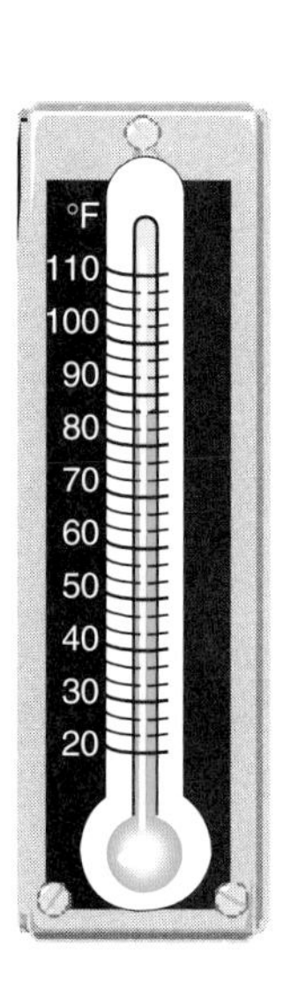

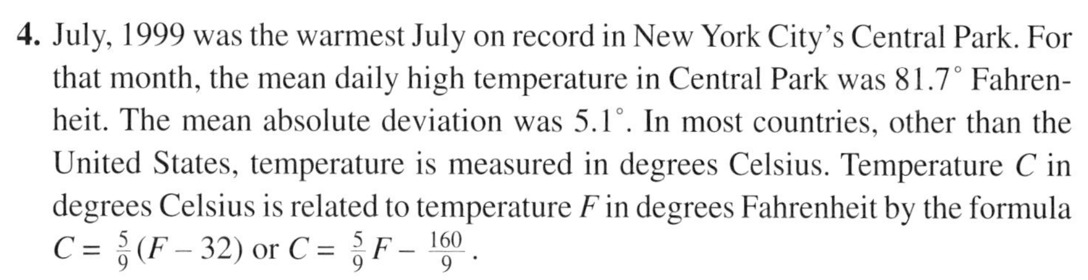

4. July, 1999 was the warmest July on record in New York City's Central Park. For that month, the mean daily high temperature in Central Park was 81.7° Fahrenheit. The mean absolute deviation was 5.1°. In most countries, other than the United States, temperature is measured in degrees Celsius. Temperature C in degrees Celsius is related to temperature F in degrees Fahrenheit by the formula $C = \frac{5}{9}(F - 32)$ or $C = \frac{5}{9}F - \frac{160}{9}$.

a. The temperature in Central Park on July 6 reached a high of 92° Fahrenheit. Express this temperature in degrees Celsius so that it would be better understood by a visitor from Canada.

b. Suppose you want to calculate the mean daily high temperature for July and the mean absolute deviation in degrees Celsius. Do you need to know the July daily high temperatures in Central Park? Explain your reasoning.

c. What are the mean daily high temperature and the mean absolute deviation in degrees Celsius?

Checkpoint

In this investigation, you discovered that transformations of data affect the mean and mean absolute deviation in predictable ways.

a What is the effect on the mean of transforming a set of data by adding or subtracting the same number to each value? What is the effect on the mean absolute deviation? Explain why this is the case.

b What is the effect on the mean of transforming a set of data by multiplying or dividing each value by the same number? What is the effect on the mean absolute deviation?

Be prepared to share your group's thinking with the class.

On Your Own

In the Amar family, the mean age is 30 with a MAD of 15 years.

a. What will be their mean age in 5 years? Their MAD in 5 years?

b. What is their mean age now in months? Their MAD in months?

c. Find possible ages for the six people in the Amar family. The six ages should have a mean of 30 and a MAD of 15.

- Find the mean and MAD of your six possible ages *in 5 years*. Was your prediction in Part a correct?
- Find the mean and MAD of your six *current* ages in months. Was your prediction in Part b correct?

MORE
Modeling • Organizing • Reflecting • Extending

Modeling

1. Find a thick book, preferably one that other members of your class are using also. Measure the thickness of a piece of paper in this book in the following way. Compress more than a hundred pages from the middle of the book and measure the thickness in millimeters to the nearest tenth of a millimeter. Divide by the number of sheets of paper. Round to four decimal places.

a. How can you determine the number of sheets of paper from the page numbers?

b. Repeat your measurement ten times, taking a different number of pages each time. Place your measurements on a number line plot.

c. What is the mean of your measurements? What is the mean absolute deviation of the measurements?

d. What would you report as the measurement of the thickness of a piece of paper?

2. D'Lynn and Seung are members of their high school track team. Their times in seconds for the last seven 50-meter races are given in the table.

D'Lynn	5.8	6.4	6.3	6.3	6.1	6.0	6.6
Seung	5.6	6.7	6.3	5.7	6.1	6.8	5.5

a. Make a plot of the times that allows you to compare the two runners.

b. Using what you have learned in this investigation, summarize the times for D'Lynn and for Seung.

c. If only one of them can be entered in the final conference race, give an argument for choosing D'Lynn. For choosing Seung.

3. News accounts suggested that the first inauguration of President Clinton marked the beginning of government by an unusually young president. Refer to the table below of the ages of U.S. Presidents at inauguration and at death.

Ages of U.S. Presidents

President	Age at Inauguration	Age at Death	President	Age at Inauguration	Age at Death
George Washington	57	67	Benjamin Harrison	55	67
John Adams	61	90	Grover Cleveland	55	71
Thomas Jefferson	57	83	William McKinley	54	58
James Madison	57	85	Theodore Roosevelt	42	60
James Monroe	58	73	William H. Taft	51	72
John Quincy Adams	57	80	Woodrow Wilson	56	67
Andrew Jackson	61	78	Warren G. Harding	55	57
Martin Van Buren	54	79	Calvin Coolidge	51	60
William H. Harrison	68	68	Herbert C. Hoover	54	90
John Tyler	51	71	Franklin D. Roosevelt	51	63
James K. Polk	49	53	Harry S Truman	60	88
Zachary Taylor	64	65	Dwight D. Eisenhower	62	78
Millard Fillmore	50	74	John F. Kennedy	43	46
Franklin Pierce	48	64	Lyndon B. Johnson	55	64
James Buchanan	65	77	Richard M. Nixon	56	81
Abraham Lincoln	52	56	Gerald R. Ford	61	
Andrew Johnson	56	66	James E. Carter, Jr.	52	
Ulysses S. Grant	46	63	Ronald W. Reagan	69	
Rutherford B. Hayes	54	70	George H.W. Bush	64	
James A. Garfield	49	49	William J. Clinton	46	
Chester A. Arthur	50	56	George W. Bush	54	
Grover Cleveland	47	71			

Source: *The World Almanac and Book of Facts 1996*. Mahwah, NJ: World Almanac, 1995 and www.georgewbush.com—biography.

a. Select the measure that is the most appropriate to use to examine the media's description of the Clinton presidency prior to the inauguration of George W. Bush: the interquartile range, the mean absolute deviation, or Clinton's percentile. Compute that measure.

b. Make an appropriate graphical display of the ages of all U.S. Presidents at inauguration and another for ages at death. Which distribution appears to have greater variability?

c. Find the mean and mean absolute deviation of the age of U.S. Presidents at inauguration and at death.

d. Which distribution has the greater mean absolute deviation? Is this finding consistent with your answer to Part b?

e. How many mean absolute deviations is President Clinton's age at inauguration from the mean?

4. The table below shows the total points scored during the first eight years of the NBA careers of Kareem Abdul-Jabbar and Michael Jordan.

Two Shooting Stars

Kareem Abdul-Jabbar		Michael Jordan	
Year	**Points Scored**	**Year**	**Points Scored**
1970	2,361	1985	2,313
1971	2,596	1986	408
1972	2,822	1987	3,041
1973	2,292	1988	2,868
1974	2,191	1989	2,633
1975	1,949	1990	2,752
1976	2,275	1991	2,580
1977	2,152	1992	2,541

a. Which player had the higher mean number of points per year?

b. What summary statistic could you use to measure consistency in a player? Which player was more consistent according to your measure?

c. On the basis of mean points scored per year and consistency, who was the better scorer? Explain.

d. Jordan had an injury in 1986. If you ignore his performance for that year, would you change your answer to Part c? Explain.

e. Is the MAD sensitive to outliers? Explain why or why not.

Organizing

1. Refer back to "On Your Own" in Investigation 3 of this lesson.
   a. Subtract the mean from each of the times on Jenny's clocks, giving the difference as a positive or negative number.
   b. Find the average of the differences. What happened? What might explain this fact?
   c. Explain how the calculation of the mean absolute deviation overcomes the difficulty in Part b.

2. Refer to Investigation 4 (page 66) for your measurement work on transformed data.
   a. Find the median and interquartile range for the original measurement data.
   b. Find the median and interquartile range after each measurement is transformed to inches.
   - How do the median and interquartile range of the transformed data compare to those of the original data?
   - In general, what is the effect on the median and interquartile range if you divide each value in a data set by the same number?

   c. Investigate the effect that multiplying each value in a data set by the same number has on the median and interquartile range. What can you conclude?
   d. Find the median and interquartile range after adding 2 cm to each original measurement.
   - How do the median and the interquartile range of the transformed data compare to those of the original data?
   - In general, what is the effect on the median and interquartile range of adding the same number to each value of a data set? Explain your reasoning.

   e. Suppose the same number is subtracted from each value in a data set. How do you think the median and interquartile range of the transformed data compare to those of the original data? Explain your reasoning.

3. Find two different sets of values so that the interquartile range (IQR) of the first set is larger than the IQR of the second, and the mean absolute deviation (MAD) of the second is larger than the MAD of the first.

4. For Parts a through d, use the set of values 1, 2, 3, 6, 23.
   a. Compute the mean absolute deviation from the mean.
   b. Compute the *median* absolute deviation from the mean.
   c. Compute the mean absolute deviation from the *median*.
   d. Compute the median absolute deviation from the median.
   e. Which is larger, the mean absolute deviation from the median or the mean absolute deviation from the mean? Test other sets of values to see if your answer holds for all of them.

Reflecting

1. Describe a situation in which the interquartile range would be a better measure of variability than the MAD. Describe a situation in which the mean absolute deviation would be preferable.
2. Comment on the statement, "If two sets of data have the same mean and the same mean absolute deviation, they have the same distribution."
3. Describe a situation in a science class for which the concept of mean absolute deviation might be important.
4. At this point, what are the things you understand about variability? What do you feel you do not understand?

Extending

1. The **standard deviation** is another way of measuring the amount of variation about the mean. It is found by computing the average of the *square* of the distance between each data point and the mean. Then, the *square root* ($\sqrt{\ }$) of this number is calculated. Here are the ratings for the Toyota MR2 convertible reported in Lesson 2:

   9, 8, 7, 8, 6, 9, 8, 7, 7, 6, 8, 8, 7, 7, 7, 6, 1, 6, 8, 8

   These ratings have a mean of 7.05.

   a. List the distance from the mean of 7.05 for each rating.

   b. Square each distance. (That is, multiply each distance by itself.) Find the total of these squared distances, then divide by the number of ratings (20) to find the average of the squared distances.

   c. Finally, calculate the square root of this number using your calculator or computer software. You have calculated the standard deviation of the ratings for the Toyota MR2.

   d. Compute the standard deviation for the car that appears to have the least variability in its ratings. Compare it with the standard deviation for the MR2.

   e. Compute the mean absolute deviation of the ratings for the MR2 and for the car you chose in Part d. Is there much difference between the standard deviation and the mean absolute deviation in these cases?

2. Suppose the AAA raters changed their minds and gave the Toyota MR2 a "5" on Cargo Space instead of the "1" it originally received.

a. How do you think the new mean and the new standard deviation will compare to the original mean and standard deviation?

b. You can find the mean and standard deviation quickly using technology. In the calculator display at the left, the mean is $\bar{x}$. The standard deviation is σx (σ is the lower case Greek letter "sigma"). The letter n indicates the number of values used in the calculations. These are standard mathematical notations. Use your calculator or computer to find the mean and standard deviation of the adjusted MR2 data and check your prediction from Part a.

```
1-Var Stats
x̄=7.05
Σx=141
Σx²=1049
Sx=1.700619082
σx=1.657558445
↓n=20
```

3. The symbol $\sum$ (upper case Greek letter "sigma") means to add the values indicated. $\sum x$ is a mathematical shorthand for indicating the *sum of all the values x*.

a. The data set below represents the amount of change carried by the 28 students in a first-hour science class. Enter the data into your calculator and find the *one-variable statistics* $\sum x$ and $\sum x^2$. Explain in terms of the data what $\sum x$ and $\sum x^2$ represent.

Amount of Change

$.00	.25	.07	.00	.00	.35	.00	1.70	.00	.85
.35	.25	.00	.50	1.00	1.55	.50	.00	.00	.85
1.30	.00	.65	.25	.15	.90	3.35	.00		

b. Write an expression using $\sum$ notation which represents the mean of a data set.

c. Which of the expressions below represents the standard deviation for this set of data?

i. $\sqrt{\frac{\sum(x-\bar{x}^2)}{28}}$ **ii.** $\sqrt{\frac{\sum(x-\bar{x})^2}{28}}$ **iii.** $\frac{\sqrt{\sum(x-\bar{x})^2}}{28}$

d. Which of the following represents the mean absolute deviation for this set of data? (Note: $|x|$ means the absolute value of x; for example, $|-5|=5$ and $|3|=3$.)

i. $\frac{\sum|x-\bar{x}|}{28}$ **ii.** $\frac{|\sum(x-\bar{x})|}{28}$ **iii.** $\frac{\sum|x|-\bar{x}}{28}$

4. If the data are available, get the heights of the players on sports teams at your school. Analyze these data with appropriate graphical displays, measures of center, and measures of variability. Write a report on the differences in the teams.

Lesson 4 Relationships and Trends

The article that accompanied the guide to the American Automobile Association's auto test ratings as reported in Lesson 1 stated:

> *Each vehicle must make compromises between power and fuel efficiency, ride and handling, cargo space and interior room, price and value.*

If you want to find a car rated highly on a combination of two characteristics, it may be necessary for you to compromise.

Car	Interior Space	Cargo Space
A. Buick LeSabre	9	9
B. Chevrolet Camaro	5	5
C. Chrysler Concorde	9	7
D. Chrysler PT Cruiser	8	9
E. Dodge Intrepid	9	5
F. Ford Mustang GT	5	5
G. Mazda Miata	6	3
H. Toyota MR2	6	1
I. Toyota Prius	7	5
J. Volvo V40 Wagon	7	10
K. Mazda 626	7	7

A rating of 1 is the lowest. A rating of 10 is the highest.

Think About This Situation

Suppose you are interested primarily in interior space and cargo space for the cars in the $20,000–$25,000 range tested by the AAA. Examine the ratings of the tested cars above on these two characteristics.

a According to the table above, do the cars with the best interior space also have the best cargo space?

b Do the cars with the worst interior space also have the worst cargo space?

c In general, does it look like cars with better interior space have better cargo space and cars with poorer interior space have poorer cargo space?

d Does there have to be a compromise between interior space and cargo space?

INVESTIGATION 1 Scatterplots

A **scatterplot** helps make comparisons like those in the "Think About This Situation." On the following scatterplot, the *horizontal axis* (called the ***x*-axis**) represents the ratings for interior space, and the *vertical axis* (called the ***y*-axis**) represents the ratings for cargo space. (It doesn't really make any difference in this case which characteristic goes on which axis. However, sometimes it will matter.)

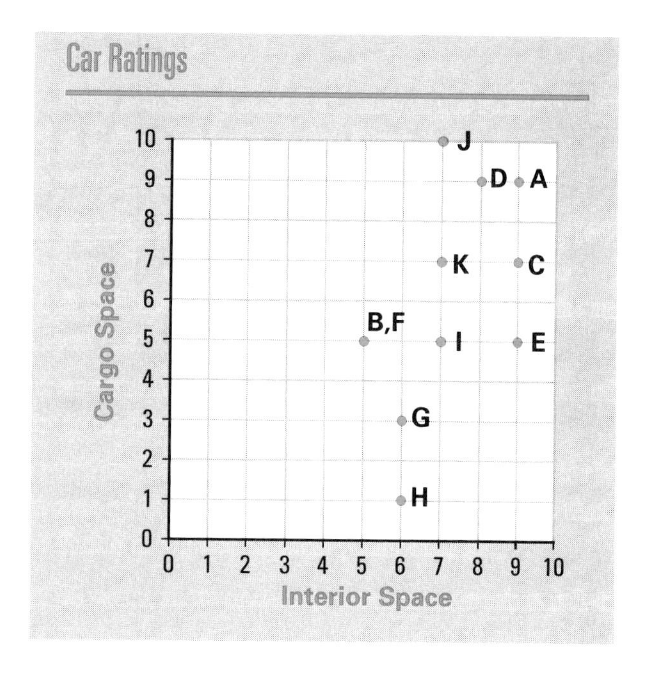

1. Explain the scatterplot above.

a. What does the point D in the plot represent? What does the point G represent?

b. How many cars had a 7 rating for interior space? Name the cars and indicate which points represent them on the plot. How could you verify your answers?

c. How many cars had a 9 rating for cargo space?

d. How many cars had the same ratings for both characteristics? Where are the points that represent them located on the plot?

e. What point would represent a car with the highest possible ratings on both characteristics? What point would represent a car with the lowest possible rating on each characteristic?

f. In an **ordered pair**, such as (6, 3), the **first coordinate**, 6, represents the value on the x-axis. The **second coordinate**, 3, represents the value on the y-axis. Which car is represented by the point (6, 3)? By the point (7, 10)?

g. What ordered pair would represent the ratings of the Toyota MR2? The Intrepid?

2. Now look at the line on the scatterplot below.

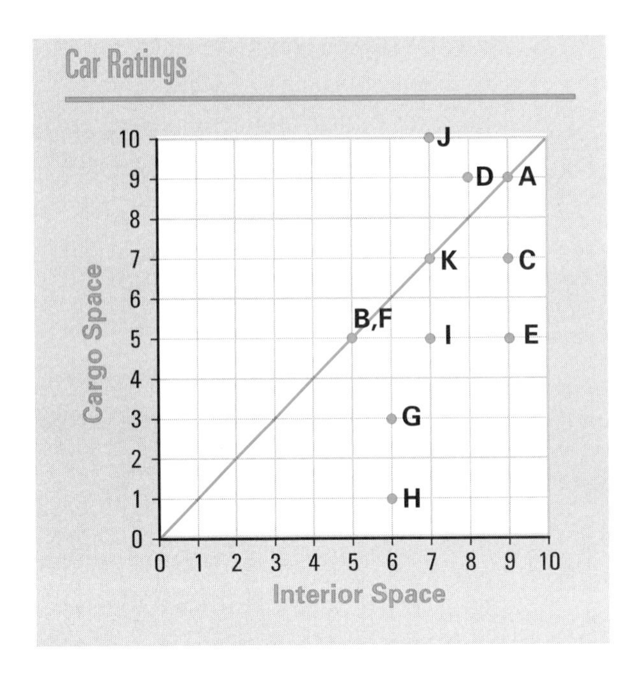

a. What is true of every ordered pair on the line? If x represents interior space ratings and y represents cargo space ratings, what equation could be used to describe all ordered pairs on the line?

b. How many cars have a *higher* rating for interior space than they do for cargo space? Where are the points that represent these cars located on the plot?

c. Describe the location of the points that represent the cars for which $x < y$.

d. Which car or cars are the best considering *both* interior space and cargo space? Explain why you chose the car or cars you did.

e. Which car ranks lowest considering *both* characteristics? Where is the point that represents that car on the plot? Explain how you selected this car.

Chrysler Concorde

Dodge Intrepid

Both the Concorde and Intrepid are made by the Chrysler Corporation. They are manufactured on the same chassis (frame, wheels, and machinery). You might expect the ratings of the two cars across all tested characteristics to be similar. Are they?

Chrysler Comparison

Car	Acceleration	Transmission	Braking	Steering	Ride	Handling	Driveability	Fuel Economy	Comfort	Interior Space	Driving Position	Instruments	Controls	Visibility	Entry/Exit	Quietness	Cargo Space	Exterior	Interior	Value
Concorde	8	9	6	7	9	8	8	6	9	9	8	7	8	7	9	9	7	7	8	9
Intrepid	8	9	6	8	9	8	9	6	9	9	8	7	8	7	9	9	5	7	8	9

3. Use your calculator or computer to make a scatterplot to help you compare the ratings.

a. Examine the scatterplot for the Concorde and Intrepid. Does your scatterplot show only 7 points? If so, why? If not, describe what your calculator or computer does to display more than one point in the same location.

b. Trace the plot to display the coordinates of the points as they were entered in the lists.

- For how many categories were the two cars rated exactly the same?
- Is it easier to answer this question by using the plot or the table? Why?

c. Now produce the graph of the line $y = x$ on your calculator or computer. Trace along the line and observe the readout of the coordinates. What do you notice?

d. Which car had more ratings that were greater than the other car's ratings? How can you tell from the plot?

e. Where are all of the points located for which $x > y$? In words, describe what $x > y$ means for this situation.

f. Interpret $x \leq y$ for this situation.

g. Are the ratings of the Concorde and Intrepid similar? Is it easier to answer this question by using the plot or the table? Explain your reasoning.

4. Now refer to the resting pulse rates and exercising pulse rates for your class as recorded in the Class Data Table.

a. On graph paper or using your calculator, make a scatterplot with the x-axis representing resting pulse rates and the y-axis representing exercising pulse rates. Draw in the line $y = x$ to help you compare the pulse rates.

b. Write down at least four observations about your scatterplot.

c. What can you say about the student who is the farthest vertical distance from the line $y = x$? Is this student also the farthest horizontal distance?

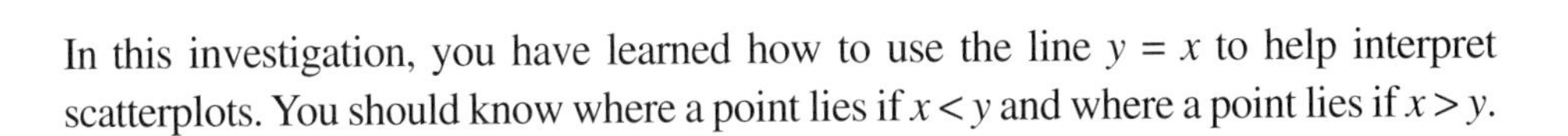

Checkpoint

Suppose a scatterplot is made of the average 1990 and 2000 temperatures for 50 major world cities. The horizontal axis represents the average temperature in 1990 and the vertical axis represents the average temperature in 2000.

a What does a point on the plot represent?

b Where are the points located that represent cities that had the same average temperature in 1990 and 2000?

c How could you use the line $y = x$ to determine if temperatures generally increased from 1990 to 2000?

d Is it always helpful to draw in the line $y = x$ on a scatterplot? Why or why not?

Be prepared to share your group's responses with the whole class.

Miami, FL

In this investigation, you have learned how to use the line $y = x$ to help interpret scatterplots. You should know where a point lies if $x < y$ and where a point lies if $x > y$.

Paris, France

Chicago, IL

On Your Own

The table below gives the cost of movies produced by Carolco Pictures and the total box office income from those movies.

Movie Guide

Movie	Cost (millions of dollars)	Box Office Income (millions of dollars)
Terminator 2	90.0	52.9
The Doors	29.0	35.0
L.A. Story	20.0	28.0
Jacob's Ladder	27.8	26.0
Narrow Margin	21.0	10.6
Air America	35.0	30.5
Total Recall	60.1	118.3
Mountains of the Moon	20.0	3.3
Music Box	18.0	5.4
Johnny Handsome	20.0	6.6
Deepstar Six	9.0	8.6
Iron Eagle II	15.5	10.5
Red Heat	32.0	35.0
Rambo III	58.0	53.75

Source: *The Wall Street Journal*, July 1991.

a. Make a scatterplot with cost on the x-axis and income from the box office on the y-axis. Draw in the line $y = x$.

b. For how many movies is the cost greater than the income from the box office? Where are the points that represent these movies on the plot? Which **inequality**, $x < y$ or $x > y$, describes these points?

c. Use the plot to determine which movie has the greatest income from the box office. Where is the point representing that movie located on the plot? What are its coordinates?

d. Which movie had the greatest loss? The greatest profit? How does the plot help you find the answer?

e. Does there appear to be a relation between the cost of making these movies and the income they generate? Explain.

INVESTIGATION 2 Plots Over Time

In this unit you have displayed and summarized data in a variety of ways. You explored how those different ways helped you make sense out of situations ranging from sports and entertainment to worker salaries and best buys. Sometimes people are interested in a particular situation at a particular moment in time; for example, the present win/loss record of a school team. On another occasion they may be interested in the win/loss record of the team over the last several years. There are many such situations in the world of business in which trends over time are of particular interest. For example, trends in life expectancy are very important to the insurance industry.

The plot below shows the life expectancy at birth for people born in the United States between 1920 and 1996. It is called a **plot over time**.

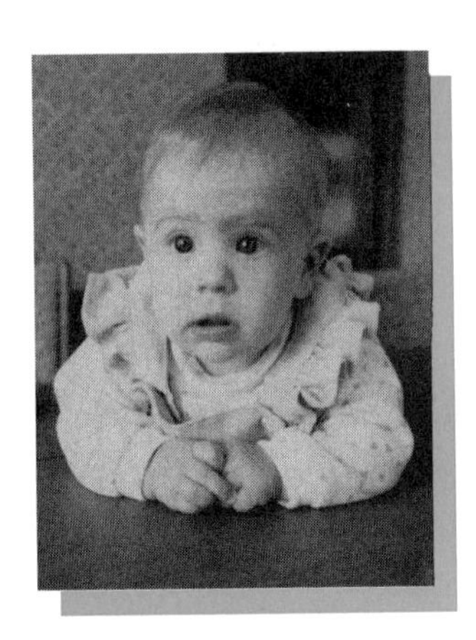

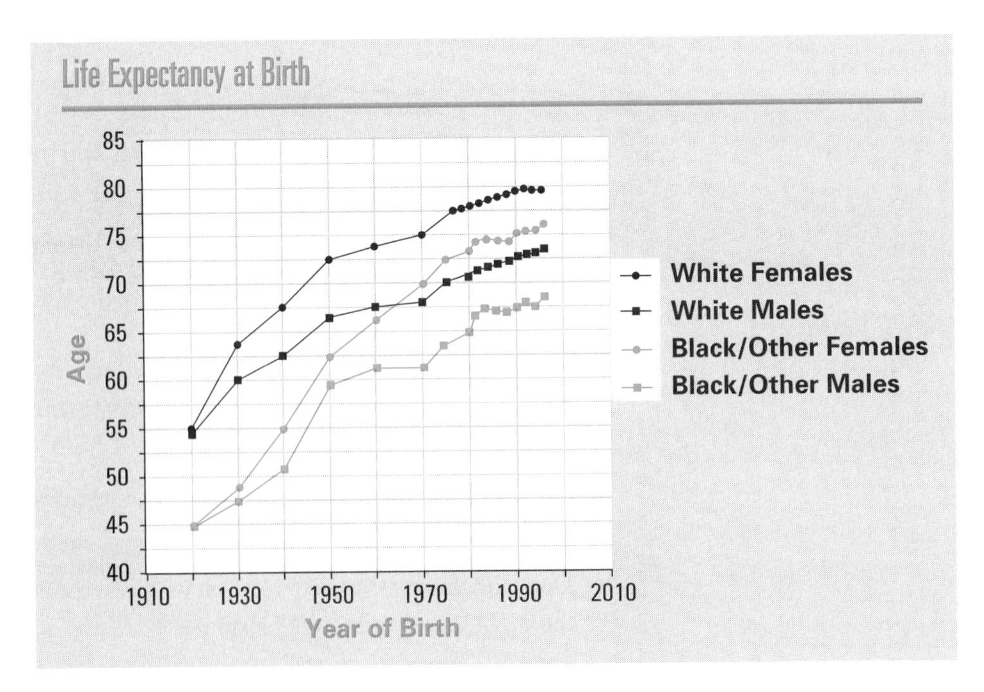

Source: *The World Almanac and Book of Facts 1997*. Mahwah, NJ: World Almanac, 1996; U.S. Census Bureau, *Statistical Abstract of the United States: 2000* (120th edition). Washington, DC, 2000.

1. From the plot of life expectancy, you can see how the life expectancy of each group has changed over time, how the groups compare at any one time, and how the changes for the groups compare.
   a. What is the life expectancy of black females born in 1920?
   b. Which group had the greatest increase between 1940 and 1950? How can you see this from the plot?
   c. What happened to the life expectancy of males born between 1960 and 1970?
   d. Where is the only place in the plot that two lines cross? Describe what is happening there.
   e. According to the plot, what is your life expectancy?
   f. Describe the trends in the plot. From the evidence in the plot, how high do you think life expectancies will be for people born in the year 2010?

Analyzing the track record of appliances or automobiles over time can help consumers make wise decisions. For example, when purchasing a used car, information on frequency and kinds of repair on earlier models together with ratings of cars can help you decide on the model and year.

2. Each year, *Consumer Reports* provides a rating of automobile cumulative repair records. The table below contains those ratings for several cars. A rating of 5 represents few problems, and 1 represents many problems. As you would expect, older cars have more problems. The ratings include categories such as paint, rust, brakes, clutch, ignition system, suspension, and transmission.

Average Auto Repair Ratings

	Year							
Model	93	94	95	96	97	98	99	00
Concorde	2.07	2.57	2.71	2.79	3.21	3.71	4.43	4.71
Intrepid	1.86	2.07	1.93	2.64	3.29	3.57	4.29	4.86
LeSabre	2.36	2.64	3.29	3.43	3.71	3.93	4.43	4.71
Mustang		3.14	3.14	3.57	3.64	4.07	4.14	4.50
Mazda 626	2.29	2.29	3.14	3.57	3.86	4.50	4.71	
Camaro		2.71	2.93	2.43	2.79	3.57	4.00	
Miata	3.79	4.07	4.21	4.07	4.64		4.64	

   a. What does a rating of 4.07 for the 1998 Mustang indicate about the repair records for that Mustang?
   b. Which 1994 car had the fewest problems?

3. Plotting an event over time allows you to see trends and make comparisons. The horizontal axis usually represents time. In this plot, the horizontal axis indicates the model year and the vertical axis gives the repair ratings for the Ford Mustang.

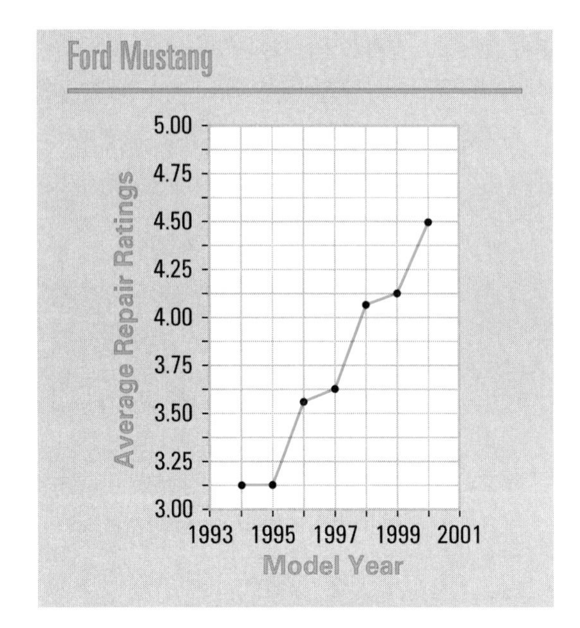

a. Use the plot shown here to estimate the average repair rating for the model year 1996 Ford Mustang.

b. Between which model years was there the greatest change in the ratings? How does the plot help you see that change?

c. Describe the change in the ratings for the Mustang over the 7-year span.

d. If you were considering buying a Mustang, how could you use this information to help make your decision?

4. As you might expect by now, you can use your calculator or computer to make a plot over time. Produce a plot over time of the Mustang repair ratings for the model years from 1994 to 2000.

a. How does the plot on your calculator or computer compare to the one in Activity 3? Trace along the plot and check the coordinate readouts against the data table.

b. Change Ymin to 0 and Ymax to 5.

- Does the new plot change your perception (visual impression) of the amount of change in the ratings?
- What happens if you set the x-axis to begin at 0 and end at 105, with a scale of 5?
- Use the automatic scaling option for statistical plots, if your graphing tool has one. How did this alter your plot?

c. Compare the average repair ratings for the Mustang and the Intrepid by graphing them on the same plot over time.

d. Change Ymin to 0 and Ymax to 10, with Yscl = 1. How does this change your perception of the two sets of ratings?

e. What would happen to the plots and your perception of the change in ratings over the years if you changed the minimum value of x to 60 and the maximum value of x to 105? Do you think this is a reasonable way to represent the data? Why or why not?

Checkpoint

In this investigation, you explored how to interpret and make a plot over time.

a Describe a plot over time.

b What information can you learn from a plot over time?

c How can you use a plot over time to find the time period when the least change occurs? The most?

d How can the scales on the axes affect your interpretation of a plot over time?

Be prepared to share your group's thinking on the interpretation of plots over time.

On Your Own

The data below give the number of cassette tapes, record albums (LPs), and compact discs (CDs) that were sold for years from 1983, when CDs became generally available, until 1999.

Sound Purchases

Year	Cassette Sales (in thousands)	LP Sales (in thousands)	CD Sales (in thousands)
1983	236,800	209,600	800
1985	339,100	167,000	22,600
1987	410,000	107,000	102,100
1989	446,200	34,600	207,200
1991	360,100	4,800	333,300
1993	339,500	1,200	495,400
1995	272,600	2,200	722,900
1997	172,600	2,700	753,100
1998	158,500	3,400	847,000
1999	123,600	2,900	939,900

Source: *The Universal Almanac 1994*. New York: Andrews and McMeel, 1993; *The World Almanac and Book of Facts 2001*. Mahwah, New Jersey: World Almanac Education Groups, 2001.

a. Produce a plot over time of the sales of the CDs, making appropriate choices for the viewing window. Describe any pattern you observe from the plot.

b. On the same display, make plots over time of the sales of cassettes and of LPs. Use different marks, if possible. Change your viewing window if necessary. Describe what happened to cassette, LP, and CD sales from 1983 to 1999.

MORE
Modeling • Organizing • Reflecting • Extending

Modeling

1. Refer to the height and armspan measurements in the Class Data Table you prepared in Lesson 1.

a. Make a scatterplot of the heights and armspans of the students in your class.

b. Draw in the line $y = x$.

- Do the points fall near this line?
- Are more points above this line or below it?
- If the point representing Wayne's measurements is below the line, what do you know about his height and armspan?
- Which inequality, $x < y$ or $x > y$, describes points that are below the line?

c. Describe the relationship shown on your scatterplot.

d. The **centroid** of any scatterplot is the point $(\bar{x}, \bar{y})$ where $\bar{x}$ is the mean of the x values and $\bar{y}$ is the mean of the y values. Find the centroid and mark it on your scatterplot.

e. Are there any points close to $(\bar{x}, \bar{y})$? If so, what can you say about the students represented by those points?

2. Refer to the thumb circumference and wrist circumference measurements in your Class Data Table.

a. Make a scatterplot of the thumb circumference (on the x-axis) and wrist circumference (on the y-axis) of the students in your class.

b. Describe the relationship shown on your scatterplot.

c. Do the points fall near the line $y = x$? Draw in a line that the points seem to fall near. What is the relationship between x and y on this line?

d. Find the centroid $(\bar{x}, \bar{y})$ and mark it on your scatterplot.

e. Are there any points close to $(\bar{x}, \bar{y})$? If so, what can you say about the students represented by those points?

3. The following table gives data about the life expectancy (at birth) for people living in the Americas, their (average) daily calorie supply, and the infant mortality rate.

Living in the Americas

Region/Country	Life Expectancy at Birth (years) 1970	1997	Daily Calorie Supply	Infant Mortality per 1,000 births
Antigua/Barbuda	—	75	2,365	17
Argentina	67	73	3,136	21
Bahamas	65	74	2,443	18
Barbados	69	76	3,207	11
Belize	—	75	2,862	35
Bolivia	46	61	2,170	69
Brazil	59	67	2,938	37
Canada	73	79	3,056	6
Chile	62	75	2,810	11
Colombia	59	70	2,800	25
Costa Rica	67	76	2,822	12
Dominica	—	74	3,093	17
Dominican Republic	59	71	2,316	44
Ecuador	58	70	2,592	30
El Salvador	58	69	2,515	31
Guatemala	53	64	2,191	43
Guyana	65	64	2,392	59
Haiti	48	54	1,855	92
Honduras	53	69	2,368	36
Jamaica	67	75	2,575	10
Mexico	62	72	3,137	29
Nicaragua	54	68	2,328	42
Panama	66	74	2,556	18
Paraguay	65	70	2,485	27
Peru	54	68	2,310	44
St. Kitts & Nevis	—	70	2,263	30
St. Lucia	62	70	2,822	24
St. Vincent & the Grenadines	63	73	2,434	18
Surinam	64	70	2,578	24
Trinidad & Tobago	66	74	2,751	15
United States	71	76	3,642	7
Uruguay	—	74	2,830	18
Venezuela	65	72	2,398	21

Source: *The New York Times Almanac 2001*. New York, NY: The New York Times Company, 2000.

a. Study the numbers carefully and write a brief summary of your observations.

b. Prepare an appropriate graphical display to compare the life expectancy in 1970 with that in 1997. Describe how life expectancy in the Americas has changed between these years.

c. Which country is farthest from the line $y = x$? What can you say about that country?

d. Make an appropriate graphical display of the infant mortality rate. Describe any patterns or clusters you see and give some possible reasons for their occurrence.

4. The following plot shows how long the average American had to work (before taxes) to earn enough to purchase a gallon of gasoline or a pound of hamburger.

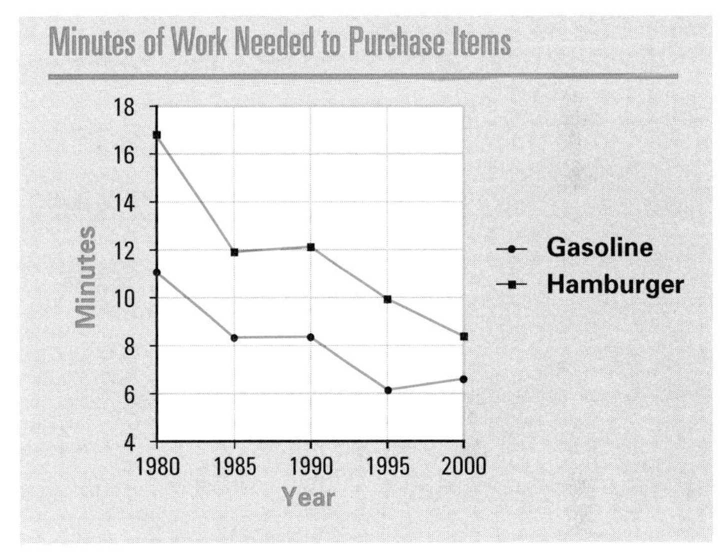

Source: U.S. Bureau of Labor Statistics

a. How many minutes did the average American have to work to buy a pound of hamburger in 1980?

b. Describe the change in the number of minutes someone had to work to buy a pound of hamburger from 1980 to 2000.

c. Starting at 1980, during what five-year period was there the greatest change in the number of minutes of work needed to buy a pound of hamburger? How can you tell this from looking at the plot?

d. How does the change over time in the minutes of work needed to buy a pound of hamburger compare with that for a gallon of gasoline?

e. Estimate the difference in the minutes of work needed in 1980 to buy the two items.

f. Do you think it ever took more minutes of work to buy a gallon of gasoline than a pound of hamburger? Explain your reasoning.

g. Copy the plot on page 87, keeping everything the same except double the distance between 4 and 18 on the y-axis. How does this change your perception of the decrease in the number of minutes of work needed to buy a gallon of gasoline?

h. Which changed more over the years—the minutes of work needed to buy a pound of hamburger or the minutes of work needed to buy a gallon of gasoline? Explain your answer.

i. Can you conclude that the price of a gallon of gasoline decreased from 1980 to 2000? Explain.

5. The plot below shows the number of female active duty officers in the Air Force and Navy from 1950 to 1998.

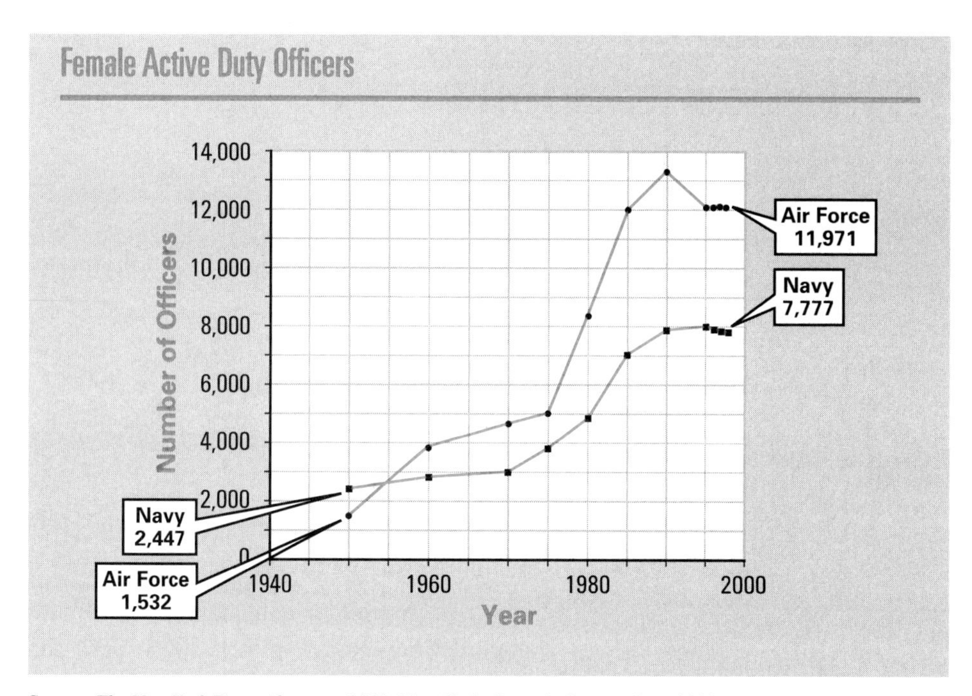

Source: *The New York Times Almanac, 2001*. New York: Penguin Putnam Inc., 2000.

a. Did the Air Force and Navy have their highest number of active duty female officers in the same year? Their lowest?

b. Estimate the number of female active duty officers in the Air Force in 1980.

c. During which five-year time period was there the greatest increase in the number of female active duty officers in the Air Force?

d. For how many years during the period 1950–1998 were there 5,000 or more female active duty officers in the Air Force?

e. Overall, what trend do you see in the number of female active duty officers in the Air Force over the period 1950–1998?

f. In what time periods did the Navy have an increase in the number of female active duty officers?

g. Overall, what trend do you see in the number of female active duty officers in the Navy over the time period 1950–1998?

h. What, if anything, can you say about the number of female active duty officers in the Navy in 2000?

i. Did the Air Force or the Navy experience the biggest gain in the number of female active duty officers during the time period 1950–1998? Explain the basis for your conclusion.

Organizing

1. The table on page 90 contains the average high and low temperatures in January and July for selected cities around the world. A scatterplot of the average maximum temperatures in January and July is shown below.

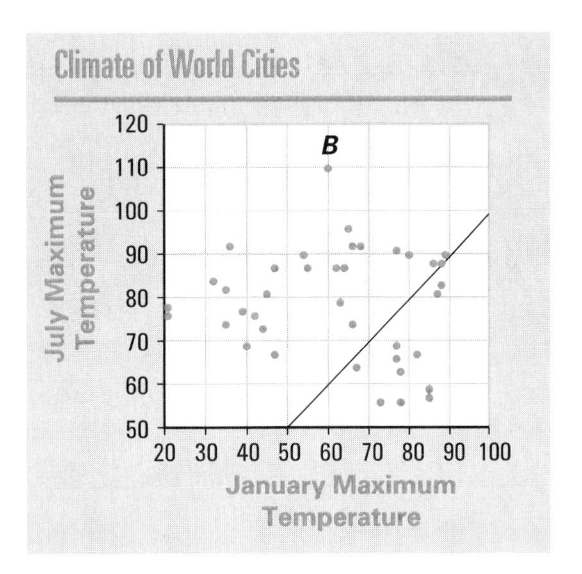

a. How many cities have nearly the same maximum temperature in January as in July? Which cities are these?

b. What is true about the cities represented by points located below the line? Where are these cities located geographically?

c. What was the temperature change for the city that is represented by point *B*? Which city is represented by point *B*?

d. Which city had the greatest change from January to July? Use the plot to find your answer and explain how you used the plot.

e. Suppose A stands for the temperature in January and U stands for the temperature in July. Which of the following describe(s) the region on the plot that contains the points representing cities warmer in January than in July?

$A > U \quad A < U \quad A = U \quad U = A \quad U > A \quad U < A$

Shade in that region on a copy of the graph.

f. Select a point in the shaded region and verify that it does represent a city that is warmer in January than in July.

Climate of Selected World Cities

City	Ave. January Temp. (in °F)		Ave. July Temp. (in °F)	
	Max.	Min.	Max.	Min.
Accra, Ghana	87	73	81	73
Amsterdam, Netherlands	40	34	69	59
Athens, Greece	54	42	90	72
Auckland, New Zealand	73	60	56	46
Baghdad, Iraq	60	39	110	76
Bangkok, Thailand	89	67	90	76
Beirut, Lebanon	62	51	87	73
Berlin, Germany	35	26	74	55
Bogota, Columbia	67	48	64	50
Bombay, India	88	62	88	75
Budapest, Hungary	35	26	82	61
Buenos Aires, Argentina	85	63	57	42
Cairo, Egypt	65	47	96	70
Calcutta, India	80	55	90	79
Cape Town, S. Africa	78	60	63	45
Casablanca, Morocco	63	45	79	65
Dublin, Ireland	47	35	67	51
Geneva, Switzerland	39	29	77	58
Hanoi, Vietnam	68	58	92	79
Hong Kong	64	56	87	78
Istanbul, Turkey	45	36	81	65
Jerusalem, Israel	55	41	87	63
Kabul, Afghanistan	36	18	92	61
Karachi, Pakistan	77	55	91	81
Lagos, Nigeria	88	74	83	74
Lima, Peru	82	66	67	57
London, UK	44	35	73	55
Madrid, Spain	47	33	87	62
Manila, Philippines	86	69	88	75
Melbourne, Australia	78	57	56	42
Mexico City, Mexico	66	42	74	54
Montreal, Canada	21	6	78	61
Moscow, Russia	21	9	76	55
Nairobi, Kenya	77	54	69	51
Osaka, Japan	47	32	87	73
Paris, France	42	32	76	55
Santiago, Chile	85	53	59	37
Sao Paulo, Brazil	77	63	66	53
Seoul, South Korea	32	15	84	70
Taipei, Taiwan	66	53	92	76

Source: *New York Times 2001 Almanac*. New York, NY: The New York Times Company, 2000.

2. Refer to the temperature chart in Organizing Task 1.

a. Make a scatterplot of the maximum (on the x-axis) and the minimum (on the y-axis) temperatures for January. Leave room to the left and below it to complete Part c.

b. Which city has the largest range in temperature for the month of January?

c. Below the scale on the x-axis, make a box plot of the maximum temperatures for January. To the left of the scale on the y-axis, make a box plot of the minimum temperatures for January.

d. Describe the variability in minimum temperature in January.

e. Describe what information the box plots add to the scatterplot.

f. Explain why you do or do not find this new type of plot to be useful.

3. For each statement, sketch a plot over time in one-year increments that illustrates the statement.

a. The cost of a computer is decreasing every year.

b. The cost of a delivered cheese pizza is increasing every year by approximately the same amount.

c. There is no predicting the cost of a stereo; some years it is up and others it is down.

d. The annual cost of electricity is increasing.

4. The table below gives the amount of daylight in minutes for Kalamazoo, Michigan, on the first day of each month in one year.

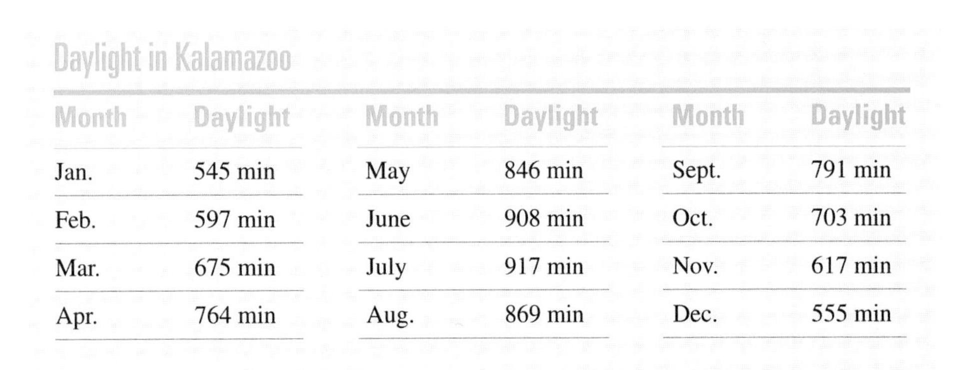

Daylight in Kalamazoo

Month	Daylight	Month	Daylight	Month	Daylight
Jan.	545 min	May	846 min	Sept.	791 min
Feb.	597 min	June	908 min	Oct.	703 min
Mar.	675 min	July	917 min	Nov.	617 min
Apr.	764 min	Aug.	869 min	Dec.	555 min

a. Prepare a plot over time of these data. Code January as 1, February as 2, and so on.

b. Describe the pattern you see in the plot.

c. Use your plot to estimate the number of minutes of daylight on October 15.

d. Explain how you could use your plot to estimate the number of minutes of daylight for any day of the year. How confident would you be of your estimate?

Reflecting

1. Suppose you were given a plot of the sales of various brands of athletic shoes, with the 1998 sales on the horizontal axis and the 2002 sales on the vertical axis. Each point represents the number of sales for a specific brand of athletic shoe. If the point representing your favorite brand lies below the line $y = x$, what can you conclude?

2. To answer a question, Toni and David each made a scatterplot of a set of data. Toni indicated that the area containing the correct points for the question was *above* the line $y = x$. David maintained the area was *below* the line $y = x$. Is there any way they could both be correct? Explain your reasoning.

3. Suppose you are studying a plot over time.

   a. How does the plot help you to see the range of the values over time?

   b. How does the plot help you to see the variability in the values over time?

   c. How can you detect from the plot where change is the greatest?

   d. How might you estimate the median y value?

   e. How might you use the plot to predict a value at a time in the future?

4. Refer to Investigation 2, Activity 3, on page 82 for the plot over time of average repair ratings for the Ford Mustang. Is the point (97.5, 3.855) on the plot? What, if anything, does this point represent?

5. How do you decide whether to use your calculator or computer or to use paper and pencil to produce a graphical display of data?

6. Look at a local newspaper over a week's time. Keep a record of the types of plots used to display data. Write a brief summary of your findings.

Extending

1. The following table gives the population (in thousands) of selected major cities in the United States for 1990 and 2000.

   a. Enter the data for the populations of the cities in 1990 and 2000 in your calculator or computer. Use List 1 for the 1990 data and List 2 for the 2000 data. Produce a scatterplot of the data.

   b. Overall, would you say the population in the cities increased or decreased? How can you tell this from the plot?

   c. Find the differences between the populations by subtracting the 1990 data from the 2000 data. Make a box plot of the differences. How can you tell from the box plot whether, overall, the population in the cities has increased or decreased?

Major U.S. Cities: 1990 and 2000 Population (in thousands)

City	1990	2000	City	1990	2000
Albuquerque, NM	385	449	Louisville, KY	269	256
Allentown, PA	105	107	Milwaukee, WI	628	597
Anaheim, CA	266	328	Newark, NJ	275	274
Anchorage, AK	226	260	Oklahoma City, OK	445	506
Atlanta, GA	394	416	Portland, OR	437	529
Aurora, CO	222	276	Reno, NV	134	180
Berkeley, CA	103	103	San Francisco, CA	724	777
Boise City, ID	126	186	St. Louis, MO	397	348
Cleveland, OH	506	478	St. Petersburg, FL	239	248
El Paso, TX	515	564	Syracuse, NY	164	147
Gary, IN	117	103	Topeka, KS	120	122
Hartford, CT	140	122	Washington, DC	607	572
Independence, MO	112	113			

Source: U.S. Census Bureau.

d. Use your calculator or computer to find the mean change. How does this relate to your earlier conclusion on population change?

2. Collect at least 20 ordered pairs of data from a situation of your choice in which the line $y = x$ has a useful interpretation. Make a scatterplot and write a paragraph describing the observations you can make from your plot. Be prepared to share your plot and observations with the class.

3. The following data are the number of immigrants to the United States, in millions, during the 1900s.

Decade	1910s	1920s	1930s	1940s	1950s	1960s	1970s	1980s
Immigrants	6.0	4.1	0.5	1.0	2.5	3.8	7.0	10.0

Source: Urban Institute. www.urban.org/pubs/immig/immig.htm

Make a plot over time of the number of immigrants since 1900. Describe the changes in the pattern of immigration. What might have caused the fluctuations?

Lesson 5 Looking Back

In this unit, you have worked together in groups to make sense of situations and solve problems involving data. You have learned how to analyze data using graphical displays such as histograms, box plots, and scatterplots. You have used visual patterns in data and summary statistics such as the median and interquartile range to describe distributions, to make decisions, to make comparisons, and to make predictions. The three tasks in this final lesson give you an opportunity to pull together the ideas you have developed in the unit.

1. In Project 3 of the first lesson, you collected data about the shoe length and stride length of the members of your class. Refer to those measurements in the Class Data Table. Working together as a group, construct appropriate plots and use the statistics you have learned to answer the following questions.

a. How do the distributions of shoe lengths and stride lengths compare?

b. Is there more variability in shoe length or in stride length?

c. Is there a relationship between shoe length and stride length?

d. Write a paragraph describing what you have learned about shoe lengths and stride lengths of the students in your class. (You may need to compute some additional summary statistics or make additional plots to illustrate your points.)

e. Compare your work here to your work in the first lesson. What are the most important things you have learned about analyzing data?

2. Every several years the *Places Rated Almanac* provides ratings of metropolitan areas in the United States and Canada. Each area is rated in several different categories on a scale from 0 to 100, with 100 being the highest rating. People can use these ratings as they consider relocating to a new city.

At the top of the next page are ratings from the *Places Rated Almanac* in six categories for 12 different major metropolitan areas. Using what you have learned from this unit, write a report on which of these twelve cities you think should be rated highest. Include suitable plots and statistics to support your choice.

Rating Cities

City	Cost of Living	Jobs	Education	Climate	Crime	Recreation
Atlanta	39	99	83	70	12	76
Boston	1	95	100	59	47	77
Chicago	9	87	99	16	2	97
Dallas	49	95	90	48	9	89
Detroit	15	81	71	30	17	97
Houston	60	95	71	80	25	85
Los Angeles	6	48	77	99	1	93
Miami	35	84	74	91	0	90
New York	0	42	95	71	1	92
Philadelphia	10	78	90	57	45	75
San Francisco	0	67	98	97	18	93
Washington, DC	5	93	98	56	38	89

Source: Savegeau, David and D'Agostino, Ralph. *Places Rated Almanac, Millennium Edition.* New York: Macmillan, 2000.

3. Each year the Recording Industry Association of America analyzes music sales for the year. The table below gives the break down of music sales by type of music.

Types of Music as Percent of Total Music Sales

Genre	87	90	95	96	97	98	99
Rock	45.5%	36.1%	33.5%	32.6%	32.5%	25.7%	25.2%
Country	10.6	9.6	16.7	14.7	14.4	14.1	10.8
R&B	9.0	11.6	11.3	12.1	11.2	12.8	10.5
Pop	13.5	13.7	10.1	9.3	9.4	10.0	10.3
Rap	3.8	8.5	6.7	8.9	10.1	9.7	10.8
Gospel	2.9	2.5	3.1	4.3	4.5	6.3	5.1
Classical	3.9	3.1	2.9	3.4	2.8	3.3	3.5
Jazz	3.8	4.8	3.0	3.3	2.8	1.9	3.0
Other[1]	6.2	8.8	10.1	8.2	9.4	11.3	9.1

1 Includes soundtracks, children's music, and other categories not shown separately. Totals may not add to 100 percent due to "Don't know/No answer" responses.

Source: Recording Industry Association of America, *1999 Consumer Profile.*

Use an appropriate plot to help investigate the change in percentage of sales during this period for three genres of your choice.

a. Which of your three music types showed the greatest change in percentage of sales over the period 1987–1999?

b. Which of your three music types had the greatest market share during the period? What was the percentage and when did it occur? Did you use the table or the plot?

c. Were there any years in the period 1987–1999 for which two of your three music types had the same percentage of sales? If so, what were the music types and what were the percentage of sales?

d. For which of your three music genres did the market share fluctuate the most? Explain your reasoning.

e. For which of your three genres would you feel *least* confident estimating the market share for the current year? Explain why you picked that genre.

Checkpoint

Patterns in data can be seen in graphical displays of the distribution and can be summarized using measures of center and variability.

a Describe the kinds of information you can get by examining

- a stem-and-leaf plot;
- a number line plot or histogram;
- a box plot;
- a scatterplot;
- a plot over time.

b Describe the kinds of situations for which each plot is most useful.

c How do you decide which measure of center to use to provide a summary of a distribution? How do you decide on which measure of variability to report?

d What is the effect on measures of center of transforming a set of data by adding a constant to each value or multiplying each value by a positive constant? On measures of variation?

Be prepared to share your group's descriptions and thinking with the entire class.

Own Your Own

Write, in outline form, a summary of the important mathematical concepts and methods developed in this unit. Organize your summary so that it can be used as a quick reference in future units and courses.

Patterns of Change

Unit 2

Lesson 1

Related Variables

Some things in life are always changing. The earth rotates on its axis and is in orbit around the sun, turning day into night and summer into winter. The prices you pay for food, clothing, and entertainment change from month to month. The cars on streets and highways change speed as traffic conditions change. Even the time you spend on homework assignments in mathematics changes.

Some changes can be observed; other changes can be experienced. There are changes that come and go quickly, called fads. Fads occur in games, clothing, music, and even in sports.

For example, one new sport began when some young daredevils in New Zealand found a bridge over a deep river gorge. They tied one end of a strong elastic cord (called a bungee) to the bridge, and then took turns tying the other end around their waists or feet. They began jumping off the bridge and bouncing up and down at the end of the cord to entertain tourists. Soon word of this new pastime got back to the United States. It wasn't long before some Americans tried bungee jumping on their own.

As you can imagine, bungee jumping is a very risky sport, especially if the jumper doesn't plan ahead very carefully. If the apparatus isn't right, the consequences could be fatal!

Some amusement parks around the world are installing bungee jumps to attract daredevils who want a thrilling ride. Those parks have important planning to do before opening for business. First, they need to make sure their bungee apparatus is safe. Then they also want to set prices in a way that maximizes profit.

Think About This Situation

Suppose the Five Star Amusement Park intends to set up a bungee jump.

a How could they design the bungee jump so that people of different weights could all have safe but exciting jumps?

b What patterns would you expect in a table or graph showing the expected stretch of a 50-foot bungee cord with different weights?

Jumper Weight (in pounds)	Cord Stretch (in feet)
50	?
100	
150	
200	
250	

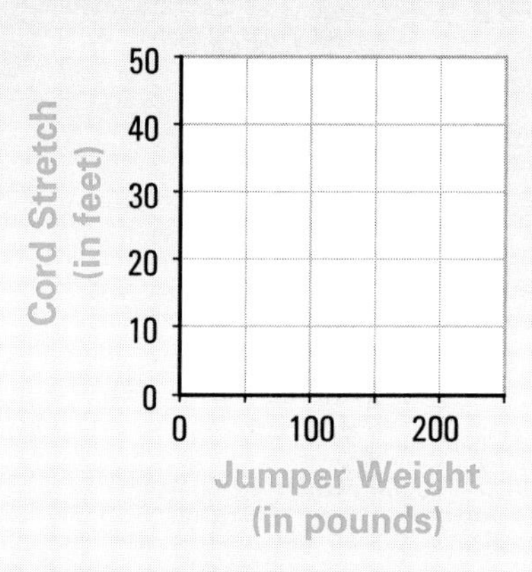

c How could they find the price to charge for each jump so the park could maximize profit?

d What other safety and business problems would Five Star have to consider to set up and operate the new bungee attraction safely and profitably?

INVESTIGATION 1 Modeling a Bungee Apparatus

The distance that a bungee jumper falls before bouncing back upward seems likely to depend on the jumper's weight. In designing the apparatus, it is necessary to know how far the elastic cord will stretch for different weights. It makes sense to do some testing before anyone takes a real jump.

1. In your group, make and test a simple model of a *bungee apparatus* using some rubber bands and small weights, such as fishing weights.

a. Loop together several rubber bands of the same size to make an elastic rope and then attach a weight as shown.

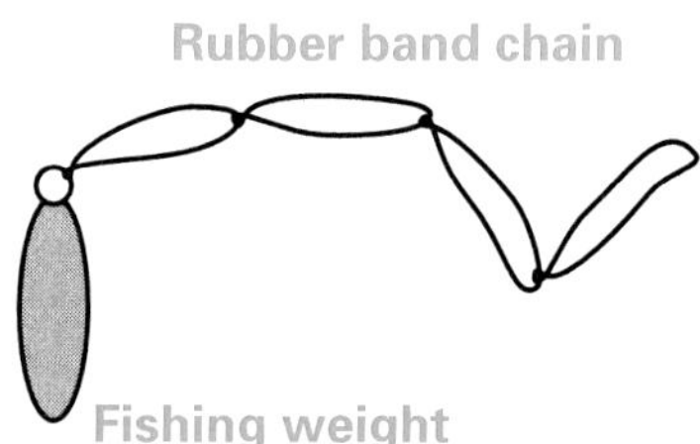

b. Use your model to collect test data for at least five different weights. Record your data in a table like this:

Bungee Test

Weight	Amount of Stretch

2. Now study the data from your experiment and write a short summary explaining, as carefully as possible, your answers to these questions:

a. How does the weight relate to the amount of stretch of the rubber bands?

b. How does the amount of stretch change as the weight changes?

3. Compare your results to those of other groups who might have used more, fewer, or different rubber bands; different weights; or different methods for collecting the data pairs.

The data from tests of your model bungee apparatus are ordered pairs of numbers in the form (*weight*, *stretch*). Looking for patterns in these kinds of data is often easier if the data pairs are presented in a graph like the one shown here.

The sample graph shows specific information (*weight*, *stretch*) and also a pattern in the relation between the two *variables*.

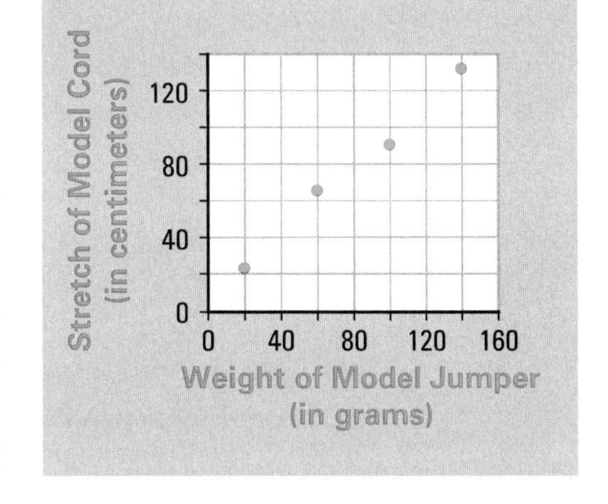

4. Make a coordinate graph using your group's experimental data. An example of such a graph is shown above. Use your graph to answer the following questions.

a. What does the pattern of points on your graph say about the pattern of change in the amount of stretch as the weight changes?

b. Is it reasonable to connect the points on your graph and use the resulting graph to predict the amount of stretch for weights other than those you tested? Try it and run some further tests to check your predictions.

c. How does the pattern in your group's graph compare to those of other groups that might have used different rubber bands, different cord lengths, or different weights?

Checkpoint

Data from experiments often suggest possible relationships between variables.

a What are the important variables in the design of a bungee apparatus?

b When change in one variable is related to change in another variable, the pattern of that relation can be described in *words*, with a *table* of sample data, or with a *graph* of sample data. What are the advantages and disadvantages of describing patterns of change in related variables by each of these methods?

Be prepared to share your group's thinking with the whole class.

In mathematical models of situations, the quantities that change are called **variables**. In many cases, we describe the relation between two variables by saying that one variable **is a function of** the other, especially if the value of one variable **depends on** the value of the other.

On Your Own

Suppose that in designing a bungee jump, Five Star Amusement Park began by considering the weight of a typical customer. Bungee cords of different lengths were tested.

a. Assuming a fixed weight, do you think the amount of stretch in a bungee cord is a function of its length? If so, write an explanation of the change in stretch you would expect as cord length changes.

b. Make a table like the one below for data in the form (*cord length*, *amount of stretch*). Complete the table showing a pattern that you think might occur.

Cord Length (in meters)	5	10	15	20	25	30
Amount of Stretch (in meters)	1					

c. Sketch a graph of the (*cord length*, *amount of stretch*) data showing the pattern of change that you would expect to occur.

d. Describe an experiment, similar to the one in the investigation, to find the likely relation between *cord length* and *amount of stretch* for some fixed amount of weight.

MORE

Modeling • Organizing • Reflecting • Extending

Modeling

Each of the following situations gives you some information about variables and relations among those variables. Then you are asked to make tables or graphs of the given information and to answer questions about the relationships. Keep an eye out for interesting patterns in the changes of related variables.

1. Before the amusement park installs the bungee jump apparatus, some business decisions must be made.

a. Make a list of all the things you can think of that need to be done or decided before a park is ready to open a bungee attraction.

b. In the list of business decisions for Part a, which involve numbers that can be determined? Keep in mind that every business tries to make a profit, and the profit is equal to the total money collected minus the expenses. What aspects of the business will be affected by the decisions?

c. One key decision is the price to charge each jumper. That price will certainly influence the total number of customers each day. Data from a survey of park customers were used to predict the following numbers of bungee jumpers:

Price Charged	\$20	\$30	\$40	\$50	\$60
Daily Customers	100	70	40	20	10

- Describe the pattern relating price and predicted number of customers.
- Make a coordinate graph of the (*price*, *customers*) data and explain how the pattern of that graph matches the pattern in the data. Remember to plot *price* on the horizontal axis and *customers* on the vertical axis.
- Predict the number of customers if the price is set at \$25, at \$45, and at \$100. Explain the reasoning that led to your predictions.

d. The data on possible prices and the number of customers can be used to predict daily income for the bungee jump. If the price is set at \$20 and the predicted 100 customers come, what would be the park's daily income from the jump?

Use the data in Part c to make a table of (*price, income*) predictions. A sample table is at the top of page 103.

Price Charged	\$20	\$30	\$40	\$50	\$60
Daily Income					

- Make a graph of the (*price*, *income*) data.
- Write a description of the relation between price and daily income that is shown by your table and graph.
- What ticket price seems likely to give the greatest daily income?

2. In one test of the bungee apparatus at the amusement park, a radar gun was used to study the drop of a test jump. The radar was connected directly to a computer that produced the graph shown here.

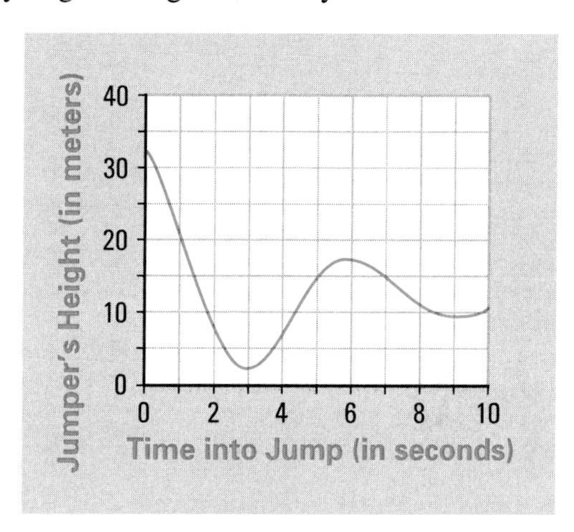

a. What does the pattern of the graph tell you about the jumper's motion?

b. According to the graph, approximately what was the jumper's height above the ground

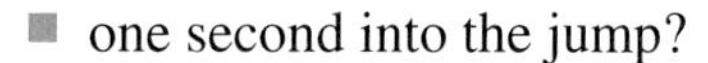

- one second into the jump?
- two seconds into the jump?
- five seconds into the jump?

c. When did the jumper come closest to the ground, and how close was that?

d. How high did the jumper bounce on the first rebound? When did the jumper reach that height?

e. When and at what height was the jumper falling fastest?

f. When and at what height did the cord begin to slow the jumper's rate of fall?

3. The test team using the radar gun took the gun for a ride on the amusement park Ferris wheel. They aimed it at the ground during two nonstop trips around on the wheel, giving a graph relating height above the ground to time into the trip. What is the highest the rider will be above the ground?

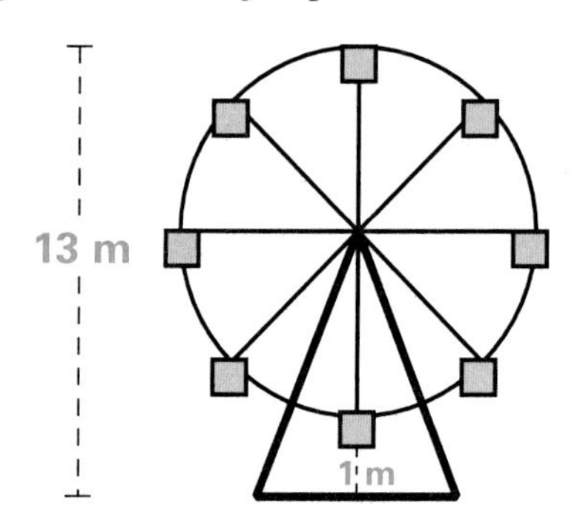

a. On a set of axes like the one at the right, sketch a graph that you believe would fit the pattern relating *height* above the ground to *time* during the Ferris wheel ride. The total time for the trip was 100 seconds. Write an explanation of the pattern in your graph. **Hint:** You might experiment with a bicycle wheel as a model of a Ferris wheel; as you turn the wheel, how does the height above the ground of the air valve stem change?

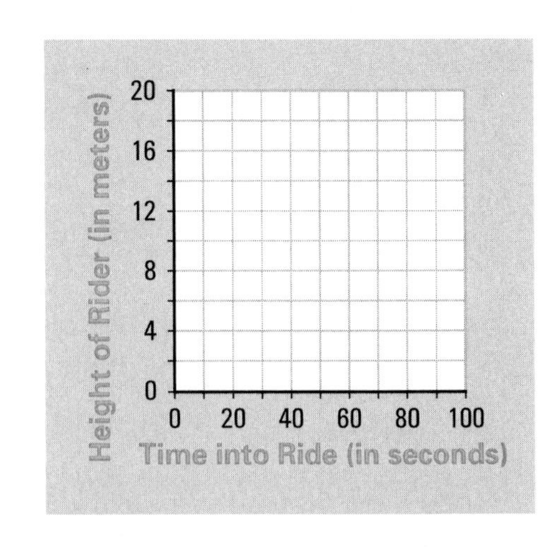

b. Given below is the graph of (*time*, *height*) data for one Ferris wheel test ride. Write an explanation of what the graph tells about that test ride.

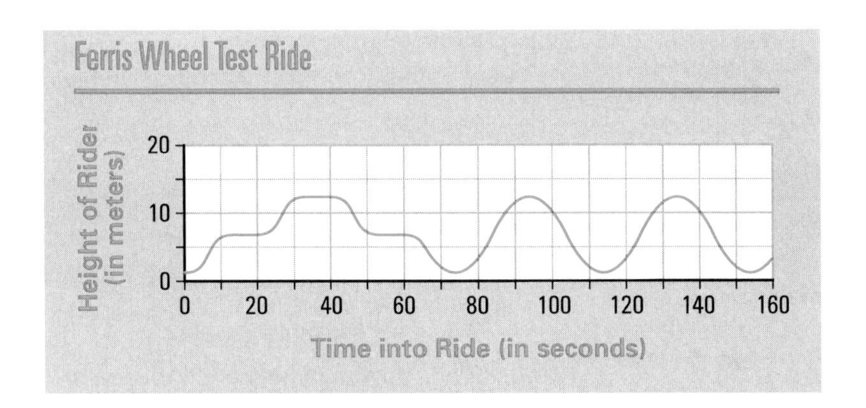

c. Given below are some (*time*, *height*) data for a second Ferris wheel test ride. Write a short description of the pattern of this test ride.

Time (in seconds)	0	2	5	10	15	20	22.5	25	30	35	40	42.5	45	50	55	60	62.5
Height (in meters)	1	1	3	3	11	11	13	11	11	3	3	1	3	3	11	11	13

d. Sketch a graph of the relation shown in the (*time*, *height*) data in Part c. Do you think the table or the graph better shows the pattern of change in height as a function of time?

4. When new motion pictures are released, they first appear in theaters all over the country. Then video cassettes, DVDs, and laser discs are made and sold to video rental companies which rent the movies to people for home viewing. Suppose that a video rental store near you buys 30 copies of a new release and begins renting them.

a. Stores keep records of the number of times that movies rent each week. How do you think the number of rentals per week of a new release will change over time?

b. Make a table of (*week*, *rentals*) data showing what you believe is a reasonable pattern relating *time, in weeks, since the video became available* and *number of rentals of that video per week* for weeks 1 to 20.

c. Sketch a graph of the relation between *time since release* and *number of rentals per week* that matches your data in Part b.

5. The Wild World Amusement Park has a huge swimming pool with a wave machine that makes you feel like you are swimming in an ocean. Unfortunately, the pool is uncovered and unheated, so the temperature forecast for a day affects the number of people who come to Wild World. On a typical summer day, when the forecast called for a high temperature of 32°C (or 90°F), about 3,000 people visited the park. On another day, when the forecast called for a high temperature of 20°C (or 68°F), only 250 people came for the ocean-wave swimming.

a. Make a table of (*temperature*, *swimmers*) data that you think shows the most likely pattern relating the two variables.

b. Sketch a graph of the sample data in Part a.

c. Describe in words the patterns of your data table and your graph.

Organizing

1. As you worked on the investigations of Unit 1, "Patterns in Data," you studied many different variables such as stride length, calories in fast foods, and production costs of movies. Examine the following data from a class survey on student time spent watching television, talking on the telephone, and working on homework.

Student Time (in number of hours per week)

TV	Phone	Homework	TV	Phone	Homework	TV	Phone	Homework
10	4	10	9	10	16	10	5	9
3.5	13	25	7	8	15	8	8	18
8	4	10.5	12	4.5	8	18	4	6
5	8	15	7	5.5	13	2	10	19
14	1	5	4	9	16	13	4	9
12	4	6	6	6.5	12	1	7	15
21	5	5	10	7.5	14			

a. Prepare a box plot and a five-number summary of the survey data for each of the three variables. Explain what these graphs and statistics tell about the variables.

b. What relation would you expect between number of hours per week spent watching television and number of hours doing homework? Complete the following sentence: *As time watching television increases, time doing homework ....*

c. What relation would you expect between number of hours per week spent talking on the telephone and number of hours doing homework? Complete the following sentence: *As time talking on the telephone increases, time doing homework ....*

d. Sketch graphs showing the patterns you described in answer to Parts b and c. Then make scatterplots of the data given in the table and comment on the match between your expected patterns of change and the patterns given by the survey data.

2. In Unit 1, "Patterns in Data," you saw that changing each value of a data set by adding a constant number changed the mean and mean absolute deviation in a predictable way. Suppose a set of test scores has a mean $\bar{x}$.

a. If 10 points are added to each test score, write an expression for the new mean. Describe the new mean absolute deviation.

b. If a number c is added to each test score, write an expression for the new mean. Describe the new mean absolute deviation.

c. Make up a list of 10 test scores and then find the median and interquartile range. Add 5 points to each score and then find the new median and interquartile range.

d. How do the median and interquartile range of a set of data change if a constant number is added to each data point?

3. Shown at the right is a pattern of "growing" squares made from toothpicks.

a. Study the pattern and draw a picture of the next likely shape in the pattern.

b. Make a table similar to the one below. A toothpick is 1 unit long. The small square has 1 square unit of **area**.

Number of Toothpicks on One Side of the Shape	1	2	3	4	5	6
Area of the Shape (in square units)		4				

- Sketch a graph of the relation shown in the (*side length*, *area*) data.
- Describe in words the relation between the area of a shape and the number of toothpicks on a side.

c. Complete a table similar to the one below. Recall that the distance around a shape is called its **perimeter**.

Number of Toothpicks on One Side of the Shape	1	2	3	4	5	6
Perimeter of the Shape (in units)		8				

- Sketch a graph of the relation shown in the (*side length*, *perimeter*) data.
- Describe in words the relation between the perimeter of a shape and the number of toothpicks on a side.

d. Describe similarities and differences in the patterns of change in Parts b and c.

4. A manatee is a large sea mammal native to Florida waters that is listed as endangered. The chart below gives the number of manatees killed in watercraft collisions near the Gulf Coast of Florida every year from 1977 through 2000.

Manatee/Watercraft Mortalities

Year	Number of Manatees Killed	Year	Number of Manatees Killed	Year	Number of Manatees Killed
1977	13	1985	33	1993	35
1978	21	1986	33	1994	49
1979	24	1987	39	1995	42
1980	16	1988	43	1996	60
1981	24	1989	50	1997	54
1982	20	1990	47	1998	66
1983	15	1991	53	1999	82
1984	34	1992	38	2000	78

Source: www.savethemanatee.org/76-00%20Mortality%20Table.htm

a. Prepare a plot over time of the number of manatees killed. Describe the pattern you see in the plot.

b. During what one-year period was there the greatest change in number of manatees killed? How can you tell this from the graph? From the data table?

c. Do you think it is the calendar year or some other factor or factors that are causing the pattern of change in manatees killed by watercraft? Explain your reasoning.

Reflecting

1. Do you think that all variables are related to each other? For each of the following pairs of variables, determine whether changes in one variable are related to changes in the other and describe the pattern, if one exists.

   a. the *height* of an adult and the *cost* of a movie ticket

   b. amount of *time* spent on homework and amount of *money* spent on lunch

   c. calendar *date* and number of *daylight hours*

   d. TV *channel* setting and TV *volume* setting

   e. the *pitch* played by plucking a string and the *tension* on that string

2. Weather reports give forecasts of many important variables—temperature, humidity, rainfall, wind-chill, atmospheric pressure, and so on. Explain why it is useful to be able to predict patterns of change in those variables as time passes.

3. Notice some of the variables that are around you every day. Think about how changes in one variable seem to cause changes in others, or how changes in one variable occur as time passes in a day, week, month, or year. For example, as meal time gets closer, most people get more hungry, but after a meal the hunger decreases! Describe two variables that you notice every day that seem related. Explain how changes in one of those variables seem related to changes in the other. Sketch a graph or make up a table of possible values illustrating the pattern you notice.

4. In this lesson's investigation you used words, tables, and graphs to describe patterns of change. Do you prefer one of these forms over the others? Why?

Extending

Complex business decisions often involve patterns relating several variables. The following case studies will test your ability to think like a good businessperson.

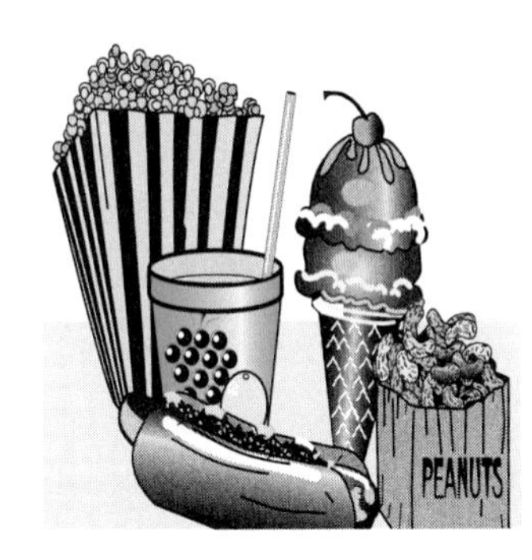

1. At most amusement parks, the sales of food and drinks bring in a great deal of money to the park operators. A typical park might sell dozens of kinds of snacks, sandwiches, and drinks. Pick just one food item—such as popcorn, hot dogs, ice cream, or soft drinks—and consider the variables that affect *profit* from sales of that item.

   a. List several factors that will affect profit from sales of your food item.

   b. Which of the factors you listed in Part a can be quantified or expressed as numerical variables, and which of those numerical variables can be changed by the park management?

c. Pick at least three variables that affect profit from your food item and sketch graphs showing the patterns relating each variable to profit. For example, you might consider the relation between the park's cost of buying hot dogs and the profit from sales of those hot dogs. What pattern would you expect in typical (cost, profit) data pairs?

d. Explain in words the pattern of each relation graphed in Part c.

e. Make up sample tables of values showing the likely relations between the variables you identified and food sales profit.

f. Given at the right is a graph of (hot dogs sold, hot dog profit) data from one amusement park. Find the coordinates of each labeled point on the graph and explain what those coordinates tell about the hot dog business at the park.

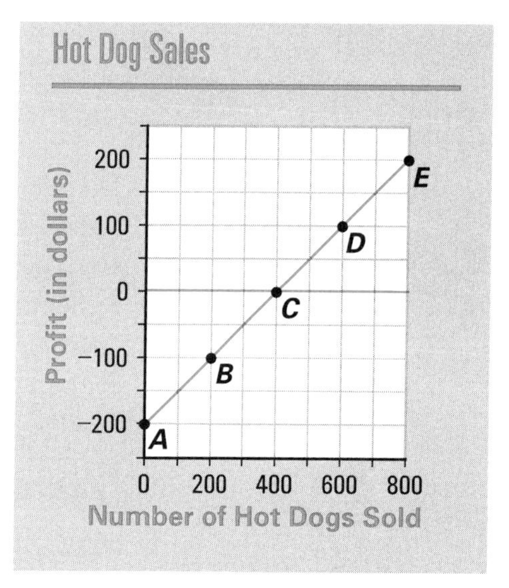

2. Suppose that for a fund-raising event, your school can rent a climbing wall. Students will be charged a small fee to borrow safety equipment required to climb the wall. Complete the following activities to help find the likely income from the climbing wall.

a. Do a survey of your class to find out how many customers you might expect for various possible prices. Complete a table like the one shown here.

Climb Price (in dollars)	1.00	2.00	3.00	4.00	5.00	6.00	7.00	8.00
Number of Customers								

b. Use your class survey to estimate the number of customers from your entire school for various possible prices. Make a table similar to the one in Part a.

c. Sketch a graph of the *(climb price, number of customers)* data. Describe the relation between those two variables. How will the number of customers change if the price is changed?

d. Use your market survey information displayed in the *(climb price, number of customers)* table to estimate the income that would be earned by the climbing wall for various possible prices. Display that *(climb price, income)* data in a table and in a graph and describe the pattern relating the two variables.

e. What does your work in Part d suggest as the best possible price to charge?

Lesson 2

What's Next?

The United States government—as well as local, state, and other national governments around the world—does lots of counting. Government agencies make hundreds of census counts every year. They keep track of jobs and unemployment, automobiles, accidents, animals, illnesses, forests, and farmlands. Every ten years, the U.S. Census Bureau counts every American citizen.

By the year 2000, the world's population was estimated to be about 6.1 billion. Some projections estimate it will increase by 50% by 2050.

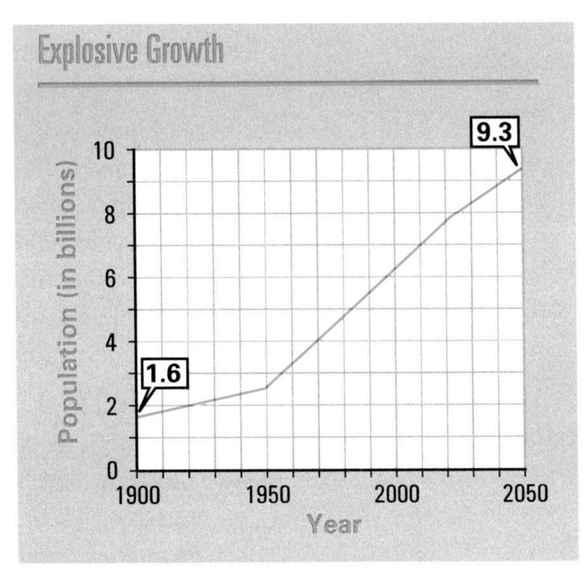

Source: *World Population Prospects*, The 2000 Revision Highlights. United Nations, 2001. www.un.org/esa/population/wpp2000.htm

Think About This Situation

The population of our world changes rapidly—in 15 years it is expected to grow by over one billion people.

a What are some of the major factors that might influence population change in a country?

b Why would it be important to know year-to-year population changes, and how you could estimate those changes without a full census survey?

INVESTIGATION 1 People-Watching

Brazil is the largest country in South America, and its 2000 population was about 170 million people. Census statisticians estimated the population of Brazil from one year to the next using small surveys and the following facts about patterns of change:

Population Change in Brazil

- Based on recent trends, births every year equal about 2.1% of the total population.
- Deaths every year equal about 0.6% of the total population.
- The net change due to births and deaths is an increase of 1.5% each year.

Source: *2000 World Population Data Sheet*, Population Reference Bureau. http://www.prb.org/pubs/wpds2000

1. Working with your group, use these facts to estimate Brazil's population for 2001.

a. What was the estimated change in Brazil's population by 2001 due to
- births?
- deaths?
- both causes combined?

b. Explain why the net population change due to births and deaths was an increase of about 1.5% each year.

c. What was the estimated total Brazilian population for 2001?

2. Using your group's estimation for the 2001 population as a basis, calculate estimates for the change and total population in 2002; then for 2003, 2004, 2005, and 2006. Record your estimates in a table like the one below.

Population Estimates

Year	Change (in millions)	New Total (in millions)
2002	_____	_____
2003	_____	_____
2004	_____	_____
2005	_____	_____
2006	_____	_____

The 2000 U.S. census report said that there were 276 million residents in the fifty states and territories at that time. Before the ink was dry on that report, the actual population had changed. So the Census Bureau makes annual estimates of the change based on current trends.

Population Change in the U.S.

- Births every year will equal about 1.5% of the total population.
- Deaths every year will equal about 0.9% of the total population.
- Immigrants from other countries will add about 0.8 million people each year.

Source: *2000 World Population Data Sheet*, Population Reference Bureau.http://www.prb.org

Using the above statistics, work with your group on Activities 3 through 5 to see how statisticians estimated the U.S. population after 2000.

3. If the 2000 U.S. population was 276 million:

a. What was the estimated change in population by 2001 due to

- births and deaths?
- immigration?
- all causes combined?

b. What was the estimated total U.S. population in 2001?

4. Using your estimate for 2001 population as a basis, calculate estimates for the change and total population in 2002; then for 2003, 2004, 2005, and 2006.

5. Describe the pattern of change in the U.S. population estimates as time passed, and compare that pattern to the change in Brazil's population estimates.

Checkpoint

In population studies of Brazil and the United States, you made estimates for several years, based on growth trends from the past.

a What calculations are needed to estimate population growth from one year to the next in the two different countries?

b Using the word *NOW* to stand for the population of the United States in any year, write an expression that shows how to calculate the population in the *NEXT* year.

Be prepared to compare your expression relating* NOW *and* NEXT *with those of other groups.

On Your Own

Complete these tasks using the population statistics for Brazil and the United States given in the investigation.

a. Write an expression using *NOW* and *NEXT* that shows how to use the population of Brazil in one year to calculate the population in the next year.

b. If the birth rate in the United States increased to 2.1%, like Brazil, how would that change the population increase from one year to the next for the United States? Estimate the 2001 population using the larger birth rate.

INVESTIGATION 2 The Whale Tale

Human beings belong to a class of the animal kingdom called *mammals*. There are hundreds of different kinds of mammals, from tiny mice and flying bats to giant elephants and whales.

Commercial hunting has pushed some large mammals close to extinction. For example, in 1986 the International Whaling Commission declared a ban on commercial whale hunting to protect the small remaining stocks of different types of whales.

Scientists make census counts of whales to see if the numbers are increasing. They do this to help determine if endangered populations should be downlisted to threatened or removed entirely from the list of endangered and threatened wildlife. It's not easy to count whales accurately. A research report about the bowhead whales of Alaska said that the 1993 population of that stock was somewhere between 6,900 and 9,200. It also said that the difference between births and natural deaths gives an increase of about 3.1% per year. Alaskan Inuits are allowed to hunt and take about 50 bowhead whales each year for their livelihood.

Source: nmml.afsc.noaa.gov/CetaceanAssessment/bowhead/bmsos.htm

1. Assume the 1993 bowhead whale population was 6,900.

a. What one-year change in that population would be due to the difference between births and natural deaths?

b. What one-year change in that population would be due to the Alaskan Inuit hunting?

c. What would you predict for the total 1994 population?

2. Using the word *NOW* to stand for the whale population in one year, write an expression that shows how to calculate the population in the *NEXT* year.

In studies of population increase and decrease, we often want to predict change over many years, not simply from one year to the next. Some calculators and computer software can use the answer to a calculation in the next calculation. This can be very helpful when repeating *NOW* and *NEXT* calculations. With a calculator, this often is done using a "last answer" function or key. In a spreadsheet, just refer to the cell that holds the calculation for the previous year.

The following calculator procedure produces estimates of the bowhead whale population up to the year 2005 using the 1993 population estimate of 6,900 as the *starting number*. (Some calculators use an EXECUTE key rather than the ENTER key.) A sample response is given, as well. Modify the procedure, if necessary, to work with your technology. The display mode is set so no digits to the right of the decimal point are displayed.

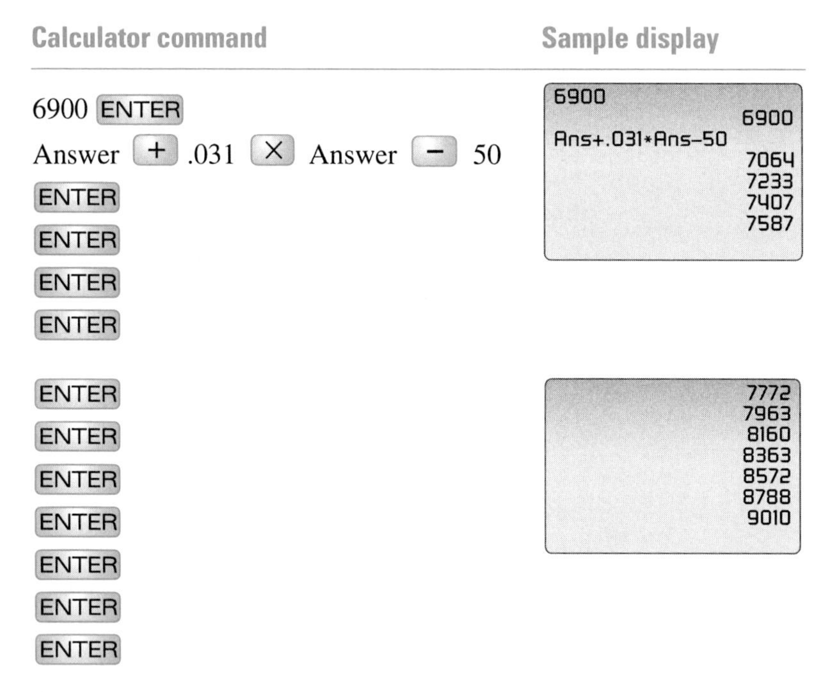

3. Explain the purpose of each keystroke or command in this procedure.

4. Modify the given steps to find whale population estimates starting from the high figure for 1993 of 9,200. Make a table showing the estimates for the years up to 2005.

5. In the 1990s, Canadians and Russians requested and received permission to harvest a limited number of bowhead whales. Test the effects of a larger hunt on the bowhead whale population. For example, suppose a total of 100 whales are hunted each year, instead of 50. Compare the results until the year 2005 in this case with those for a hunt of 50 whales, starting from the high figure for 1993 of 9,200.

6. How are the calculations for predicting whale populations similar to, and how are they different from, those for predicting human populations of the United States and Brazil?

Checkpoint

In this study of whale populations, you again made estimates for several years, based on growth trends from the past.

a What calculations must you do to estimate the change in number of whales from one year to the next?

b Explain how to use your technology's "last answer" function to calculate the total population in the next year.

c Explain how your calculator or computer software can be used to predict the total population many years ahead.

Be prepared to share your procedures with the entire class.

On Your Own

Recall that the Census Bureau estimates U.S. population growth based on birth, death, and immigration data. The 2000 U.S. population was 276 million, with a birth rate of 1.5%, a death rate of 0.9%, and about 0.8 million people immigrating to the United States each year.

a. Modify this calculator procedure to work with your technology. Use your procedure to estimate the population in the year 2010 and the year 2020. Compare the population estimates for the years 2000, 2010, and 2020.

276 ENTER

Answer + .006 × Answer + .8 ENTER

ENTER

ENTER

⋮

b. Explain what calculations are being done at each step of the procedure.

c. Suppose the U.S. immigration rate increased to 2 million per year. Calculate the population in the years 2010 and 2020 based on this assumption.

d. How would the U.S. population change if 2 million people each year *leave*, rather than enter, the country? Compare the results for the years 2010 and 2020 in this case to those in Part c.

MORE
Modeling • Organizing • Reflecting • Extending

Modeling

1. Recall that data on population growth showed a U.S. population of 276 million in 2000, a growth rate of 0.6%, and 0.8 million immigrants annually. Use your calculator or computer software and this information to answer the following questions.
   a. When will the population reach 300 million?
   b. When will the population double?

2. The People's Republic of China is the country with the largest population in the world. According to the 2000 *World Population Data Sheet,* approximately 1.2 billion people lived in China in 2000. Although families are encouraged to have only one child, in 2000 the population of China was growing at a rate of 0.9% per year.
   a. Predict the population of China for each of the next 10 years and record your predictions in a data table.
   b. When will the population of China reach the 2 billion mark?
   c. Using the word *NOW* to stand for the population in any year, write an expression that shows how to calculate the population in the *NEXT* year.
   d. Suppose that 7 million people leave China each year for other countries. How would this estimate affect the growth of the population of China over the next 10 years? (7 million = 0.007 billion)
   e. Using *NOW* to stand for the population in any year, write an expression that shows how to calculate the population in the *NEXT* year under the condition that 7 million Chinese leave annually.
   f. Experiment with different values for the growth rate and the number of people leaving China each year. Search for a balance that will lead to *zero population growth* in China.

3. According to the 2000 *World Population Data Sheet,* the country with the second largest population in the world is India, with approximately 1 billion people in 2000. The birth rate in India is about 2.7% of the total population each year, and the death rate is approximately 0.9% of the total population each year. Also, about 4.5 million people leave the country each year.
   a. Estimate the population of India for each of the 10 years from 2000 to 2010. Record your estimates in a table.
   b. When will the estimate of India's population reach 2 billion?

c. Using *NOW* to stand for the population in any year, write an expression that shows how to calculate the population in the *NEXT* year.

d. What combinations of growth rate and number of people leaving the country could lead to *zero population growth* in India?

4. Suppose that beginning in 1993, the hunting of bowhead whales was allowed to increase to 200 whales per year.

a. If the 1993 population was 6,900, what effect would that change have on the population over the next 10 years? Record your estimates in a data table. What if the 1993 population was actually 9,200?

b. If the 1993 population was actually 8,000, what hunting limit would produce a stable population?

c. Scientists studying whales have also tried to estimate populations in earlier years, before good census data were available. For example, commercial bowhead whale hunting was banned in 1914 because the stock was hunted to such a low figure.

- Assume the 1993 population was 8,000 (about midway between low and high estimates); that growth has been about 3% for some time; and that hunting is so small that it can be ignored. You can estimate the population for earlier years using a procedure similar to the one here.

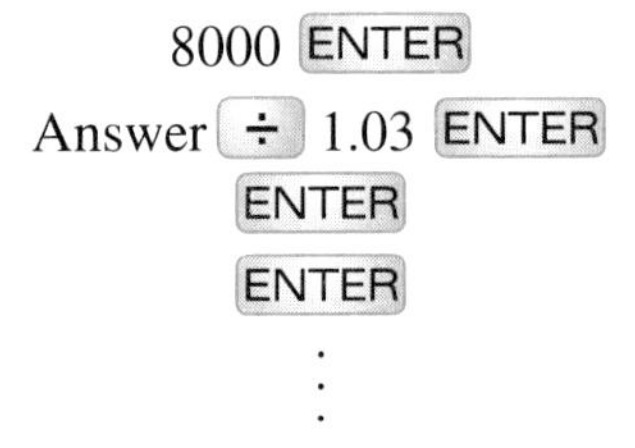

Report estimates for 1983, 1973, 1963, 1953, 1943, 1933, 1923, and 1913.

- Using *NOW* to stand for the whale population in any year and *THEN* for one year earlier, write an expression showing how to calculate *THEN* from *NOW*.

5. If money is invested in a savings account or a business or real estate, its value usually increases by some percentage each year. Suppose that when a child is born the parents set aside \$500 in a special savings account that earns interest at the rate of 4% each year.

a. What will the value of that account be after 1 year? After 2 years? After 5 years? After 18 years, when the child is ready to graduate from high school?

b. Using *NOW* to stand for the savings account value at the end of any year, write an expression for calculating the value of the account at the end of the *NEXT* year.

c. Suppose the interest rate is 8% instead of 4%. Calculate the value of the savings account after 1, 2, 5, and 18 years.

- How do these values differ from those calculated in Part a?
- How would the *NOW-NEXT* expression for calculating the value of the savings account in this situation differ from that in Part b?

d. How will your answers to Parts a and b change if the account continues to earn 4% interest, but the family deposits an additional $500 each year?

Organizing

1. The studies of populations changing over time can be represented with graphs if you form ordered pairs of (*year*, *population*) data. Recall that the 2000 population of Brazil was 170 million people and that the growth rate each year is about 1.5%. Use your calculator to plot the (*year*, *population*) data for each ten-year period from 2000 to 2050.

a. Make a sketch of the plot and write a brief description summarizing the pattern of the plotted data.

Note: Make the sketches for Parts b and c on the same set of axes.

b. Sketch the pattern of (*year*, *population*) data you would expect in Brazil if the birth rates increased.

c. Sketch the pattern of (*year*, *population*) data you would expect in Brazil if the birth and death rates were equal.

d. Explain how the pattern of points on each graph shows the *NOW* to *NEXT* change in the population.

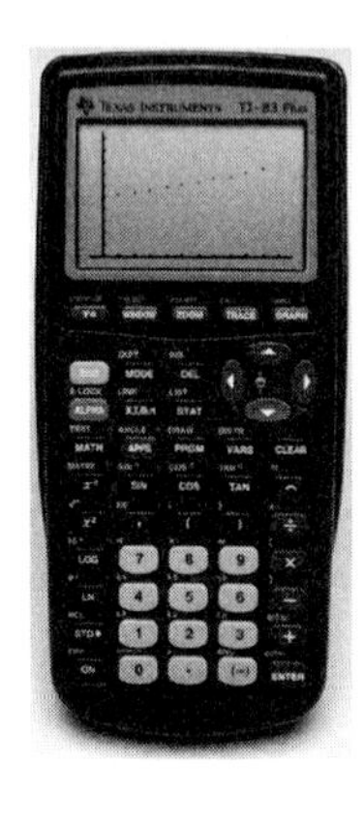

2. Calculate how a $500 bank balance would change over 10 years with 4% annual interest and no withdrawals. Record your results in a data table, and then use your calculator to plot the (*year*, *bank balance*) data.

a. Make a sketch of the plot and write a brief description of the pattern in the plotted data.

Note: Make the sketches for Parts b, c, and d on the same set of axes.

b. Sketch the pattern of (*year*, *bank balance*) data you would expect if the interest rate was reduced to 3%.

c. Sketch the pattern of (*year*, *bank balance*) data you would expect if the interest rate was doubled to 8%.

d. Sketch the pattern of (*year*, *bank balance*) data you would expect if withdrawals exactly equaled interest earned each year.

e. Explain how the pattern of points on each graph shows the *NOW* to *NEXT* change in each bank account.

3. Explain what the calculator procedure does here. Predict what the result will be if this procedure is used with starting numbers different from 10. Check your ideas by modifying the procedure, if necessary, and then executing the procedure.

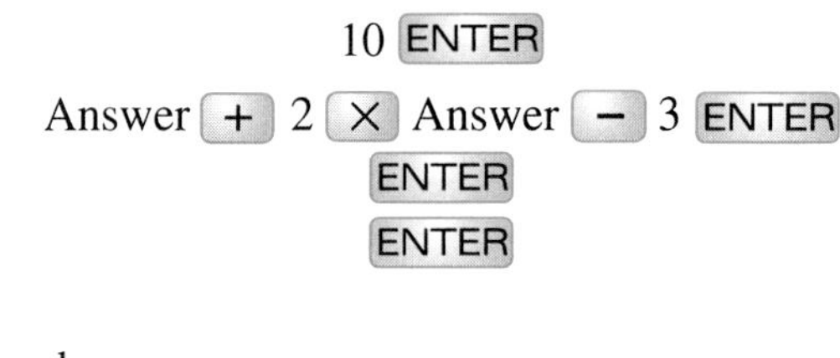

4. Investigate the following two expressions relating *NOW* and *NEXT* using your calculator or computer. Try several different possible choices for starting values. Use the same starting values with each expression.

$NEXT = NOW + 0.05 \times NOW \qquad NEXT = 1.05 \times NOW$

What appears to be true about these two expressions? Write an explanation for your findings.

Reflecting

1. The models of population change studied in this lesson are somewhat different from those involved in the bungee modeling investigation. Look back over the examples of the investigations in this lesson and see if you can figure out the meaning of the lesson title "What's Next?" What do the prediction models have in common that involves the word *NEXT*? How does the word *NOW* get involved in describing these situations?

2. Compare the use of a "last answer" function on your calculator or computer to the equation relating *NOW* and *NEXT* for calculating the total population at a particular point in time.

3. In the population model of this lesson, you made estimations based on given assumptions about how the populations would change. Do you know anything about differences between Brazil and the United States that would help explain the differences in assumptions?

4. Invent a situation that could be modeled as follows:

Starting value = 7,000

$NEXT = NOW - 0.12 \times NOW + 600$

Extending

1. The kinds of models of change used in studying populations are sometimes quite different from the ones you have investigated so far. For example, many psychologists study the way people learn and remember information. Suppose that when school closes in June you know the meaning of 500 Spanish words, but you don't study or speak Spanish during the summer vacation.

a. One model of memory suggests that during each week of the summer you will forget 5% of the words you know at the beginning of that week. Make a table showing the (*weeks*, *words remembered*) data pairs for 10 weeks and describe the pattern of data in the table.

b. A second model suggests that you will forget 20 words each week. Make a table showing the (*weeks*, *words remembered*) data pairs for 10 weeks following this model and describe the pattern of data in that table.

c. Graph the data from the two models and describe the patterns of data in those graphs.

d. How would answers to Parts a through c be different if the number of Spanish words you knew at the start of summer was 300?

e. Which model do you think best represents memory loss? Explain your reasoning.

f. Suppose again that you know the meaning of 500 Spanish words when school closed, and 10 weeks of summer are gone. You decide to do an intensive vocabulary review for the remaining 2 weeks before school starts. If you are able to regain 20% of your missing vocabulary each week, how many words will you know when school begins? Which model of memory loss did you assume for the first 10 weeks?

2. The International Whaling Commission considers three factors in its population estimates that influence hunting policy: current population, percent increase by natural means, and number of kills from hunting allowed.

a. Experiment with the numbers for bowhead whales in Alaska. Try different population estimates, and then for each estimate, choose different growth rates and different numbers of hunting limits to see the long-run effects on the whale population. In a table, record your choices for the three factors and the corresponding population changes over time. What different results can you get over a period of 20 years?

b. What different graph patterns can occur in the various cases?

3. Study the calculator procedure in Task 4 Part c of the Modeling section. Explain why the division operation (÷ key) is the proper choice for estimating populations in previous years.

Lesson 3

Variables and Rules

Important problems often involve variables that are related by regular patterns. Those patterns make it easy to calculate the value of one variable if the values of others are known. For example, many American teenagers have part-time jobs after school, on weekends, and in the summer. A first job might pay as low as the minimum wage. Even at a low rate, however, the money you earn can add up if you are a good saver. Suppose you earn $5.25 per hour. Here is a table showing some (*time worked*, *money earned*) data.

Time Worked (in hours)	0	1	2	3	4	5	10	15
Money Earned (in dollars)	0	5.25	10.50	15.75	21.00	26.25	52.50	78.75

Think About This Situation

The table shows money earned for a sample of hours worked. But the payroll computer for a company with many workers will have to be able to calculate pay for any number of hours.

a In this case, pay increases by $5.25 from one hour to the next. Using the words *NOW* and *NEXT*, how could you write that rule with an equation that begins *NEXT* = __________?

b Using the letters *H* (for number of hours worked) and *E* (for number of dollars earned), how could you write a rule that gives earnings for *any* number of hours worked with an equation that begins *E* = __________?

c How could you calculate the money earned for any number of hours worked, such as 23 or 42, and for fractions of hours, such as 8.25 hours or 12.5 hours?

d Why might rules like those called for in Parts a or b be more useful than tables or graphs of the (*time worked*, *pay earned*) data?

INVESTIGATION 1 Money Matters

Earning and spending money is an important part of daily life for most Americans. We spend a lot of time thinking about ways to earn, to save, or to shop. Unfortunately, that quite often means figuring out how to pay off bills or loans.

Installment Buying Typical American shoppers borrow money for large purchases like houses, cars, furniture, or stereo systems. Suppose your family finds a special deal for a new $1,250 home entertainment system: no interest will be charged for one year, and the family can pay the loan back at the rate of $120 per month.

1. Make a table showing the relation between *number of payments* and *unpaid balance*.

Number of Payments	0	1	2	3	...	8	9	10
Unpaid Balance (in dollars)					...			

   a. Make a plot of (*number of payments*, *unpaid balance*) data pairs.

   b. In your group, discuss how the pattern in the graph matches the pattern in the data table.

2. Suppose that another store offers a different deal: a price of only $1,100, with payments of $130 per month.

   a. What will the unpaid balance be after 3 payments?

   b. How many payments will be required to reduce the unpaid balance to $450?

3. For the two payment plans offered in this situation, write rules showing how the unpaid balance changes from one month to the next, using the words *NOW* and *NEXT* in equations that begin *NEXT* = __________.

4. Now consider rules relating the number of payments to the unpaid balance.

   a. Complete the following sentence to give rules relating the number of payments and unpaid balance variables in the two payment plans:

   To calculate the unpaid balance after any given number of payments, you...

   b. Use the letter N for *the number of payments made* and U for *the number of dollars unpaid* to write rules, for both payment plans, that give the unpaid balance after any number of payments. For each plan, use an equation that begins $U =$ __________.

Profits and Losses Typical businesses watch patterns of change in their costs, income, and profit from operations. For example, the Palace Theater shows only second-run movies and charges a single low price of $2.50 for all shows all day. The income from ticket sales depends on the number of tickets sold.

5. Make a table of (*number of tickets sold, income in dollars*) data like the one below.

Palace Theater Income Data

Number of Tickets Sold	Income in Dollars
0	
50	
100	
150	
200	
250	
300	

a. Plot these data on a graph and explain what the pattern of the plotted points tells you about the way that income changes as ticket sales increase.

b. Use the letter T to stand for the variable *number of tickets sold* and the letter I to stand for the variable *daily income in dollars*. Write a rule for calculating income from ticket sales with an equation that begins $I =$ __________.

6. Ignoring costs for operating the concessions stand, the operating expenses for the Palace Theater average $450 per day. Assume there are no other expenses nor other sources of income. How could you determine the theater's daily profit? Use this relation to make a table of (*number of tickets sold, profit in dollars*) data like the one below.

Palace Theater Profit Data

Number of Tickets Sold	Profit in Dollars
0	
50	
100	
150	
200	
250	
300	

a. Plot these data on a graph. Explain what the pattern of the plotted points tells you about the way that profit changes as ticket sales increase.

b. Have each group member write a question about Palace Theater ticket sales and profit that can be answered using information from the data table or the graph. Share your questions with each other. Decide on the answers and discuss how the answers are revealed by the table or the graph.

c. Let P represent *daily profit in dollars* and T represent the *number of tickets sold.* Write a rule for finding profit as a function of tickets sold with an equation that begins $P =$ __________.

7. Suppose the theater offers a discount ticket of $2.00 for children and senior citizens.

a. Calculate income for these combinations of ticket sales:

Tickets and Income

Regular Tickets	Discount Tickets	Income (in dollars)
100	100	
200	100	
100	200	
150	250	
250	150	

b. Use the letter R for the *number of regular tickets* and D for the *number of discount tickets.* Write a rule for calculating income I from ticket sales as an equation that begins $I =$ __________.

c. Calculate profit for these combinations of ticket sales:

Tickets and Profit

Regular Tickets	Discount Tickets	Profit (in dollars)
100	100	
200	100	
100	200	
150	250	
250	150	

d. Write a rule for calculating *profit* from ticket sales as a function of *numbers of discount and regular tickets* sold: $P =$ __________.

Checkpoint

The earnings, installment buying, and Palace Theater situations all involved variables that changed in relation to each other.

a Describe the variables involved and the patterns of change in each case.

b Explain how the patterns are similar and how they are different.

c How are the graphs similar and how are they different?

d Write at least two rules that you determined in this investigation. What is the relationship between patterns and rules?

Be prepared to share your group's descriptions and observations with the class.

On Your Own

Rules relating variables occur in many situations. For example, here are some data about the costs of food items at a fast food restaurant.

Soft Drink	Fries	Burger	Sundae
\$0.79	\$0.89	\$1.79	\$1.25

a. What is the bill (before tax) for an order for 15 members of a school team, if each orders a soft drink, fries, burger, and sundae?

b. Write in words a rule for calculating the total cost of meals for *any number of students*, if each student gets one soft drink, one order of fries, one burger, and one sundae.

c. Use letters to write a rule that gives the total cost for any number of students if each student gets one of each item.

d. Use the words *NOW* and *NEXT* to write a rule showing how the cost changes for each additional student with an equation that begins *NEXT* = __________.

e. Make a table and a graph of (*number of students*, *total cost*) data for any number of students from 0 to 10.

f. Write one question that can be answered by the data in your table or graph. Answer the question and explain how the table or graph provided the answer.

INVESTIGATION 2 Quick Tables and Graphs

Often the first step in modeling a situation is to discover patterns and then write relations among the key variables using symbolic rules. Once this is done, you can study the situation further by using your graphing calculator or computer software to calculate values, make tables, and display graphs of relationships between the variables.

Making Tables Recall that the Palace Theater took in \$2.50 for each ticket sold and had daily expenses of \$450 (ignoring concessions). Below are two tables of sample (*number of tickets sold, profit*) data.

Table 1

Number of Tickets Sold (T)	Profit P (in dollars)
0	–450.00
1	–447.50
2	–445.00
3	–442.50
4	–440.00
5	–437.50
6	–435.00
7	–432.50
8	–430.00
9	–427.50
10	–425.00

Table 2

Number of Tickets Sold (T)	Profit P (in dollars)
0	–450.00
20	–400.00
40	–350.00
60	–300.00
80	–250.00
100	–200.00
120	–150.00
140	–100.00
160	–50.00
180	0.00
200	50.00

You can use computer software or a graphing calculator to produce tables like these quickly. Producing a table generally requires the following three steps.

Enter the rule. The function rule usually must be entered in a "Y =" form. For the rule $P = 2.50T - 450$, which gives profit P as a function of the number T of tickets sold, replace T (the input variable) with X and replace P (the output variable) with Y. This gives Y = 2.50X – 450.

Set up the table. Specify the beginning or minimum value for the input or *independent* variable X. Also specify the size of each step by which the variable changes. Some technological tools allow you to specify the ending or maximum value as well.

Display the table. With most tools, displaying the table is just a matter of pressing one or two keys or giving a single command.

Examples of the screens involved are shown here. Your calculator or software may look different.

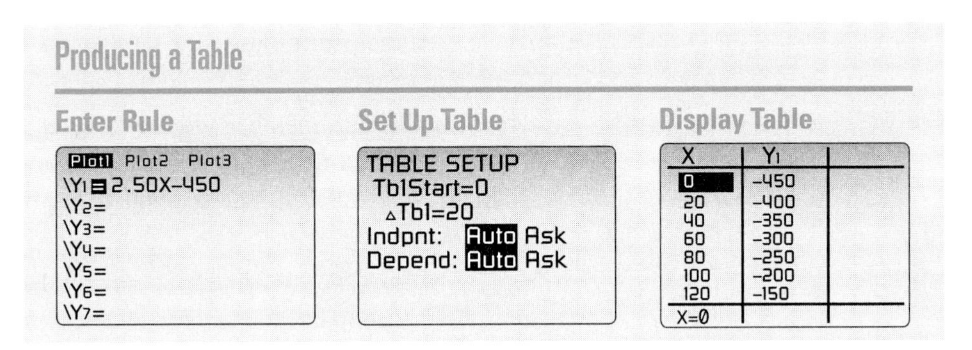

1. Examine the tables on the previous page.

a. What is the minimum X value for Table 1? For Table 2?

b. What is the step size for Table 1? For Table 2?

c. Produce each of these tables on your calculator or computer.

2. Use a table to find:

a. the profit if 280 tickets are sold;

b. the number of tickets that must be sold for the theater to break even;

c. the number of tickets that must be sold to make a profit of $800.

d. How would you set up a technology-produced table to quickly find the profit if 317 tickets are sold?

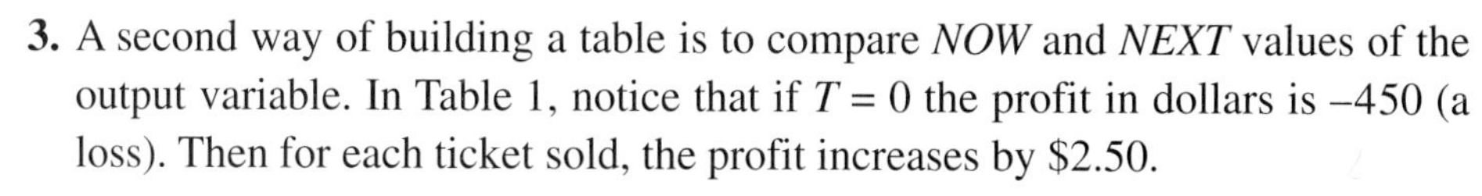

3. A second way of building a table is to compare *NOW* and *NEXT* values of the output variable. In Table 1, notice that if $T = 0$ the profit in dollars is –450 (a loss). Then for each ticket sold, the profit increases by $2.50.

   a. Use the words *NOW* and *NEXT* to write a rule describing this pattern of change.

   b. Use the "last answer" function on your calculator or computer software to build a table of profit values as in Table 1. In a data table, record each profit value and the corresponding number of tickets sold.

   c. Use the method in Parts a and b to build a table of profit values as in Table 2.

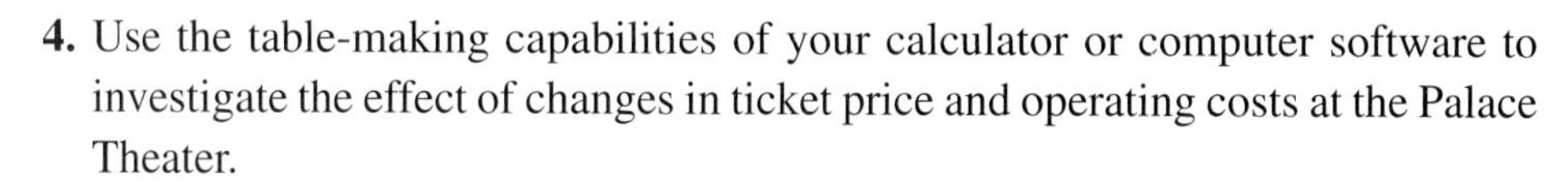

4. Use the table-making capabilities of your calculator or computer software to investigate the effect of changes in ticket price and operating costs at the Palace Theater.

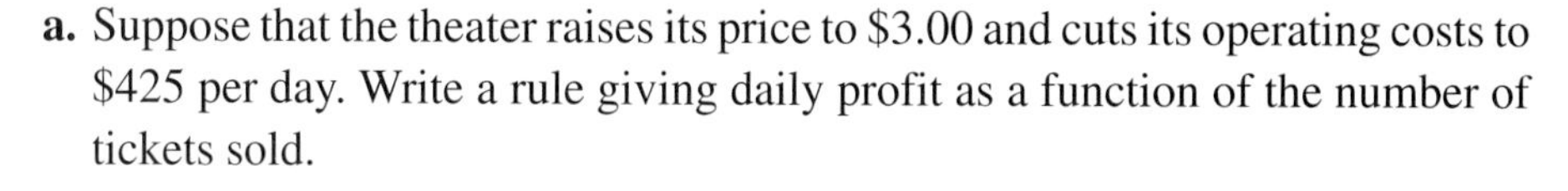

   a. Suppose that the theater raises its price to $3.00 and cuts its operating costs to $425 per day. Write a rule giving daily profit as a function of the number of tickets sold.

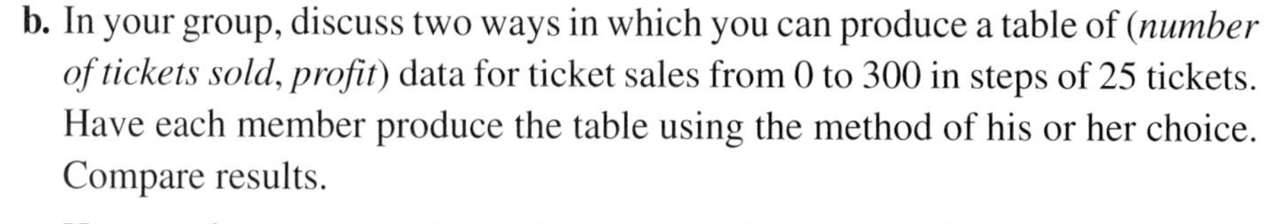

   b. In your group, discuss two ways in which you can produce a table of (*number of tickets sold*, *profit*) data for ticket sales from 0 to 300 in steps of 25 tickets. Have each member produce the table using the method of his or her choice. Compare results.

   c. Have each group member write two questions about the relation between ticket sales and theater profit that can be answered from her or his table. Then share the questions with the group and explain how the table can be used to get the answers.

   d. Individually, make another table that would help you to answer the following questions. Compare your tables and responses with those of other group members.

   - How does profit change when ticket sales increase? Use steps of 15 tickets.
   - What is the theater's profit if a blockbuster movie is shown and 600 tickets are sold?

   e. What is the fewest number of tickets that can be sold in a day and still make a profit?

Making Graphs You already have used your calculator or computer software to make graphs of data and relations between data. The graph in Activity 5 is a plot of several of the pairs (T, P) from Table 2, entered as data points and graphed as a scatterplot.

5. Carefully compare Table 2 (reproduced below) and the scatterplot.

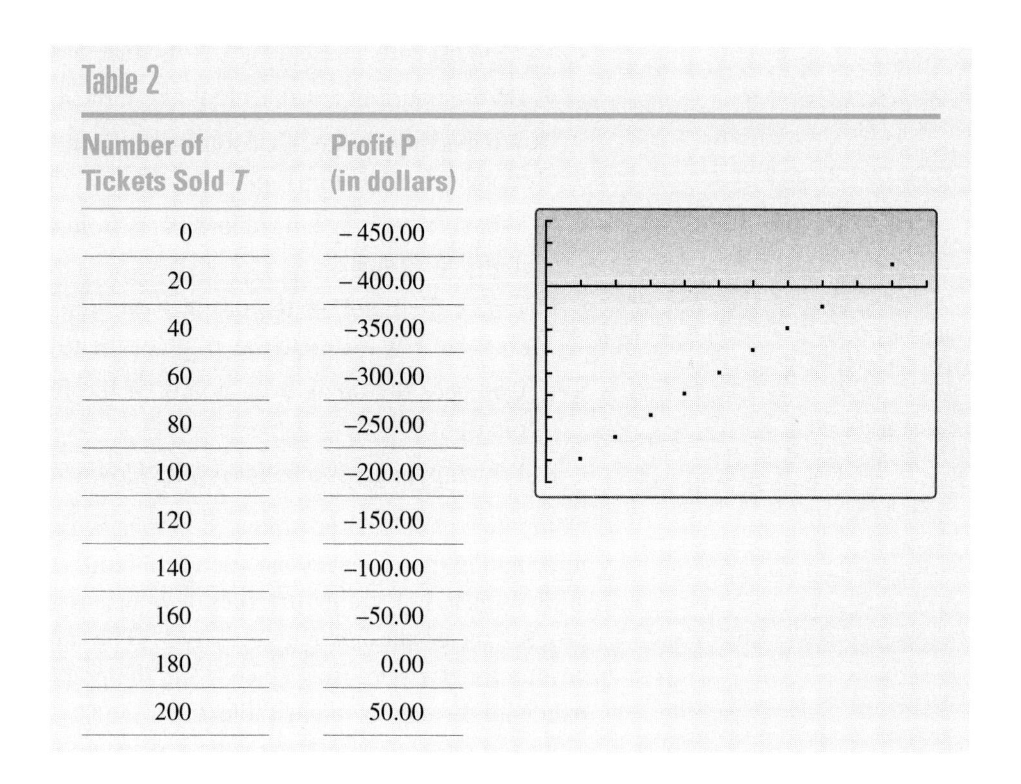

Table 2

Number of Tickets Sold T	Profit P (in dollars)
0	- 450.00
20	- 400.00
40	-350.00
60	-300.00
80	-250.00
100	-200.00
120	-150.00
140	-100.00
160	-50.00
180	0.00
200	50.00

a. What do you think the minimum and maximum values are for the viewing window (Xmin, Xmax, Ymin, and Ymax)? What do you think the scales are on the x- and y-axes? Explain your answers.

b. Check your answer by producing the scatterplot using your calculator or computer software. Adjust your answer to Part a if needed.

The graph shown here was made on a graphing calculator by entering the rule relating profit and ticket sales ($y = 2.50x - 450$). The viewing window is the same as the one in Activity 5. You can use a trace feature to read coordinates of any point on the graph. (Be sure any plots are turned off.)

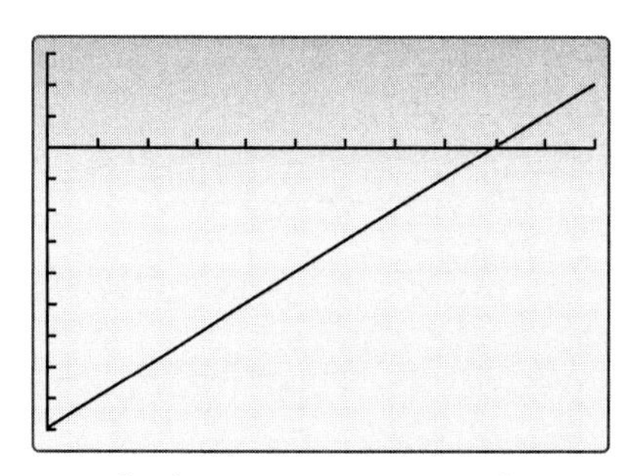

6. Experiment with the graphing capabilities of your calculator or computer software.

a. Use your calculator or computer software to produce the graph shown above. Trace along the graph and compare coordinate readouts with the data given in Table 2. Do the coordinate values look the same?

b. Re-plot the graph of this profit function using a viewing window with x values ranging from 12 to 200 with a scale of 20 and y values ranging from –475 to 60 with a scale of 50. Now trace along this new graph.

- Do the coordinate values look the same as the data given in Table 2?
- What is the change in x-coordinates from one point to the next as you trace along the graph this time?
- What is the change in y-coordinates from one point to the next as you trace along the graph?

c. Compare the graph of the original rule relating profit to number of tickets sold ($y = 2.50x - 450$) to the graph of the rule you developed in Activity 4.

- Compare the break-even points of the two ticket-pricing/operating-cost plans.
- Which plan gives a better profit? How is this seen in the graphs?

7. Computer- or calculator-produced graphs of function rules often provide more information than is necessary for a problem situation. Some graphing software allows you to link graphs to tables of values. Use software with this capability to produce the graph shown in Activity 5. (The calculator program TBLPLOT was written for this purpose.) Trace along the graph and compare coordinate read-outs with the data given in Table 2.

Here are the basic steps in producing a graph of a function rule.

- Enter the rule.
- Set the viewing window.
- Display the graph.

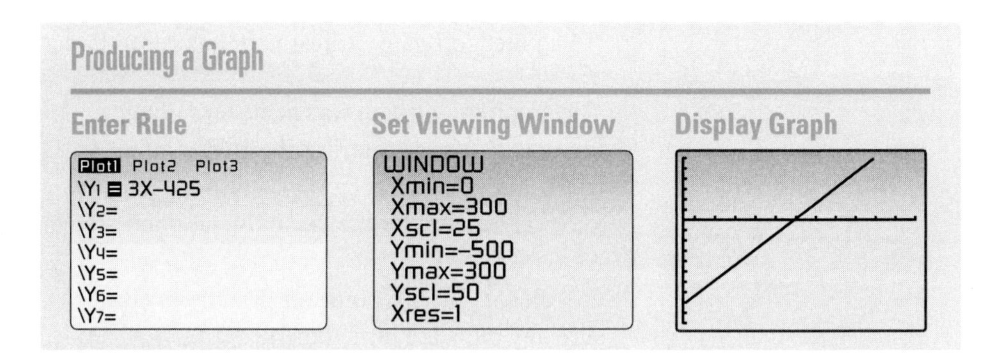

You may have to re-set the viewing window several times before getting a view of the graph that is most useful for the situation you are studying. Producing a quick table of sample data often helps.

8. Suppose you make a deal with your parents. They buy you a $125 stereo CD-player and you promise to repay $3.50 per week.

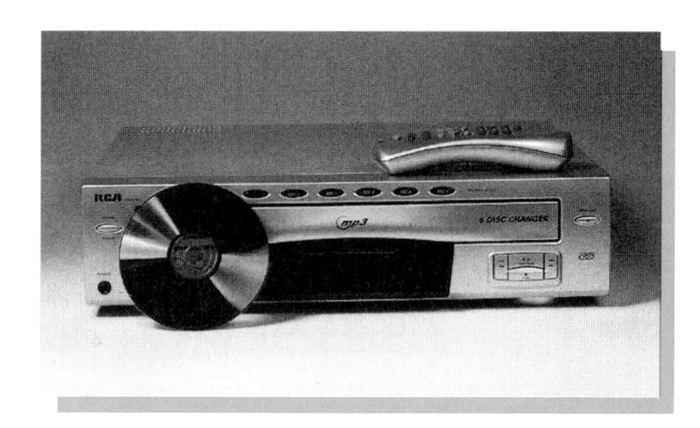

a. Write a rule giving what you still owe as a function of time since repayment began.

b. Use your calculator or computer software to make a graph of the rule in Part a. Sketch the graph on grid paper. Then label points whose coordinates answer the questions below. Be sure your responses make sense.

- How much will you owe after 10 weeks?
- When will you owe only $55?
- When will your loan be paid in full?

c. Explain how you could produce and use a table to answer Part b.

Checkpoint

Suppose a cross-country bus travels at an average speed of 50 miles per hour.

a Describe two ways to use a calculator or computer to produce a table of values showing how far that bus travels as a function of time during the trip. What are the advantages or disadvantages of each method?

b Describe two ways to use a calculator or computer to graph the relation between time and distance. What are the advantages or disadvantages of each method?

c Write and answer three different questions that can be answered using the table or the graph.

Be prepared to share your questions and methods with the class.

On Your Own

The *Concorde* is a type of supersonic airliner that is used to carry passengers from London and Paris to distant locations around the world. It cruises at a speed of about 2,200 kilometers per hour.

a. Write a rule giving the *distance* the Concorde travels as a function of *flight time*.

b. Make a table of (*flight time*, *distance*) values for a trip of 8 hours and a graph showing the same data. Then use the table and graph to answer the following questions:

- How far will the Concorde travel in 5 hours?
- How long will it take the Concorde to make the 5,780 kilometer trip from Paris to New York?

c. What patterns do you see in the table and graph that describe the way the distance traveled changes as time passes?

MORE

Modeling • Organizing • Reflecting • Extending

Modeling

1. The freshman class officers at Banneker High School ordered 1,200 fruit bars. They paid $0.25 per bar at the time the order was placed. They plan to sell the fruit bars at school football and basketball games for $0.60 per bar. No returns of unsold bars are possible.

a. How much money has the class already invested in this project?

b. What income will be earned if the class sells 100 bars? 400 bars?

c. What profit will be earned if the class sells 100 bars? 400 bars? 800 bars? All 1,200 bars?

d. Write two rules for calculating profit as a function of the number of fruit bars sold.

- In the first rule, use the words *NOW* and *NEXT* to show how profit changes for each additional fruit bar sold.
- In the second rule, use variable names P and B to show how to calculate the profit P for any given number B of fruit bars sold.

e. Use one of the rules you have written to produce a table of (*number sold*, *profit*) data from 0 to 1,200 bars in steps of 100 bars sold.

f. Produce a graph of the relation between *number sold* and *profit* earned.

g. Find the break-even point in sales—the number that must be sold to be sure that the class does not lose money. Show where that point appears in the table of data and on the graph.

h. Describe the overall relation of fruit bar sales to class profits and explain how that pattern is shown in the table and the graph.

2. Summer thunder-and-lightning storms occur in most parts of the United States. It is very dangerous to be out in the open when lightning hits.

You can estimate your distance from a storm center. Count the time in seconds between a lightning flash and the clap of thunder produced by that lightning. You see the lightning flash almost instantly, but the thunder moves at the speed of sound (about 330 meters per second).

a. How far away is a lightning strike if the thunder it produced arrives

- 2 seconds after you see the lightning?
- 3 seconds after the lightning flash?
- 4.5 seconds after the lightning flash?

b. Write two rules for calculating distance of a lightning flash from you and the time it takes the sound of that flash (the thunder) to reach your ears.

- In the first rule, use the words *NOW* and *NEXT* to show the change in the distance estimate for each additional second of time counted.
- In the second rule, use the variable names T and D to show how to find the distance for any given time counted.

c. Make a table of (*time*, *distance*) data and a graph for times from 0 seconds to 10 seconds.

d. Complete each of the following sentences to describe the relation between the distance between the lightning and you and the time it takes thunder to reach your ears.

- *As the time increases, the distance ... .*
- *As the distance increases, the time ... .*

e. Use your rule, table, and graph to estimate the time required for thunder to reach your ears from a lightning strike 2,500 meters away.

3. Janitorial assistants at Woodward Mall start out earning \$5.50 per hour and are paid weekly. However, the \$55 cost of uniforms is deducted from their first paycheck.

a. What pay would a new employee receive if she worked 5 hours in her first week? If she worked 10 hours? If she worked 20 hours?

b. Write two rules for calculating first-week's pay as a function of time worked.

- In the first rule, use the words *NOW* and *NEXT* to show the change in pay for each hour of work.
- In the second rule, use variable names *P* and *T* to show how to find the pay for any given time worked.

c. Produce tables and graphs of the relation between hours worked and pay showing the relation for 0 to 40 hours of work.

d. Write two questions about the relation between time worked and pay earned. Show how those questions can be answered with data in the table and on the graph of Part c.

e. Assuming the employee worked more than 10 hours the first week, how would you change your rules in Part b to calculate weekly pay after the first week?

4. Commercial jet airliners can fly at an average speed of 15 kilometers per minute. Thus, the distance they travel is a function of flight time.

a. Choose letters to stand for the variables *time* in flight and *distance* covered and then write a rule for calculating distance as a function of time.

b. Write two questions about the relation between time in flight and distance covered that would be of interest to an airline or a private pilot. Then show how the answers to those questions can be found in tables of (*time*, *distance*) data and on a graph of the relation between time and distance.

c. Suppose that the speed of the airliner is increased by a strong tailwind of 2 kilometers per minute. Write a rule for calculating distance as a function of time under this condition. Write a different rule for the case when the airliner is traveling against a headwind of 2 kilometers per minute.

d. The trip from New York to San Francisco is about 5,000 kilometers. Write an equation for a rule that will calculate *distance remaining to be covered* as a function of *elapsed flight time* at a speed of 15 kilometers per minute.

e. Write two questions about the relation between time in flight and distance remaining in the trip to San Francisco. Then show how the answers to those questions can be found in tables of (*time*, *distance*) data and on a graph of the relation between time and distance.

5. NASA space shuttles travel about 30,000 kilometers per hour.

a. Write two rules for calculating distance traveled as a function of time in flight.

- In the first rule, use the words *NOW* and *NEXT* to show the change in distance for each hour of flight.
- In the second rule, use variable names *T* and *D* to show how to find the distance for any given time in flight.

b. Produce tables and graphs of the relation between time in flight and distance, showing the relation for 0 to 20 hours of flight time.

c. Write two questions about the relation between time in flight and distance traveled. Show how those questions can be answered with data in the table and on the graph of Part b.

d. In July, 1969, the Apollo 11 mission put an astronaut on the moon, almost 400,000 km from the Earth. (The path taken from launch to lunar orbit was not a straight line, but assume the Apollo 11 traveled 400,000 km.) The speed of the vehicle varied widely, moving as fast as 39,000 km per hour. However, the average speed was only 5,500 km per hour. Use the average speed to write a rule estimating *distance still to be traveled* as a function of *time in flight*.

e. Produce a table and a graph of the relation in Part d and write two questions that can be answered from information in the table and graph.

f. Compare the patterns in the relations between time and distance traveled and time and distance still to be traveled.

Organizing

1. *Formulas* are equations which relate two or more variables according to rules. For example, you are probably familiar with formulas for area and perimeter of geometric figures. The size and shape of any rectangle can be described by two variables: *length* and *width*.

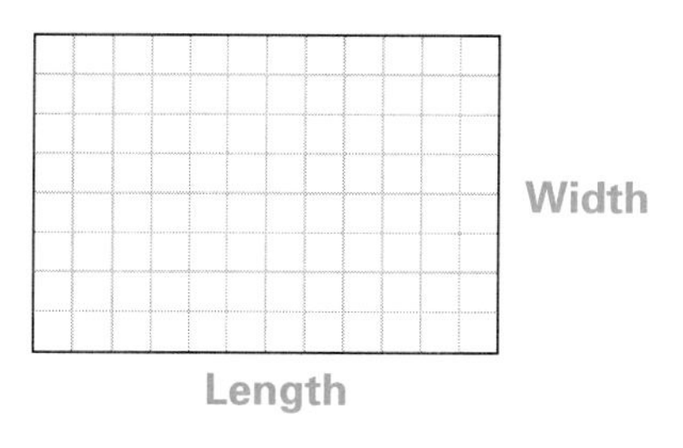

a. If each of the small squares in the rectangle above represents one square meter, what are the area and perimeter of the rectangle represented by the figure?

b. Using single-letter variable names, write rules for calculating the area and perimeter for any rectangle.

c. Suppose the rectangle models a grid of solar power cells, each of which produces 0.03 watts of power.

- How much power is produced by the entire grid of power cells?
- Using single-letter variable names, write a rule for calculating the power produced by a rectangular solar grid of any size given in terms of the *length* and *width* in meters.

d. The area of any triangle can be calculated from the formula $A = \frac{1}{2}bh$, where b and h stand for the *base* and *height* of the triangle.

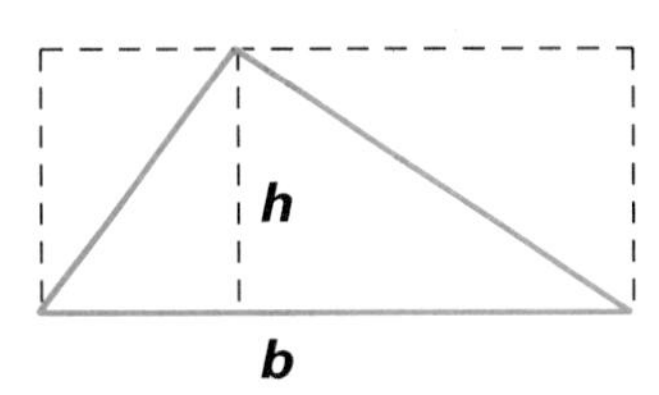

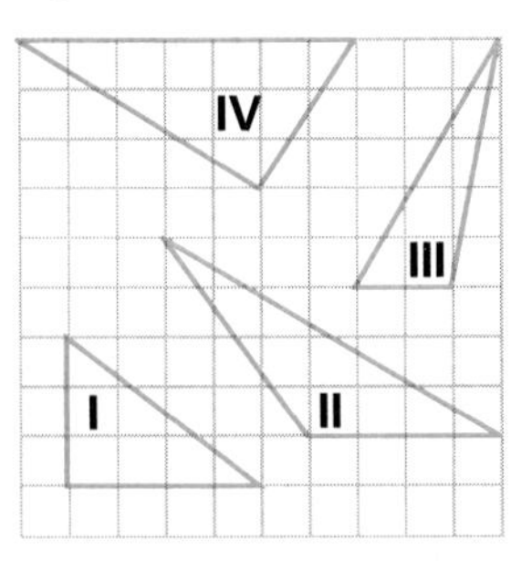

- Use the diagram above at the left to explain how the formula for the area of a triangle is related to the one for a rectangle.
- Use the formula for the area of a triangle to find areas of the triangles at the right.

2. The diagram at the right shows a circle with one *radius* and one *diameter* drawn.

diameter
radius

The circumference C and area A of a circle with diameter d and radius r are given by the formulas:

$C = \pi d$ $\qquad\qquad$ $A = \pi r^2$

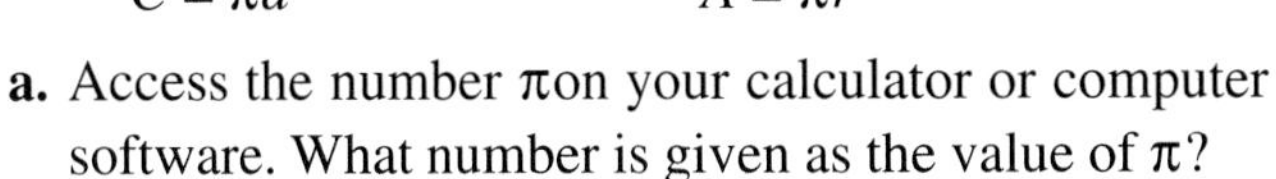

a. Access the number π on your calculator or computer software. What number is given as the value of π?

b. Calculate the circumference of a circle with diameter 3 meters.

c. Calculate the area of a circle with radius 1.5 meters.

d. Tony's Pizza Place is advertising 2-item, 10-inch pizzas for $7.95 and 2-item, 12-inch pizzas for $9.95. Which pizza is the better buy? Explain your reasoning.

e. Write formulas giving rules for finding:

- the circumference of a circle of any given *radius*;
- the area of a circle of any given *diameter*.

3. When an object is dropped, the distance it falls is related to the time it has been falling by the equation $d = 4.9t^2$, where t is time in seconds and d is distance in meters. Suppose a ball falls 250 meters down a mineshaft.

a. Use the table-building capability of your calculator or computer software to find, to the nearest second, how long it takes to fall this distance.

b. Find, to the nearest 0.01 second, the time the ball must have been falling.

c. Produce a graph showing distance as a function of time. Answer Part b by using the trace feature and adjusting the viewing window.

d. How are the procedures you used in Parts b and c similar?

4. Write rules relating X and Y in each of the following tables of numbers.

a.

X	Y
1	2
2	4
3	6
4	8
5	10

$Y =$ ___________

b.

X	Y
1	1
2	3
3	5
4	7
5	9

$Y =$ ___________

c.

X	Y
1	7
2	14
3	21
4	28
5	35

$Y =$ ___________

d.

X	Y
1	6
2	11
3	16
4	21
5	26

$Y =$ ___________

e.

X	Y
1	2
2	5
3	10
4	17
5	26

$Y =$ ___________

f.

X	Y
1	98
2	96
3	94
4	92
5	90

$Y =$ ___________

5. The graph at the right shows the pattern of growth in the fish population of a lake that has been damaged by acid rain and then cleaned up and restocked with fish. Write three different questions that can be answered from the graph and explain how the answers can be found.

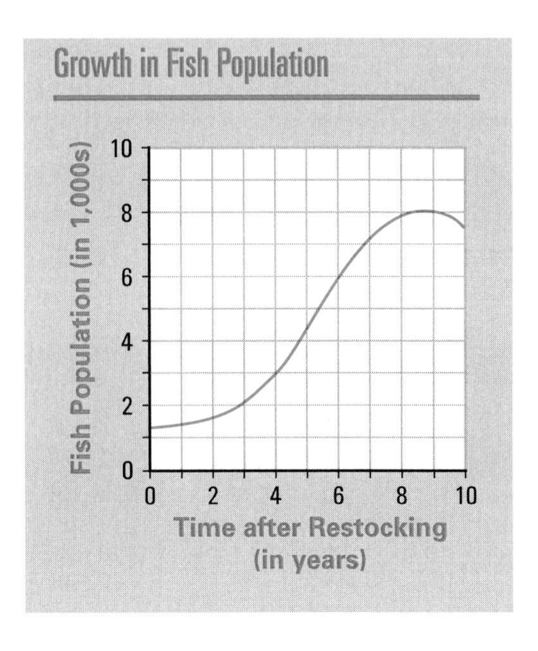

Reflecting

1. When using a graphing calculator or computer software to make tables and graphs, you have to learn the special language of the tool you have. There are usually some problems in the first few attempts.
   a. What problems did you have in building tables with your calculator or computer software? How did you solve those problems?
   b. What problems did you have in making calculator or computer graphs? How did you solve those problems?

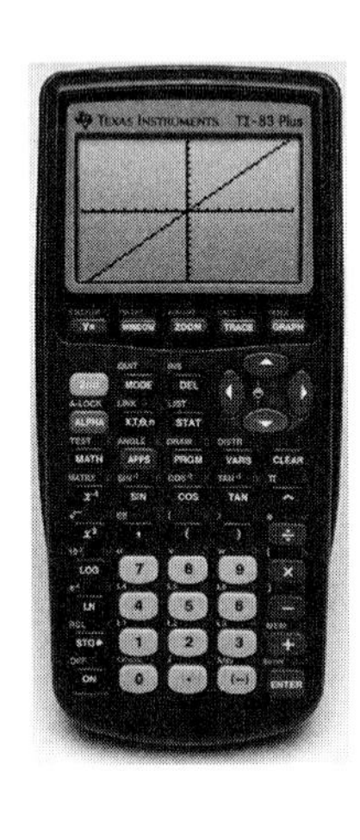

2. The relation between price P of an item in a store and the total cost C, including 5% sales tax, is given by $C = P + 0.05P$. Use your graphing calculator or computer software to produce a graph of the equation. Use a standard viewing window: Xmin = –10, Xmax = 10, Xscl = 1, Ymin = –10, Ymax = 10, and Yscl = 1. This ensures that you will view the same screen as your classmates. Using the trace feature gives you coordinates of any point on the graph, but some of those points don't make sense! Think about the situation being modeled by the equation and graph and explain why the points don't make sense.

3. Organize your class to do a survey of teachers in all the subjects, except for mathematics, taught in your school to get examples of formulas used in their fields. Science you might expect to be easy, but you should also be able to get some examples from business, art, social studies, and English.

4. Organize your class to do a survey of parents and friends who work outside of schools to get examples of formulas used in various occupations. You should be able to get examples from construction, banking, auto mechanics, and many more occupations.

Extending

1. One car rental company charges $25 per day, gives 100 free miles, and then charges 35 cents per mile for any miles beyond the first 100.

a. What will this company charge if the car is kept for one day and is driven 130 miles?

b. Use single-letter names for variables to write a rule for calculating rental charge as a function of the number of miles driven, if the car is kept only one day.

c. Use single-letter names for variables to write a rule for calculating the rental charge as a function of the number of miles driven *and* the number of days.

2. At the start of a match race for two late-model stock cars, one car stalls and has to be towed to the pits for repairs. The other car roars off at an average speed of 2.5 miles per minute. After 5 minutes of repair work, the stalled car hits the track and maintains an average speed of 2.8 miles per minute.

a. How far apart are the two cars 5 minutes after the start of the race? How far apart are they 10 minutes after the start of the race?

b. Use single-letter variable names to write a rule for calculating the distance between the two cars at any time after the start of the race.

c. On the same set of axes, sketch or produce two graphs showing the distances traveled by the two cars as time passes in the race. Discuss the patterns in the two graphs in terms of the race.

3. The price or *fare* of taxi-cab rides in many large cities is a function of the distance traveled. For example, one rule is based on distances rounded up to the nearest fifth of a mile, as follows.

Charge 90 cents for the first one-fifth mile and 20 cents for each additional one-fifth mile or part thereof.

a. Make a table of (*distance*, *fare*) data for 0 to 2 miles in steps of one-fifth mile.

b. Sketch a graph of the relation between *distance* and *fare.*

c. How far can you travel on $10 for cab fare? Explain how you found your answer.

d. Which is the better deal for a rider, the rate rule above or a rule that charges only 30 cents for each one-fifth of a mile (including the first)? Explain your reasoning.

4. Video rental stores use a variety of rules for setting the prices of their rentals. For example, one chain charges a $3.50 minimum per video for the first two days and $1.25 for each additional day.

a. Make a table of charges to rent videos from 0 days through 14 days.

b. Sketch a graph showing the pattern of the (*rental days*, *rental charge*) data in Part a.

c. Compare the rental plan given above with another that simply charges $1.50 per video per day. Which plan would you prefer and under what conditions?

5. Below is a calculator graph of the equation $y = x^2 - 4x$ for $x = -3$ to $x = 7$. Reproduce this view on your calculator or computer software.

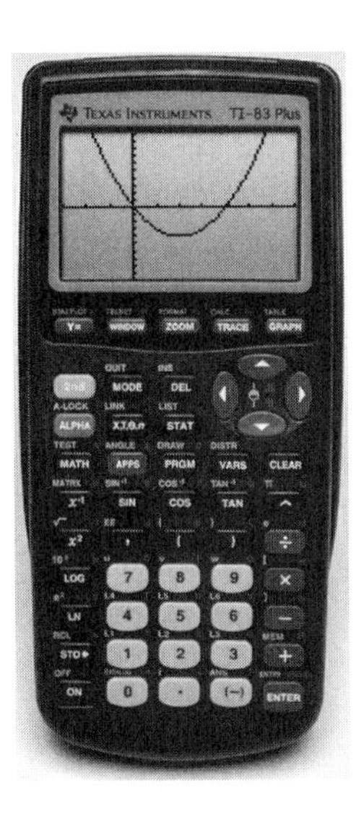

a. Use the zoom feature to take a closer look at where the graph crosses the x-axis between $x = 3$ and $x = 5$. Make sketches and describe any patterns you see.

b. Return to the original graph and zoom in to take a closer look at the graph at its lowest point. Make sketches and describe any patterns you see.

c. How can you zoom in on tables of values to check the patterns you observed in Parts a and b?

Lesson 4

Linear and Nonlinear Patterns

In nearly every situation you investigated in Lesson 3, the relations between variables were fairly easy to describe with rules. The patterns in tables most often showed constant rates of change, and graphs of those data usually gave linear patterns. The table-building and graphing capabilities of calculators and computer software make it easy to study nonlinear patterns, too. Consider the following situation in preparation for Investigation 1.

For Earth-bound humans there is something especially fascinating about things that fly through the air. You follow the flight of balls in all kinds of sports, perhaps curious about how high they go or when and where they will come down. In some sports the speed at which a ball travels is also of interest. For example, baseball teams use radar guns to "clock" the speed of a pitch.

Think About This Situation

Suppose you threw a ball straight up into the air at a velocity of 25 meters per second. (Major league pitches travel approximately 95 miles per hour, which is about 42.5 meters per second.)

a About how high do you think the ball would go before it starts falling back to the ground?

b About how many seconds do you think it would take before the ball hits the ground?

c Which of the following graphs do you think best matches the pattern of (*time*, *height*) data describing the ball's flight? Explain your choice.

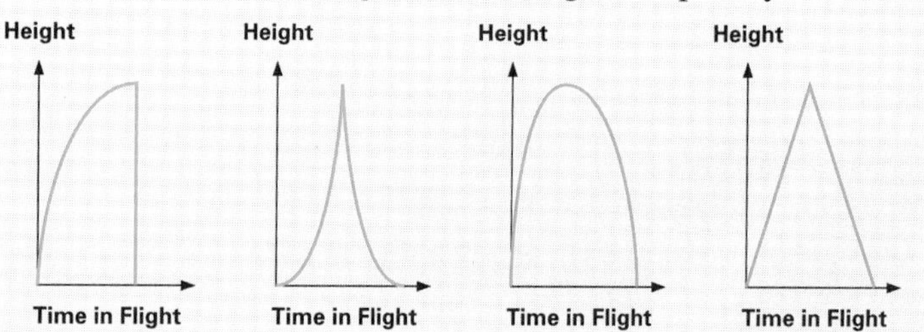

INVESTIGATION 1 What Goes Up ... Must Come Down

To study the flight of a baseball thrown straight up into the air, all you need to know is the velocity with which the ball leaves the hand of the thrower. Using principles of physics, you can predict the upward velocity of the ball as a function of time in flight with the following rule:

$$V = 25 - 9.8T$$

In this rule, the letter V stands for the upward velocity of the ball (in meters per second) at any time, and T stands for the time since the ball was thrown (in seconds).

1. Use your graphing calculator or computer software and the rule relating *velocity of the ball* and *time in flight* to study the pattern of (*time*, *velocity*) data that could be expected.

   a. Make a table of (*time*, *velocity*) data for $T = 0$ to $T = 6$ in steps of 0.5 second.
   - What is the upward velocity of the ball 0.5 second after it is thrown?
   - At what time will the velocity be about 10 meters per second? About 5 meters per second?
   - What does the pair (3, –4.4) mean?
   - Describe the pattern of change in velocity for every 0.5 second the ball is in flight.

   b. Produce a graph of the pattern of (*time*, *velocity*) data obtained in Part a. Use the table to help you choose an appropriate viewing window.

   c. Write an explanation of what the table and graph say about the pattern of the ball's velocity. Be sure to explain any points or trends in the graph that are surprising or important.

2. Looking at the rule $V = 25 - 9.8T$, can you see how the numbers and operations in the rule relate to flight of the ball? For example, does the 25 look familiar? What is the effect of the "$- 9.8T$" part of the rule?

3. Use your calculator table or graph tools to *solve these equations.* That is, find values for T and V that will make the statements true. Explain what the solutions tell about flight of the ball. Find times to the nearest 0.1 second.

   a. $10 = 25 - 9.8T$

   b. $0 = 25 - 9.8T$

   c. $V = 25 - (9.8 \times 2)$

   d. $-5 = 25 - 9.8T$

The rule relating time in flight and upward velocity of the baseball gives some information about the path of the ball, but it also leaves some interesting questions unanswered. More can be learned from a second rule that estimates the height H of the ball as a function of its time in flight, where H is in meters. Again, scientific principles suggest that this relation will be given by the following rule:

$$H = 1 + 25T - 4.9T^2$$

4. Make a table of (*time*, *height*) data for $T = 0$ to $T = 6$ seconds in steps of 0.5 second. Remember to use your table to help set your viewing window. Then plot a graph of the same data. Adjust your table or graph as necessary to estimate:

a. the height of the ball after 3 seconds; after 5 seconds;

b. the time or times, to the nearest 0.1 second, when the ball is 20 meters above the ground;

c. the maximum height of the ball and the time when the ball reaches that height;

d. the time when the ball hits the ground;

e. the upward velocity of the ball when it hits the ground.

5. Look at the patterns in the tables and graphs for your (*time*, *height*) and (*time*, *velocity*) data and try to explain how those patterns relate to each other. Look at both graphs on the same axes using a viewing window of x from 0 to 7 and y from −40 to 35.

a. What does the shape of the graph relating *time* and *height* tell about the velocity of the ball during its flight?

b. How is the shape of the graph relating *time* and *height* illustrated by the trend of numbers in the data tables for (*time*, *height*) and (*time*, *velocity*)?

6. Compare the shape of your graph relating *time* and *height* with the choice you made in answering Part c of the "Think About This Situation" on page 141.

Checkpoint

As you worked on questions about the baseball's flight, how did you use your calculator or computer to find:

a the maximum height and the time it took the ball to reach that height?

b the time the ball returned to the ground and the speed it was traveling when it hit?

Be prepared to share your group's methods with the whole class.

On Your Own

If a tennis ball is lobbed into the air with an initial upward velocity of 12 meters per second, its velocity and height will be functions of time in flight described by the following rules:

$$V = 12 - 9.8T \quad \text{and} \quad H = 1 + 12T - 4.9T^2$$

a. Find the maximum height of the tennis ball and the time it takes to reach that height.

b. What is the upward velocity of the ball at its maximum height?

c. Find when the ball will hit the ground. Round your answer to the nearest tenth of a second.

d. What is the velocity of the ball when it hits the ground?

INVESTIGATION 2 The Shape of Rules

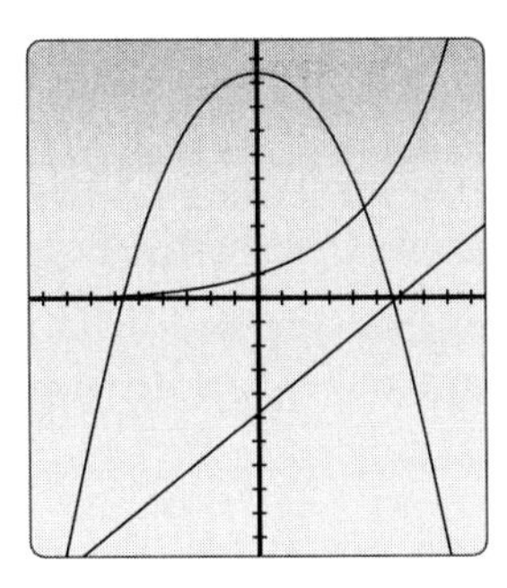

Finding symbolic rules for relations between variables is a big part of mathematical model building. It is common to start the search with some data in a table or graph. But to find a rule fitting patterns in that data, it helps to know about the match between types of rules, shapes of graphs, and patterns in tables. You can discover useful connections between rules and graphs by some experimentation with your graphing calculator or computer software.

Experiment 1

1. Do the following for each rule listed below:

- Produce a table of values and a graph. Use the standard viewing window: Xmin = −10, Ymin = −10, Xmax = 10, Ymax = 10, Xscl = 1, and Yscl = 1.
- On a separate sheet of paper, write the rule and make a sketch of the graph. How does the pattern of values in the table match the pattern in the graph?
- Have each member of your group examine how the graph appears using a different viewing window. Compare your graphs and viewing windows.

a. $y = 2x - 4$ **b.** $y = 2x + 4$ **c.** $y = 0.5x + 2$

d. $y = -0.5x + 2$ **e.** $y = 10 - 1.5x$ **f.** $y = x^2 - 4$

2. As a group, discuss similarities and differences in the graphs you produced.

3. As a group, make some conjectures about the shapes of graphs for certain types of rules. Have individual members test various conjectures and report back to the group.

Experiment 2

1. Do the following for each rule listed below:
   - Produce a table of values and a graph. Use the standard viewing window (x and y values from –10 to 10, with a scale of 1).
   - On a separate sheet of paper, write the rule and make a sketch of the graph. How does the pattern of values in the table match the pattern in the graph?
   - Have each member of your group examine how the graph appears using a different viewing window. Compare your graphs and viewing windows.

   a. $y = x^2$ **b.** $y = x^2 - 3$ **c.** $y = -x^2$

   d. $y = -x^2 + 5$ **e.** $y = (x + 3)^2$ **f.** $y = \frac{2}{x}$

2. As a group, look for patterns in the rules that seem connected to special patterns in the graphs. Make and test some conjectures about the shapes of graphs for certain types of rules.

Experiment 3

1. Do the following for each rule listed below:
   - Produce a table of values and a graph. Modify the standard viewing window so that the x-axis shows values only from –5 to 5.
   - On a separate sheet of paper, write the rule and make a sketch of the graph. How does the pattern of values in the table match the pattern in the graph?
   - Have each member of your group examine how the graph appears using a different viewing window. Compare your graphs and viewing windows.

   a. $y = \frac{1}{x}$ **b.** $y = \frac{3}{x}$ **c.** $y = \frac{5}{x}$

   d. $y = \frac{-5}{x}$ **e.** $y = \frac{5}{x+1}$ **f.** $y = \frac{x}{3}$

2. As a group, look for patterns in the rules that seem connected to special patterns in the graphs. Make and test some conjectures about the shapes of graphs for certain types of rules.

Experiment 4

1. Do the following for each rule listed below:

- Produce a table of values and a graph. Decide as a group on a common viewing window to use.
- On a separate sheet of paper, write the rule and make a sketch of the graph. How does the pattern of values in the table match the pattern in the graph?
- Have each member of your group examine how the graph appears using a different viewing window. Compare your graphs and viewing windows.

a. $y = 2^x$ **b.** $y = (1.5)^x$ **c.** $y = 3^x$

d. $y = x^3$ **e.** $y = (0.5)^x$ **f.** $y = x^{0.5}$

2. As a group, look for patterns in the rules that seem connected to special patterns in the graphs. Make and test some conjectures about the shapes of graphs for certain types of rules.

Checkpoint

As a result of the experiments, your group probably has some hunches about matches between the form of a symbolic rule, the pattern of data in tables, and the shape of graphs. Summarize your ideas in statements like this:

If we see a rule like ... , we expect to get a table like ... .

If we see a rule like ... , we expect to get a graph like ... .

Be prepared to compare your summary statements with those of other groups.

On Your Own

Without using your calculator, match as best you can each table or graph with one of the given rules. In each case, explain the reason for your choice.

I

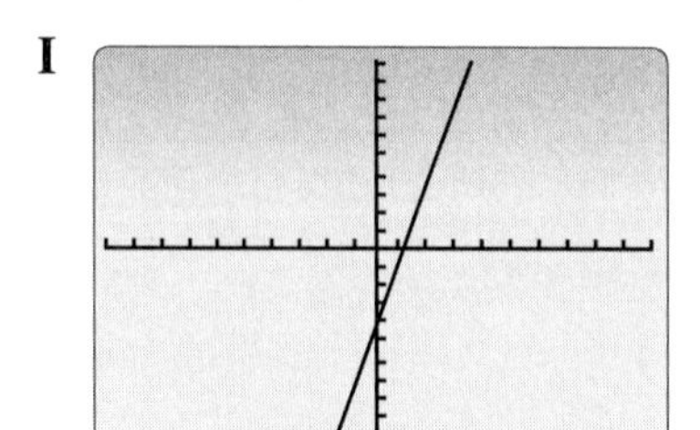

II

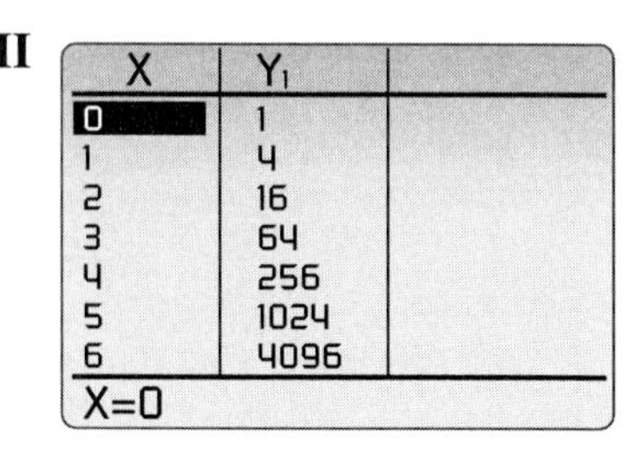

X	Y_1	
0	1	
1	4	
2	16	
3	64	
4	256	
5	1024	
6	4096	

X=0

III

X	Y_1
1	5
2	2.5
3	1.6667
4	1.25
5	1
6	.83333
7	.71429

X=1

IV

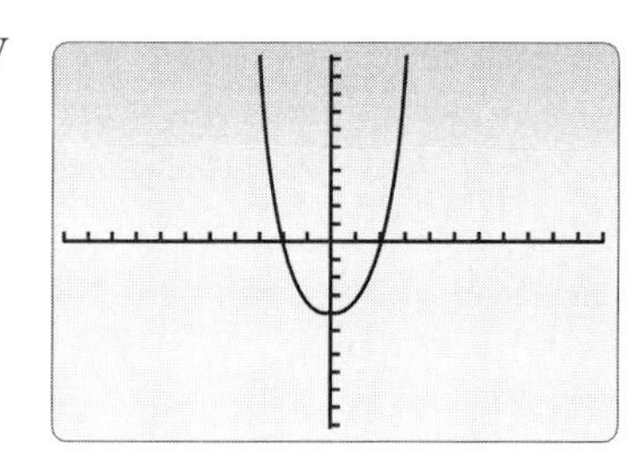

a. $y = x^2 - 4$ **b.** $y = \frac{5}{x}$ **c.** $y = -4x - 3$

d. $y = -x^2 + 4$ **e.** $y = 4^x$ **f.** $y = 3x - 4$

In this investigation you have explored only a few of many possible types of equations that can be used to model relations between variables. You will continue to study the match between rule types, patterns in tables, and graphs of sample data in the "MORE" section, and it will be an important task in the units ahead.

Modeling • Organizing • Reflecting • Extending

Modeling

1. On a clear summer day, the temperature of the air around you might be warm, perhaps 25°C. But if you flew up into the atmosphere in an airplane, a balloon, or a glider, the temperature would decrease. A scatterplot of some sample data obtained by a weather balloon is shown at the right.

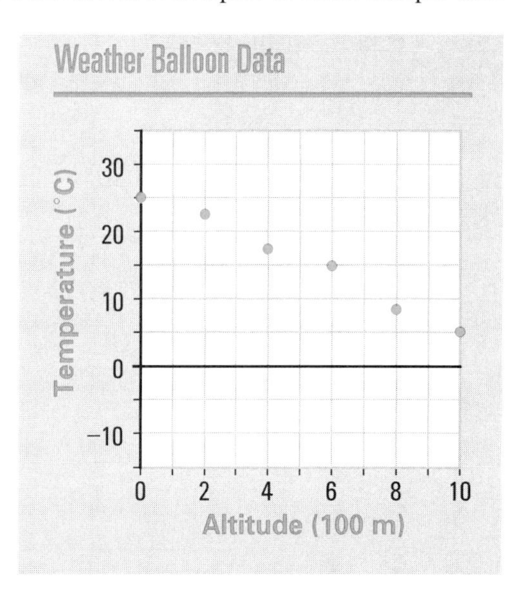

a. Write a sentence describing the way temperature changes as altitude changes.

b. Which of the following rules seems to fit the data in the graph?

$T = 25 - 2A$

$T = 25 + A^2$

$T = \frac{25}{A}$

$T = 25(0.8^A)$

c. Use what you believe is the best modeling rule to find the temperature at 300 meters; at 700 meters; at 2,000 meters.

d. To the nearest 10 meters, what altitude would correspond with a temperature of 0°C?

2. The cost of a long-distance phone call depends on the time of day and the length of the call. To save money, one family always makes calls to their relatives and friends in the evening. The following table shows (*call length*, *cost*) data for a sample of their calls to one city.

Call Length L (in minutes)	1	2	3	4	5	10	15	20
Cost C (in dollars)	3.00	3.50	4.00	4.50	5.00	7.50	10.00	12.50

a. Plot the data on a graph.

b. Write a sentence describing the way cost changes as call length changes.

c. Which of the following rules seems to fit the data in the table?

$C = 3L$ $\quad$ $C = 3L + 0.50$

$C = L + 0.50$ $\quad$ $C = 2.50 + 0.50L$

d. Use what you believe is the correct rule to find the cost of calls of these lengths: 8 minutes; 17 minutes; 30 minutes.

e. If you have $15 to make an evening call to a friend in this city, how long would you be able to talk?

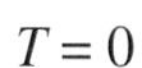

3. Suppose an experimental golf ball is hit high into the air with an initial upward velocity of 18 meters per second. The upward velocity V and height H of the ball will vary as time T passes. Those changes can be modeled by the rules $V = 18 - 9.8T$ and $H = 18T - 4.9T^2$.

a. For each value of T, find V and H. Then explain what those values tell you about the flight of the ball.

$T = 0$

$T = 1$

$T = 2$

$T = 3$

b. Use your calculator or computer software to make tables and graphs of the (*time*, *velocity*) and (*time*, *height*) data.

c. Use the graphs or tables of upward velocity and height data to estimate, to the nearest tenth of a unit, answers for these questions:

- When will the ball reach its maximum height, and what is that height?
- What is the upward velocity of the ball when it reaches its maximum height?
- When will the ball return to Earth?
- How fast will the ball be falling when it hits the ground?

d. Describe the change in velocity of the ball for each 1 second change in time. Describe the change in height of the ball for each 1 second change in time.

4. When businesses do market research for new products, they are interested in the relation between the prices they charge and the income they will receive. For example, suppose that the owners of Video City are trying to set the best rental price for video game CDs. Using information on operation costs and survey data from potential customers, their market research staff might produce a recommendation that says the *profit P (in dollars per week)* for game rentals depends on *charge per rental C (in dollars)* according to this rule:

$$P = -750 + 900C - 150C^2$$

a. Calculate these values and explain what each tells you about prices and profits for Video City game rentals.

P when $C = 1$	P when $C = 7$
P when $C = 4$	P when $C = 2.50$

b. Make a table and a graph of (*price charged, profits earned*) data for charges varying from \$0 to \$8.

c. Use the table and graph from Part b to answer the following questions:

- For what prices charged will Video City have a positive profit from game rentals?
- For what prices charged will they lose money?
- Is there a price that will give maximum profit? If so, what is it?

d. What is the weekly operation cost for video game rentals? How can you tell by looking at the rule? By looking at the graph?

e. Explain what the shape of the graph relating *price charged* and *profits earned* tells about the relation between those two variables. Why does that relation seem reasonable?

f. What other tables and graphs have you already worked with that have the same shape as the graph relating *price charged* and *profits earned*? How are their rules similar to and how are they different from the rule in this problem?

5. The stopping distance D of a car depends on reaction distance R and braking distance B. Each of these variables is a function of the speed s of the car. You can calculate reaction distance and braking distance in *feet* by using the rules

$$R = 1.1s \text{ and } B = 0.05s^2$$

where s is the speed of the car in miles per hour.

a. Write a rule that gives stopping distance as a function of the speed of a car.

b. Make a table and a graph of the relation between *speed* and *stopping distance* for speeds from 0 to 80 miles per hour in steps of 5 miles per hour.

c. What is the stopping distance for a car traveling at 20 miles per hour? At 40 miles per hour?

d. Would 200 feet be a safe distance to follow a car on the highway traveling at 60 miles per hour? Explain your reasoning.

e. Produce graphs of (*speed*, *reaction distance*) and (*speed*, *braking distance*) on the same set of coordinate axes. Compare the patterns of change in distances as speed increases. How could these patterns of change be predicted by looking at the rules for reaction distance and for braking distance?

f. Suppose that an investigating officer at the scene of a car accident in front of a school finds skid marks 100 feet in length. What, if anything, can you conclude about the speed of the associated car?

Organizing

1. Look back over the various problem situations in earlier investigations. Find several examples where two variables were related in a pattern similar to that of Task 1 in the Modeling section. Explain what those examples have in common.

2. Look back over the various problem situations in earlier investigations. Find several examples where two variables were related in a pattern similar to that of Task 4 in the Modeling section. Explain what those examples have in common.

3. For each rule below, use your calculator or computer software to produce a table and then write an equation relating *NOW* and *NEXT* values of the y variable.

a. $y = 5x + 2$ **b.** $y = -10x - 3$ **c.** $y = 2^x$ **d.** $y = 4^x$

4. Use your calculator or computer software to produce side-by-side tables for each of the following rules in steps of 1 unit starting at $x = 0$.

a. $y = 2x$ **b.** $y = x^2$ **c.** $y = 2^x$

Write a short paragraph summarizing similarities in overall patterns of change and differences in *rates* of change of the three relations.

5. A cube is a three-dimensional shape with square faces.

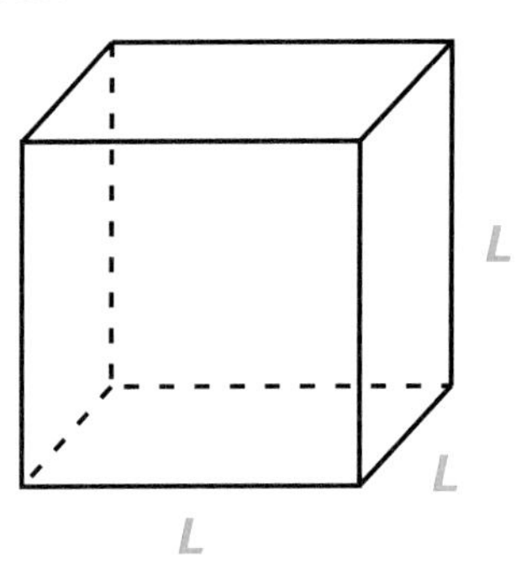

a. If the length of an edge of a cube is L, write an expression for the area of one of its faces.

b. Write a rule that gives the total surface area A of a cube as a function of the length L of an edge.

c. Suppose you wished to design a cube with surface area of 1,000 square centimeters. To the nearest 0.1 centimeter, what should be the length of the edge of the cube?

d. Did you use a graph or a table in answering Part c? Why?

e. Could you answer Part c without using a graph or a table? If so, describe your method.

Reflecting

1. Graphing calculators and computer software can be useful tools in modeling real situations with mathematics. What features of your calculator or computer software have you found to be most useful for the activities in this lesson? What calculator or computer skills do you think you need to develop further?

2. If you enter the rule Y = 5X + 100 in your calculator or computer software and press the GRAPH key or enter the graph command, you might at first find no part of the graph on your screen. The plotted points will not appear in your graphing window. Compare notes with others in your class to get some ideas about making good window choices. Write your advice as a reminder to yourself and as a help to others.

3. When you model a relation between variables by a rule written symbolically, which approach do you find most useful in finding answers to related questions—making tables, making graphs, or some other method? Explain your reasoning and give some examples from the problems you have done in this unit.

4. Drivers' education classes often give a rule-of-thumb for the space to leave between you and a car in front for every 10 miles per hour you are traveling. Find out what this rule is by talking with a drivers' education instructor or reading a manual. Judge how good the rule is in light of Task 5 in the Modeling section.

Extending

1. Consider the relation between variables given by $y = 2x + 1$. Make a table of (x, y) data for this relation from $x = -10$ to $x = 10$ in steps of 1. Using the same window settings for x, produce a graph of the rule. Use your table to help you decide on appropriate window settings for y.

a. Explore the changes in the table and graph that result when the number 2 in the rule is changed to other values, including some decimals and some negative numbers. What stays the same and what changes in the table? In the graph?

b. Next, explore the changes in the table and graph that result when the number 1 is changed in the rule to other values, including some decimals and some negative numbers. What stays the same and what changes in the table? In the graph?

c. Can you find a clue in the rule that explains why the tables and graphs fit the observed patterns and why changing the 2 and the 1 causes the observed changes in the tables and graphs?

2. Each of the following calculator graphs is in a viewing window with x and y settings from -5 to 5. Write a rule for each graph and test your prediction using your calculator or computer software. Explain whether or not you think your rule is the only possible one.

a.

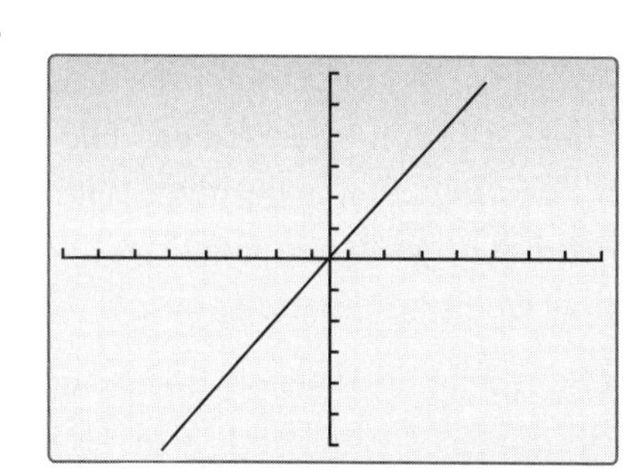

b.

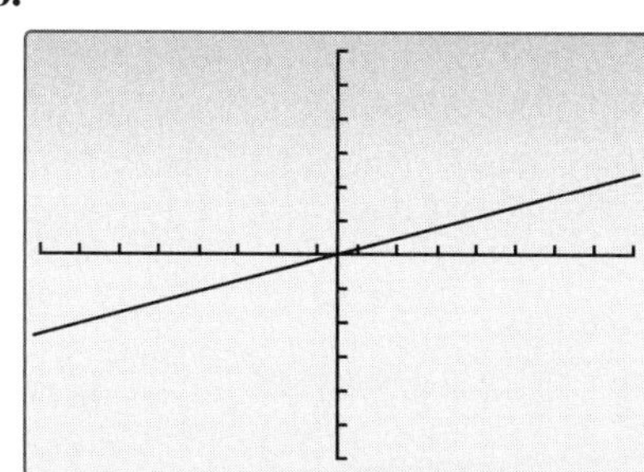

c.

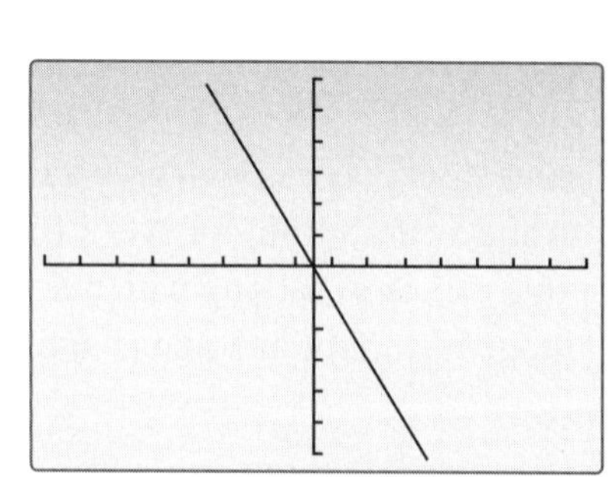

d.

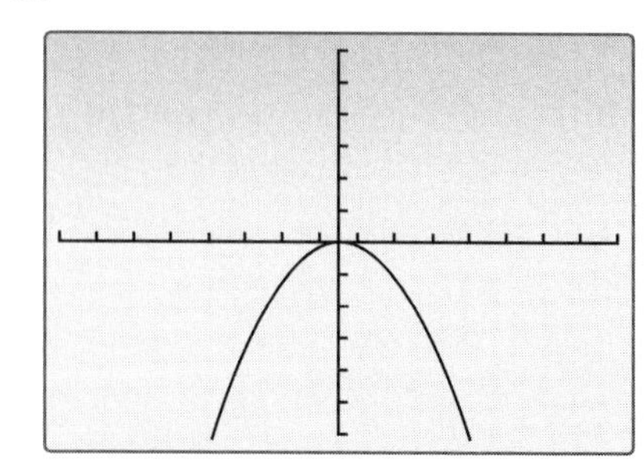

3. When a baseball is hit, you can consider its velocity toward the outfield wall and its upward velocity.

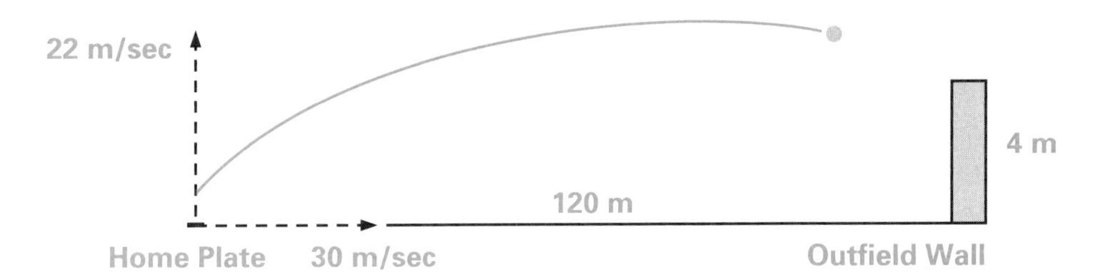

Suppose a hit toward the outfield wall has initial upward velocity of 22 meters per second and velocity toward the outfield wall of 30 meters per second. The outfield wall is 4 meters high and 120 meters from home plate. Will the ball go over the wall? Answer the following questions to help you analyze this problem.

a. The distance of the ball from home plate is a function of time since it was hit with rule $D = 30T$. When will the ball reach the outfield wall?

b. The height of the ball at any time T seconds into its flight is given by the rule $H = 1.5 + 22T - 4.9T^2$. How high will the ball be when it reaches the outfield wall? Will it be an automatic home run?

4. For Task 5 of the Modeling section you were given the rule relating reaction distance in feet to speed of a car in miles per hour: $R = 1.1s$. The reaction time for a typical driver has been found to be about 0.75 second. Explain how this experimental finding can be used to create the rule for reaction distance.

5. On some graphing calculators and with some computer software, you can display at the same time both a graph and a table, or a graph and the rule for a function entered in the "Y=" menu.

a. Examine the split-screen capability of a graphing calculator or computer software to which you have access.

b. Consider the relation between variables given by $y = 2x$. Make a table of data for this relation and a graph. Use the split-screen capability to explore changes in the table and graph when the number 2 is changed in the rule to other values, including some numbers greater than 1 and some positive numbers less than 1.

Lesson 5 Looking Back

In your work in this unit, you have studied many kinds of relations between variables. You have been asked to describe the patterns relating changes in those variables in graphs, tables of values, words, and symbolic rules. The tasks in this final lesson of the unit ask you to put it all together.

1. In some public transportation systems of U.S. cities, the fare is related to distance traveled, time of day (higher in morning and evening rush hours), or both.

The fare pattern for one city's subway system is illustrated in the following table.

Distance (in miles)	1	2	3	4	5	6	7	8	9
Fare (in dollars)	1.10	1.20	1.30	1.40	1.50	1.60	1.70	1.80	1.90

a. Write a brief description of the pattern relating distance traveled to fare charged on this subway system. How does the fare change as distance increases?

b. Write two rules for calculating fare from distance.

- In the first rule, write an equation showing how each additional mile increases the subway fare.
- In the second rule, use the variable names F (for fare) and D (for distance) to show how to calculate the fare for any given distance.

c. Make a graph of the rule relating distance and fare, using distances from 0 to 15 miles.

d. Does the fare pattern used seem fair to your group? Discuss your ideas with other groups.

e. How would the table, graph, and rules change if during rush hour each fare is increased by \$0.05?

f. How would the table, graph, and rules change if during rush hour each fare is doubled?

2. When private pilots make flight plans for their trips, they must estimate the amount of fuel required to reach their destination airport. When the plane is in flight, the pilots watch to see how much fuel they have left in their tanks. The graph at the right shows fuel remaining in the tanks as a function of time in flight under constant speed for one type of small plane.

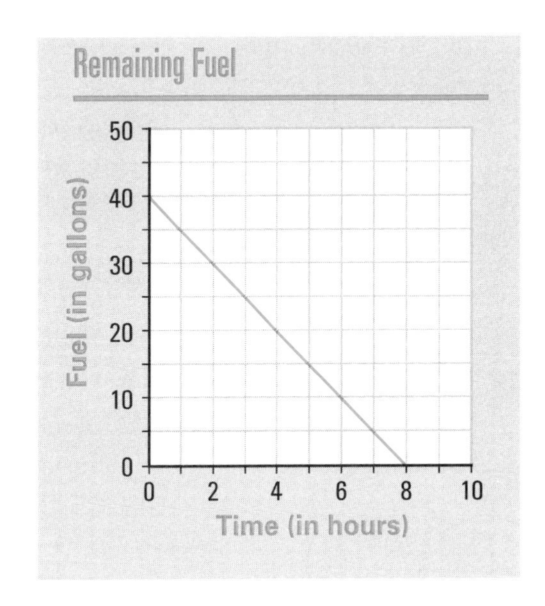

a. Make a table showing gallons of fuel left from 0 to 10 hours in steps of 1.0 hour.

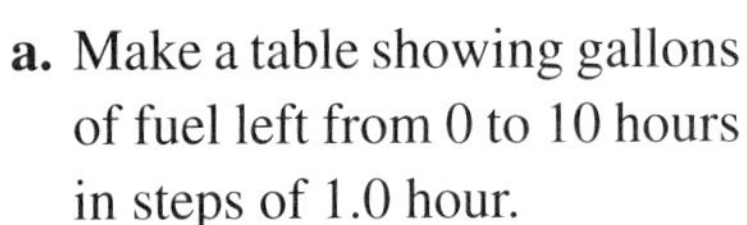

b. Describe the pattern of change in fuel remaining as time in flight passes.

c. Write two rules relating time in flight to fuel left in the tanks.
   - In the first rule, show how fuel changes from one hour to the next.
   - In the second rule, show how to find fuel left at any time in the flight.

d. Assume that the plane's tanks are full at the start of a flight.
   - How much fuel is left after 1.5 hours? After 3.75 hours?
   - How much fuel is used in a trip of 5.8 hours?

e. If the plane flies at an average speed of 125 miles per hour, how much fuel would be required for a trip of 525 miles? How is the answer shown in the graph?

f. What conditions can you imagine that would make the amount of actual fuel used different from the predictions of the graph, the table, and the rule you found?

g. Suppose two hours into the flight, the pilot discovers that the fuel attendant neglected to fill both tanks. Only 24.5 gallons of fuel remain. Assuming the same rate of fuel consumption as shown in the graph, sketch a graph representing this situation. How much flight time does the pilot have left?

3. Metal springs of various lengths and strengths are used for many important tasks—for example, absorbing shocks in car wheels and bumpers, pulling garage doors closed, weighing people and products on scales, and adjusting exercise equipment. When a spring is stretched, the length of the spring is a function of the force pulling on it.

   Suppose that for a spring used in weight training equipment, its length L (in feet) is related to the force F pulling on it (in pounds) by the rule $L = 2 + 0.01F$.

   a. Make a table of sample (F, L) data for $F = 0$ to 200 in steps of 20.

   b. Make a graph of the relation between force and length for $F = 0$ to 200.

   c. Describe the pattern that you see in the table and the graph relating change in force to change in length.

   d. To what length will that spring be stretched by a force of 50 pounds? By a force of 200 pounds?

   e. What is the length of the spring when the equipment is not in use? Describe two different ways in which you could answer this question.

   f. If the spring stretches to a length of 3.5 feet, what force is pulling on it?

   g. What do the two numbers in the rule, 2 and 0.01, tell about the pattern of change in the spring length as force is applied?

Checkpoint

When two variables change in relation to each other, the pattern of change often fits one of several common forms.

a Make sketches of at least five different graphs showing different patterns relating change in two variables.

b For each graph, write a brief explanation of the pattern of change shown in the graph and describe a real-life situation that fits the pattern.

Be prepared to share your sketches and descriptions with the whole class.

On Your Own

Write, in outline form, a summary of the important mathematical concepts and methods developed in this unit. Organize your summary so that it can be used as a quick reference in future units and courses.

Linear Models

Unit 3

Lesson 1

Predicting from Data

On Tuesday, May 10, 1994, millions of people across the United States witnessed a very rare event—a near total eclipse of the Sun. Did you see the Moon pass between our Earth and the Sun, dimming the sky in mid-day? If not, you'll have to wait until the year 2012 for the next chance!

Rare and spectacular events like eclipses get everyone's attention focused on the skies. Astronomers have been watching the movement of the Sun, Moon, Earth, and stars for thousands of years. Among their conclusions are the patterns that are used to design yearly calendars and our latitude/longitude system for locating places on the Earth.

The Sun and its shadows have been used in clever ways to find directions and take measurements. One use of sunlight and shadows gives a method of finding heights of very tall objects.

The method is based on the fact that the taller any object is, the longer its shadow will be. To make use of this relation to get actual measurements, you need to know the numerical pattern of the relation between *height* and *shadow*. You can collect that sort of data easily on a sunny day.

The graph on the next page shows data from several measurements as a flag was raised up its pole. A line has been drawn on the plot to highlight the pattern relating *length* L of the shadow and *height* H of the flag causing that shadow. That graph is called a **linear model** of the relation between the two variables.

The graph itself can be described by the equation $H = 1.5L$, giving another form of the linear model. As you know from your work in the "Patterns of Change" unit, both the graph and the equation can be used to explore the data and to answer important questions.

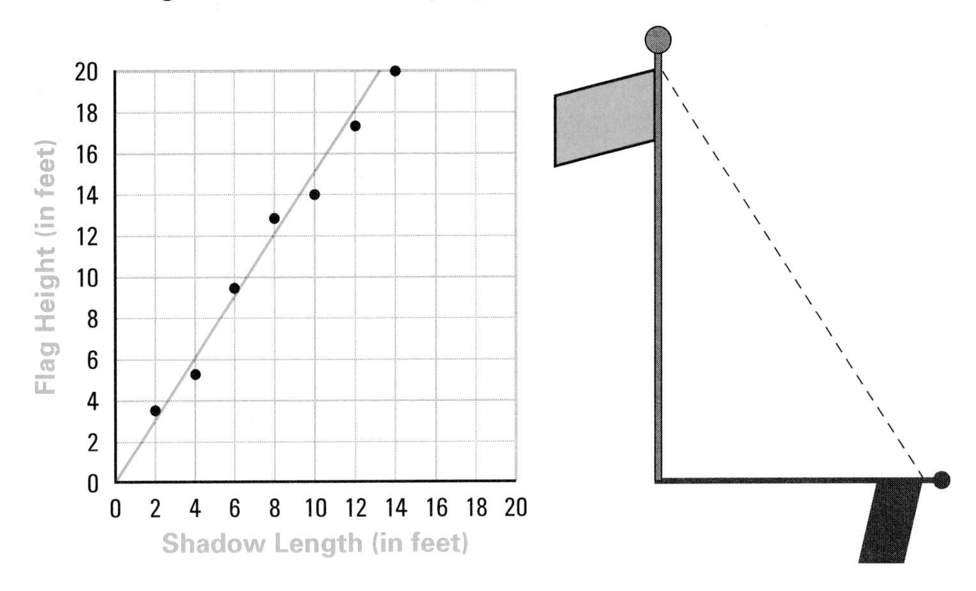

Think About This Situation

As in many experiments, the shadow data do not fall exactly on a line. No simple equation will relate all of the (*shadow length, flag height*) data pairs.

a How would you decide where to draw a line fitting the pattern in a plot like the one above?

b What predictions could you make from the given linear model?

c What kinds of equations do you expect for linear graphs?

Linear models occur in many other situations. To use linear models in solving problems, it is helpful to know the connections between the graphs, the rules that produce those graphs, and the tables of values related to the graphs and rules. Finding and using linear models is the focus of this unit.

INVESTIGATION 1 Where Should the Projector Go?

The search for a useful linear model relating variables often starts with a plot of data about those variables. Those data often come from an experiment. For example, consider this question about the many kinds of film projectors that show pictures in homes, theaters, classrooms, and auditoriums:

How can the projector be positioned so that the image will fit the screen?

Whenever such a projector is pointed directly at the screen, it enlarges images on the film into similar images on the screen. The enlargement factor depends on the distance from the projector to the screen (and special lens features of the projector). You can discover more about that relation with some simple experiments using an overhead projector.

As a class, make a test-pattern figure on an overhead transparency. Include line segments of many different lengths from 5 cm to 30 cm. In the experiment, you will be checking to see how those segments are lengthened when projected on a screen from various distances.

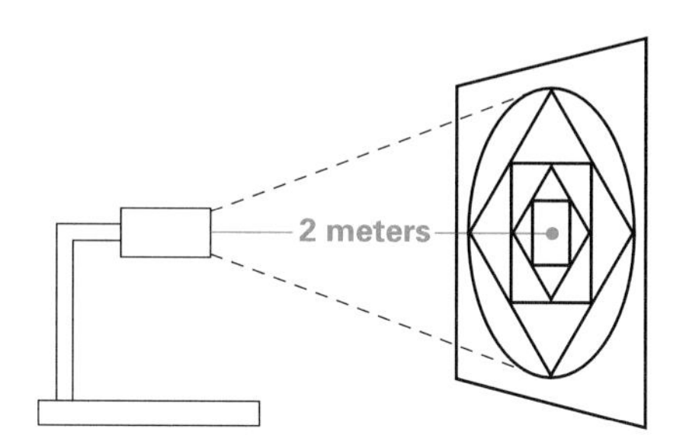

1. The first step is to collect some (*projector distance, enlargement factor*) data.

a. Place the overhead projector so that its lens is 2 meters from a screen. Focus the test pattern with the projector aimed directly at the screen. Then collect data comparing the lengths on your overhead transparency to the lengths on the screen. Estimate the enlargement factor of projection at that distance.

(**Hint:** If 5 cm on the film projects as 15 cm, 10 cm projects as 30 cm, and 15 cm projects as 45 cm, what *enlargement factor* would you record?)

b. Move the projector to a distance of 3 meters from the screen and focus the test pattern. Collect data for estimating the enlargement factor from that distance.

c. Repeat the process to find enlargement factors called for in this table:

Projector Distance (m)	1	2	3	4	5	6
Enlargement Factor						

2. Plot the (*projector distance, enlargement factor*) data on a coordinate graph. Draw a linear model that you believe is a good fit for the trend in those data.

3. Describe the pattern of change in enlargement E as the distance D changes. Explain how that pattern is shown in the data table and in the modeling line.

4. Now look for symbolic rules for the relation between distance and enlargement.

a. Find an equation using *NOW* and *NEXT* to show how the enlargement factors increase as projector distance increases in steps of 1 meter.

b. Find an equation relating D and E that matches both the linear model and the data trend.

5. Algebraic models often are used to summarize patterns in data. They are also useful in making predictions about untested values of the variables.

   a. Use your linear model or equation to predict the enlargement factor when the projector is placed 2.5 meters from the screen. Also predict the enlargement factor for a distance of 4.25 meters from the screen.

   b. Carefully make actual measurements to test both predictions. Then make a report assessing the accuracy of your predictions. If they were inaccurate, revise your model and equation.

Checkpoint

The relationship between projector distance and enlargement factor can be represented by a linear model, which can then be used to make predictions.

a Suppose you were to draw one line segment on an overhead transparency. The relation between the length of the screen image of that segment and the distance the projector lens is from the screen can be represented by a data table, a graph model, or an equation. Which representation do you think is easiest to use and most accurate for making predictions? Give reasons for your choice.

b What factors could cause inaccurate predictions from a linear model of (*projector distance, enlargement factor*) data?

Be prepared to share your group's responses with the entire class.

On Your Own

Suppose that two line segments are drawn on an overhead transparency. One segment is 4 cm long and the other is 10 cm long. An overhead projector is positioned so that the shorter segment shows up as a 12-cm segment on the screen. The longer segment shows up as a 30-cm segment on the screen.

a. Complete this table in a pattern that you would expect for other data pairs.

Sketch Length (cm)	0	2	4	6	8	10	12	14	16	18
Screen Image Length (cm)			12			30				

b. Draw a graph of the relation you would expect between *sketch length* and *screen image length.*

c. Write an equation relating *sketch length S* and *screen image length I.*

INVESTIGATION 2 The Ratings Game

Linear models can be used to summarize data patterns and make predictions in different kinds of situations. For example, consider the relation between rankings and audience size for television shows on the major networks. Those variables are important because they set advertising rates and determine the survival of the shows themselves.

The table below shows a sample from the complete list of ratings for regular television programs during the 2000–2001 season.

Television Ratings Data

Rank	Show	Average Weekly Audience*
10	Friends, NBC	10.9
20	60 Minutes, CBS	9.1
30	The Simpsons, FOX	8.4
40	Dateline, NBC	7.6
49	The Drew Cary Show, ABC	6.7
59	Ed, NBC	5.8

*in millions of households

Source: www.bayarea.com/entertainment/tv/docs/078010.htm

A scatterplot of the television ratings data is shown below.

1. Study the data and then answer these questions as a group.

a. What do the table and graph show about the relation between average weekly audience and ranking for television programs?

b. How can you predict the ranking of a show from the average weekly audience using the graph? Using the table?

c. How can you predict the average weekly audience from the rankings of a show using the graph? Using the table?

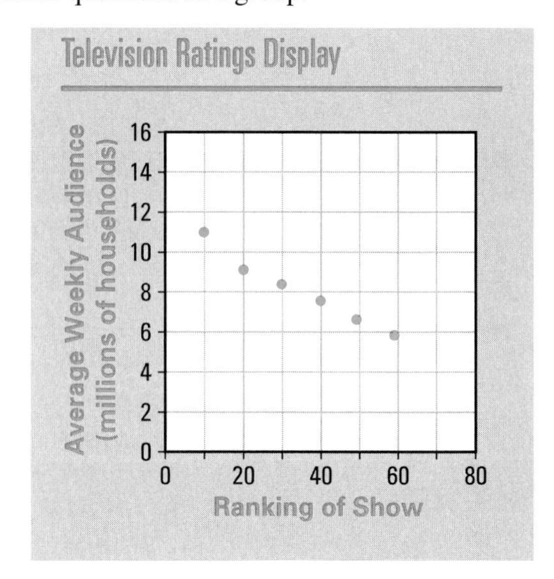

d. Which display do you find easier to use in making ranking or audience estimates?

e. Which display would be more effective in a report to another group?

f. What are the differences in average weekly audience between the shows ranked 20, 40, and 59? If you worked for a company choosing a show to sponsor, how would those differences affect your decision? What other factors would you consider?

The table and graph of television program rankings and audiences do not include data for every 2000–2001 show. They do give enough information to make some predictions. Those predictions would be made more easily and accurately if the data were modeled by a smooth graph. It looks like a straight line is a reasonable model in this case.

2. For the following activities, use a copy of the scatterplot showing ranking and audience size.

a. Draw a linear model that you believe is a good fit for the trend in those data.

b. Explain why you drew your modeling line where you did. Consider the number of points on the line and points above or below the line.

c. Use your linear model to answer the following questions.

- *The Simpsons* was the 30th ranked show for the year. What audience does your model estimate for that show?
- *Dateline* had an average weekly audience of 7.6 million households. What ranking does your model predict for that show?

d. The actual average weekly audience for *The Simpsons* was 8.4 million households. *Dateline* was ranked 40th. Compare these values to your estimates in Part c. Explain any differences between predicted and actual numbers.

3. Refer back to the sample ratings on page 162.

a. Use your graphing calculator or computer software to make a scatterplot of the sample (*ranking*, *average weekly audience*) data.

b. One equation proposed as a model of the data pattern was $y = -0.1x + 12$. Enter that equation in your calculator's or software's functions list. Test the equation's accuracy as a model. Then experiment to find another equation whose graph seems to model the data better.

c. Use your modeling equation from Part b to estimate audience size for the shows ranked 30th and 59th. Then compare those estimates to the actual data. Explain why there are differences.

Checkpoint

In Investigations 1 and 2 of this lesson, you used tables and graphs of sample (x, y) data pairs to find a line that would fit all data in a collection reasonably well.

a How do you use a modeling line to estimate y values related to any of the chosen x values?

b How do you use a modeling line to estimate the x values that will predict any chosen y value?

Be prepared to share your group's procedures with the whole class.

On Your Own

In 1975, a new Ford Mustang car had a base price of $4,906. In 1981, a comparable car had a base price of $7,900. In 1987, the base price was $9,750; in 1993, it was $13,245; and in 2001, it was $17,695.

a. On graph paper, make a plot of the sample (*years*, *base price*) data. Use a time scale where 1975 is year 0. Then draw a line that is a good fit for the pattern of the data.

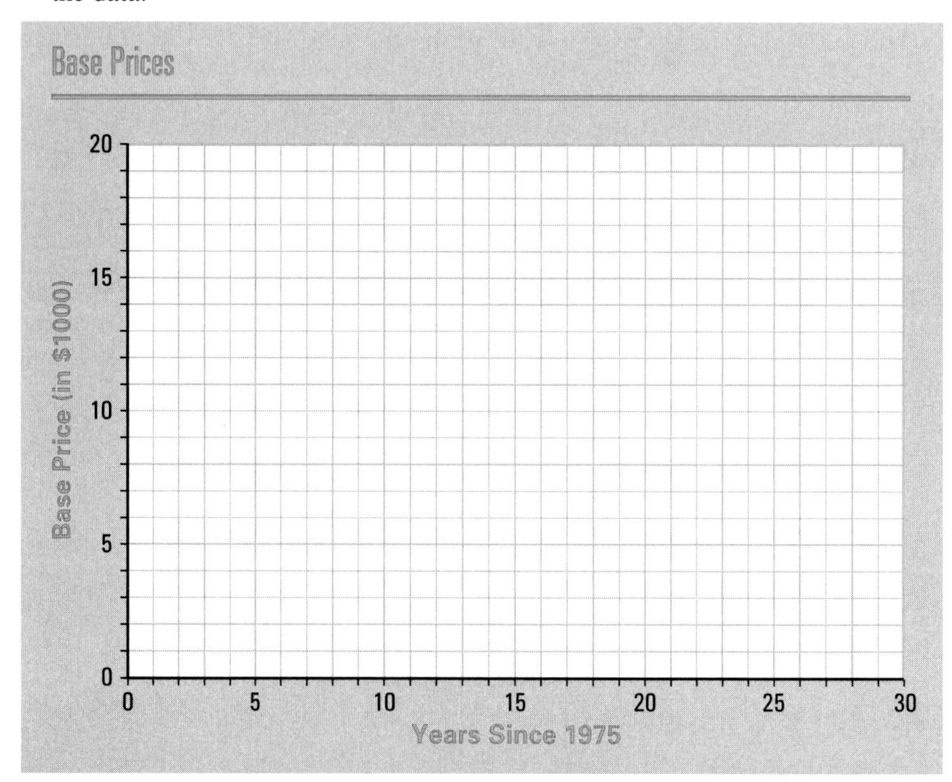

b. Extend the axes in your linear model from Part a, and estimate the base price for Ford Mustangs in 1978, 1984, 1990, 1994, and the current year.

c. Produce a scatterplot of the data with your calculator or computer software. Test the equation $y = 500x + 5{,}000$ to see if it seems a good model of the relation between years since 1975 and base price. Then experiment with variations on that equation to find what you think might be a better model.

d. Use the trace or table feature of your calculator or computer software to compare prices predicted by your revised model from Part c to the original data for 1975, 1981, 1987, 1993, and 2001. Try to explain differences between predicted and actual base prices.

INVESTIGATION 3 Choosing a Good Linear Model

When you are looking for a pattern relating two variables, it often helps to make a scatterplot of sample (x, y) data pairs. In many cases, that plot will show points that lie in a roughly linear pattern. Then you can draw a single straight line that seems to summarize the overall trend in that pattern. The challenge is finding the right place to draw that modeling line.

The table below and scatterplot on the next page give information about tuition costs at fourteen major American state universities in 1999. Public universities charge different tuition fees for in-state and out-of-state students. This is because taxes paid by state residents help support these universities. The given data show the relation between those two fees at some universities. How would you describe the relation between in-state and out-of-state tuition at these universities?

Tuition Costs

University	In-State Tuition ($)	Out-of-State Tuition ($)	University	In-State Tuition ($)	Out-of-State Tuition ($)
Delaware	12,678	21,427	Oklahoma	11,639	15,719
Florida	10,053	16,709	Penn State	13,806	20,922
Georgia	10,648	18,166	Texas	10,554	17,034
Illinois	13,276	19,740	Utah	12,867	18,572
Kansas	9,002	15,605	Washington	11,364	19,755
Maryland	13,625	20,513	West Virginia	9,088	14,440
Ohio State	11,326	19,279	Wisconsin	10,830	19,450

Source: *The College Handbook 2001*. New York: College Entrance Examination Board, 2000.

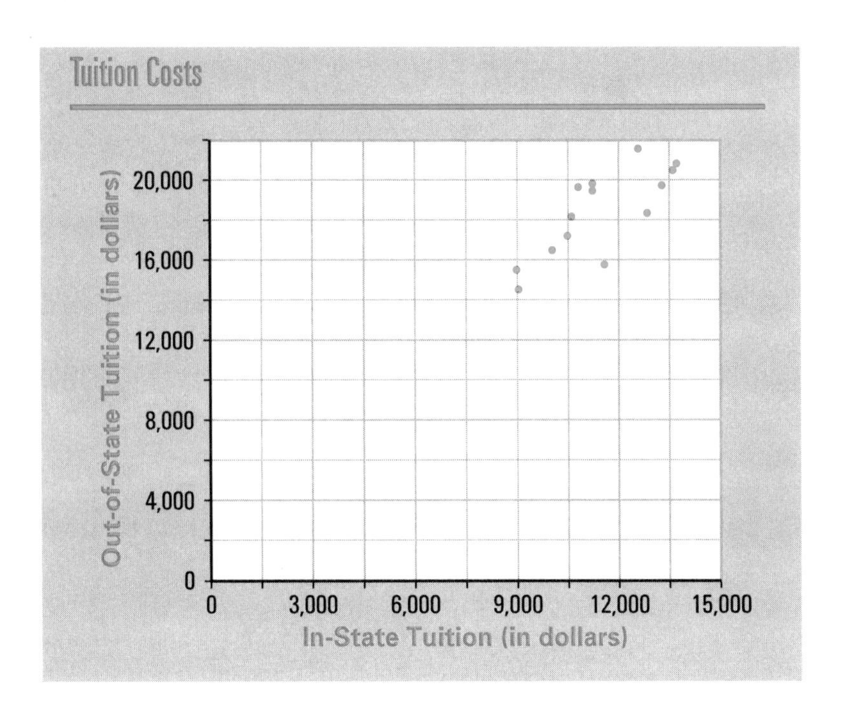

Think about drawing a straight line to model the trend in tuition data pairs. It might seem reasonable to send that line through the points that are typical of the trend. In the case of the tuition scatterplot, one such point would be (*mean in-state tuition*, *mean out-of-state tuition*) or $(\bar{x}, \bar{y})$. That point is shown with a square on the scatterplot below.

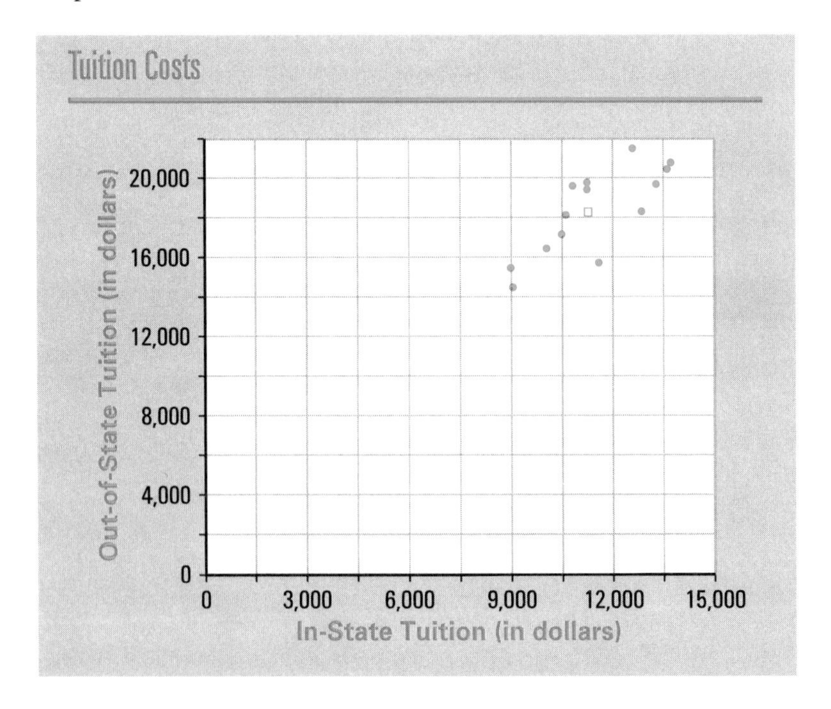

1. Calculate the mean in-state and out-of-state tuition prices for data in the table. Check your answers by comparing them to approximate coordinates of the square on the plot.

2. On a copy of the scatterplot, experiment with different positions for a straight-line model through the $(\bar{x}, \bar{y})$ point. Use a clear ruler or a long narrow object such as a piece of uncooked spaghetti. Draw the line that you believe best fits the overall trend in the data. Use that linear model to answer the questions below.

   a. What would the out-of-state tuition be for a typical university that set in-state tuition at $10,500?

   b. What would the out-of-state tuition be for a typical university that set in-state tuition at $13,000?

   c. At the University of Colorado in Boulder, in-state tuition is $11,010. Out-of-state tuition is $23,790. How does that combination compare to the pattern described by your linear model?

Even if it makes sense that a good modeling line should pass through the point $(\bar{x}, \bar{y})$, there are still many choices for a line. You might wonder if there is a way to choose the "best" among those options. Here's a simple example to explore. It uses ten data points on the scatterplot and two possible modeling lines. The point $(\bar{x}, \bar{y})$ is marked by a square. Which of the two lines shown would you choose as the better linear model for the data pattern?

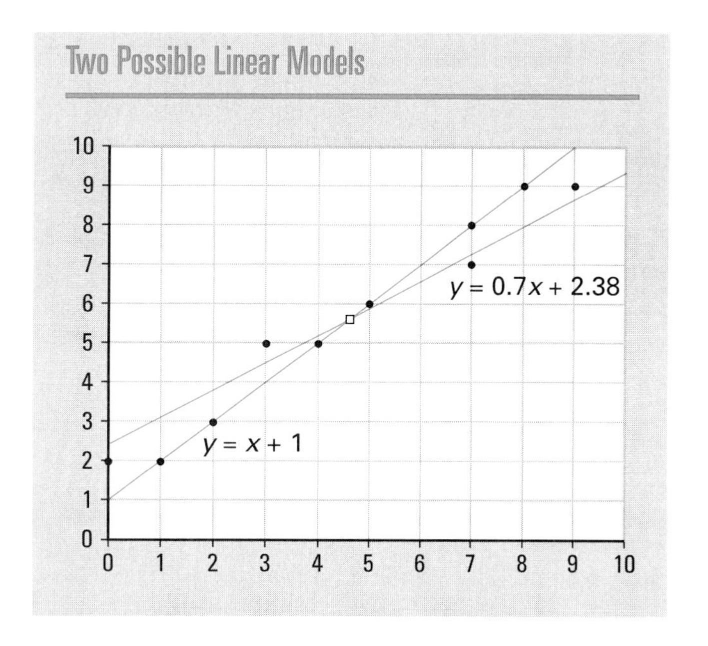

3. One way to choose between the two modeling lines is to compare their **mean absolute error**. The mean absolute error is the average difference (ignoring signs) between actual and predicted y values for the various x values.

   a. How does the absolute error for each data point and the modeling line $y = x + 1$ show up on the diagram above?

   b. Calculate the missing entries in a copy of the table on the top of page 168. Then calculate the mean absolute error for the modeling line $y = x + 1$.

Predicting from $y = x + 1$

x	Actual y	Predicted from $y = x + 1$	Error Actual – Predicted	Absolute Error
0	2	1	1	1
1	2	2	0	0
2	3			
3	5			
4	5			
5	6			
6	7			
7	8			
8	9			
9	9	10	–1	1

Mean absolute error = ________

c. Finding the mean absolute error requires many calculations. You can simplify the work by using your calculator or computer software. Complete a copy of the following table for the modeling line $y = 0.7x + 2.38$. Then calculate the mean absolute error.

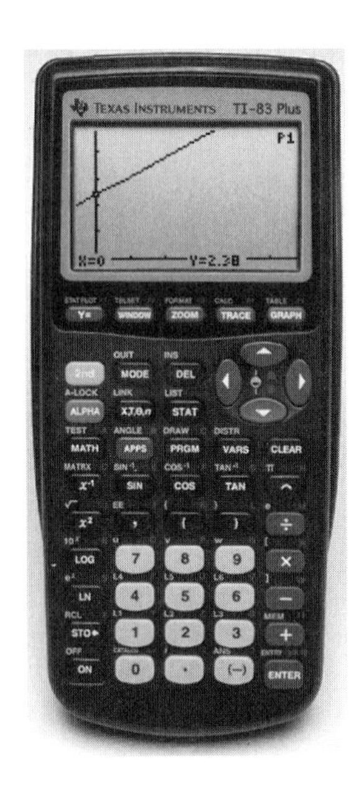

Predicting from $y = 0.7x + 2.38$

x	Actual y	Predicted from $y = 0.7x + 2.38$	Error Actual – Predicted	Absolute Error
0	2	2.38	–0.38	0.38
1	2	3.08	–1.08	1.08
2	3			
3	5			
4	5			
5	6			
6	7			
7	8			
8	9			
9	9			

Mean absolute error = ________

d. Which line would you choose as the better model? Why?

e. What are the coordinates of the point $(\bar{x}, \bar{y})$ for this scatterplot? Verify that it is on both modeling lines.

Checkpoint

In this investigation, you considered the problem of choosing the best linear models for given data patterns.

a What strategies seem sensible in finding a good linear model?

b If two different linear models are proposed, how could you compare them to see which is the better fit to the data pattern?

Be prepared to share your strategies and reasoning with the class.

The mean absolute error is the average (vertical) distance of data points from a modeling line. It provides a measure of how close a modeling line is to the data points themselves.

On Your Own

The student government at Banneker Middle School was planning a fund-raising carnival. They considered renting a "moon walk." First they wanted to see how much students would pay for five-minute walks. They collected the following data.

Price for Five-Minute Walk (in cents)	25	50	75	100	125	150
Number of Customers	100	80	55	35	20	5

a. Make a scatterplot of these data and locate the point $(\bar{x}, \bar{y})$ on the plot.

b. Draw what you believe is a good modeling line for the pattern in the data. Use that model to estimate the number of customers if the price is set at \$0.35 and at \$1.15.

c. Two possible equation models are given below. In each model, x stands for price and y stands for number of customers. Calculate the mean error of prediction for these two models.

i. $y = -0.7x + 100$

ii. $y = -0.8x + 110$

d. Use your calculator or computer to produce a scatterplot of the (*price, number of customers*) data. Then experiment to see if you can find a better-fitting model than either proposed in Part c.

MORE
Modeling • Organizing • Reflecting • Extending

Modeling

1. To find the enlargement factor of a film or slide projector at some given distance from the screen, you really only need to look at the size of one object and the size of its image. Use this fact to complete the table of data describing a slide projector. Then use that information to complete the items that follow.

Projection Size Data

Projector Distance	Size on Film	Size on Screen	Enlargement Factor
1 meter	10 mm	25 mm	
2 meters	10 mm	50 mm	
3 meters	10 mm	75 mm	
4 meters	10 mm	100 mm	

a. Make a scatterplot of the (*projector distance*, *enlargement factor*) data. Draw a linear model that fits the pattern of those data.

b. Use the model or the pattern of the table to estimate the enlargement factor when the projector is:

- 2.5 meters from the screen;
- 5 meters from the screen.

c. How would you describe the pattern of change in the enlargement factor as the projector distance from the screen increases? How is the pattern illustrated by the graph and the table?

2. The *Sun and Surf Company* rents boogie boards, beach chairs, and umbrellas at a beach in Bethany Beach, Delaware. Naturally, their business is affected by the weather. Following are some typical boogie board rental data from sixteen weekend days during July and August. Study the data to answer the following questions.

Rentals

High Temperature (°F)	Boogie Board Rentals	High Temperature (°F)	Boogie Board Rentals
72	4	92	37
78	14	94	41
87	29	85	26
89	33	87	30
87	28	71	6
94	40	79	12
78	15	95	41
83	22	94	42

a. Obtain a copy of the scatterplot of the (*temperature*, *boogie board rentals*) data. Plot the point $(\bar{x}, \bar{y})$ on the scatterplot. What does $(\bar{x}, \bar{y})$ represent?

b. Draw a linear model which contains the point $(\bar{x}, \bar{y})$ and represents the trend in the data.

c. Use your linear model to predict the number of boogie board rentals on a 60° day. Do the same for a 75° day and a 100° day.

d. On a day when the temperature reached 100° at Bethany Beach, the manager of the *Sun and Surf Company* reported rentals of 38 boogie boards. The owner was suspicious that some rentals were not being reported. What do you think?

3. The graph shows data from a market research survey of potential customers for a helicopter ride over some mountains. People were asked how much they would pay for a ten-minute ride. Based on those results, the survey company predicted likely numbers of customers each day at several typical prices.

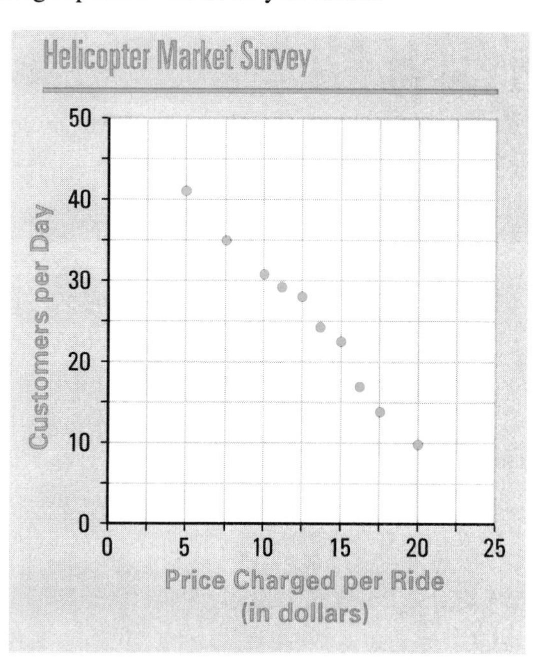

a. Describe the trend of the relation between price per ride and customers per day as shown in the data plot. How will the number of customers probably change as the price is raised higher and higher?

b. Draw a linear model that you believe is a good fit for the pattern in the data. Explain how you arrived at your model choice.

c. Use your model to estimate the number of customers when the price is set at $7.50, $12.50, and $17.50.

d. Suppose you found that the helicopter ride averaged 32 customers per day in one month. What guess would you make about the price that had been set?

4. The following shadow data were collected at 10:00 A.M. on a sunny spring day in Washington, D.C. Use them to investigate the connection between heights of objects and lengths of their shadows.

Object Height (m)	Shadow Length (m)
1.0	3.1
1.5	4.5
2.0	5.9
2.5	7.4
3.0	9.2
3.5	10.6

a. Plot the (*object height*, *shadow length*) data on a graph. Draw a line that is a good fit for the pattern of these data.

b. Use your linear model to estimate shadow lengths for:
- a flagpole that is 4 meters tall;
- a person who is 1.75 meters tall;
- the Washington Monument, which is 169 meters tall.

c. Use your linear model to estimate heights for:
- a light pole that casts a shadow 8 meters long;
- a tree whose shadow is 15 meters long;
- a water tower that casts a shadow 50 meters long.

d. Use your calculator or computer to produce a scatterplot of the data in the table. Experiment with different expressions in the function list to find an equation whose graph is a good model for the data. Compare shadow lengths predicted by the rule with those in the table.

5. Each March, many people in the United States are caught in "March Madness"—the national college basketball championship. Below are statistics from the 2001 women's championship game.

Notre Dame Fighting Irish

Player	Minutes	Points	Rebounds
Riley	35	28	13
Barksdale	5	0	2
Haney	35	13	5
Joyce	20	2	0
Ratay	25	3	4
Ivey	40	12	5
Siemon	40	10	9
Totals	200	68	38

Purdue Boilermakers

Player	Minutes	Points	Rebounds
Hurns	39	17	7
Wright	34	17	4
Cooper	23	6	6
Komara	37	8	2
Douglas	40	18	7
Hicks	1	0	0
Parks	12	0	2
Crawford	11	0	4
Noon	3	0	0
Totals	200	66	32

Source: und.fansonly.com/sports/w-baskbl/stats/040101aaa.html

a. Combine the data for the two teams to study the relation between minutes played and points scored. Use your graphing calculator or computer software to produce a scatterplot of the (*minutes played*, *points scored*) combined data.

- Describe the overall relation between those variables and any interesting deviations from that pattern.
- Would a linear graph be a good model of the pattern in these data? Explain your reasoning.

b. Repeat the analysis of Part a to study the relation between *minutes played* and *rebounds.*

c. Repeat the analysis of Part a to study the relation between *points scored* and *rebounds.*

6. The next set of data gives information about the pass statistics of some professional football quarterbacks. The table shows the relation between number of touchdown passes and number of interceptions thrown by ten of the all-time top-rated quarterbacks in the National Football League.

National Football League All-Time Quarterbacks

Name	Touchdowns	Interceptions
Steve Young	232	107
Joe Montana	273	139
Brett Favre	235	141
Dan Marino	420	252
Mark Brunell	86	52
Jim Kelly	237	175
Roger Staubach	153	109
Neil Lomax	136	90
Troy Aikman	158	127
Sonny Jurgensen	255	189

Source: *The New York Times 2001 Almanac.* New York, NY: The New York Times, 2000.

a. Use your calculator or computer to produce a scatterplot of the (*touchdowns*, *interceptions*) data.

b. Rona and Aaron each found an equation for modeling lines relating touchdown passes x to interceptions y.

Rona's equation was $y = 0.5x + 25$.

Aaron's equation was $y = 0.7x + 8$.

Do these lines contain the point $(\bar{x}, \bar{y})$?

c. Visually compare the graphs of these lines on the scatterplot. Does one appear to fit the pattern in the data better than the other? Explain why or why not.

d. Decide which of the lines is the better model by comparing their mean absolute errors.

e. Use the better model to predict the number of career interceptions to be expected from a top-rated quarterback who threw 120 touchdowns.

Organizing

1. When using coordinate graphs, mathematicians usually refer to the horizontal axis as the x-axis and the vertical axis as the y-axis. Each point on the graph has a pair of coordinates (x, y).

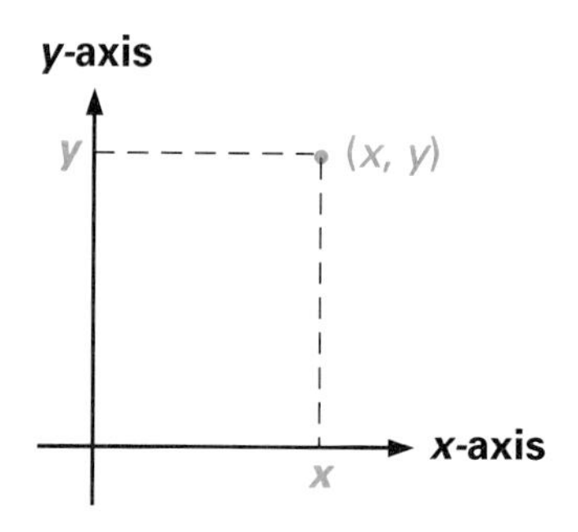

To make a scatterplot of data or a graph of an equation relating two variables, some decisions must be made. One is which variable should have its values listed first in the ordered pairs. The values of that variable are represented on the x-axis. The values that are listed second in the data pairs are represented on the y-axis.

There are two general rules for guiding this decision. First, if you are studying the way some variable changes as time passes, the time variable is usually listed first in the ordered pair. Second, in many cases one variable depends on the other. That means you can control or choose the values of one variable. The other variable responds to those changes or choices. In such a situation, the variable you can control or choose directly is called the **independent variable**. It is usually listed first in the coordinate pairs. The other variable, called the **dependent variable**, is listed second in the coordinate pairs. In some cases, there is no natural order for the variables involved.

The following list gives eight situations involving variables. In each case, name the two variables involved. Explain which you would list first and represent on the x-axis. Which would you list second and represent on the y-axis?

a. whale population in the Arctic changing over the years

b. profit from a concert and tickets sold for the concert

c. height and weight of players on a soccer team

d. shoe size and length of the wearer's foot

e. money earned and hours worked at a job

f. age and height of young people from ages 1–20

g. size of world population for the years 1900 through 2000

h. driving speed and time required for a trip

2. The graph below shows data from a market research survey of potential audience members for a local "battle of the bands." People were asked how much they would pay to attend the concert. Based on those results, the survey company predicted likely audience sizes at several typical prices.

a. Suppose you were asked to experiment with different expressions to find an equation whose graph models the data shown below. What equation would you try first?

b. Explain why you chose that equation.

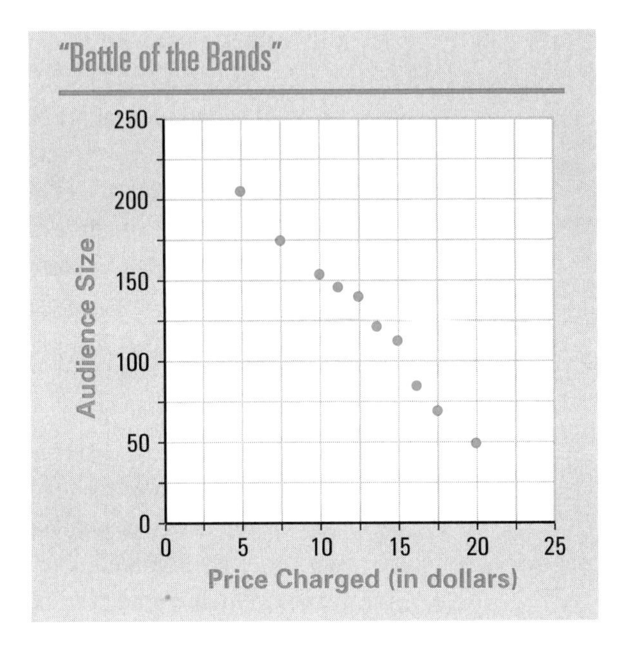

3. The enlargement factors of film projectors and overhead projectors can be studied using the arithmetic of *ratios* and *proportions*. Study the following table which gives some sample data from an experiment on projection of overhead transparencies.

Transparency Image Length (in cm)	1	3	5	7	9
Screen Image Length (in cm)	4	12	20	28	36

a. Write the ratio

$$\frac{\text{transparency image length}}{\text{screen image length}}$$

for each data pair. How are the ratios related?

b. Suppose a transparency image length was 6 cm and its screen image had length x. Use the data to complete the following proportion:

$$\frac{6}{x} = \frac{?}{?}$$

c. Solve the proportion in Part b.

d. Write and solve proportions corresponding to each of the following questions.

- What screen image length will correspond to a transparency image length of 8 cm?
- What transparency image length will correspond to a screen image length of 16 cm?

4. Refer back to the data about shadows in Modeling Task 4.

a. Write and compare the ratios of *object height* to *shadow length* for each of the data pairs.

b. Write and solve proportions to answer each of the following questions:

- What shadow length will be produced by an object of height 3 meters?
- What object height will produce a shadow of length 8 meters?

5. The graph below shows a scatterplot of data and two different possible modeling lines.

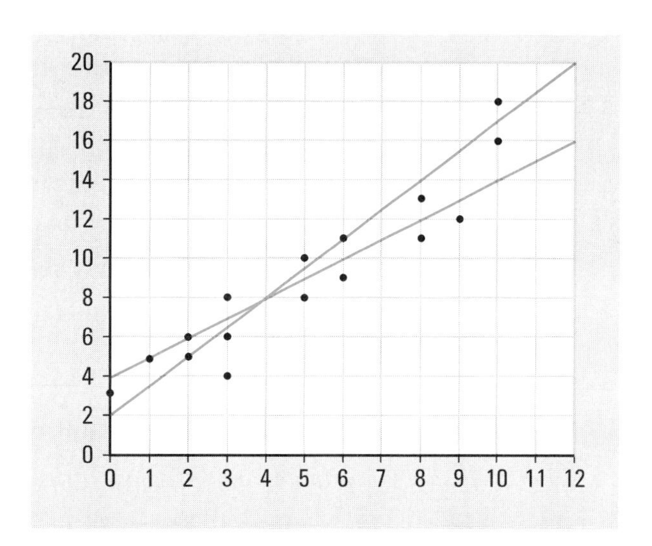

a. Which of the two modeling lines do you believe is the better fit for the data pattern?

b. How can you support your choice with a numerical measure of accuracy of model fit? Use these facts:

i. The equations of the modeling lines are $y = 1.5x + 2$ and $y = x + 4$.

ii. The data points all have whole number coordinates, beginning with (0, 3), (1, 5), (2, 5), and so on.

Reflecting

1. The sketches that follow show five ways of drawing a linear model on the same scatterplot. See if you can figure out the reasoning used to make each drawing. Then decide which linear model seems best. Explain why the others are not based on sensible strategies.

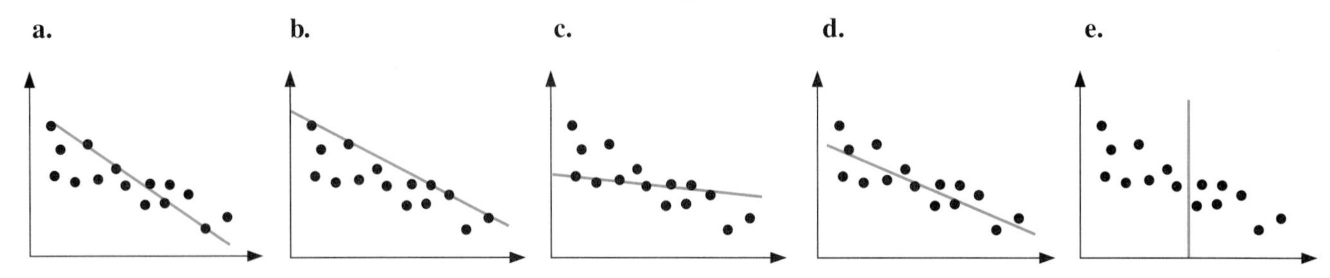

2. How are the two statistics *mean absolute deviation* and *mean absolute error* similar? How are they different?

3. Think about the variety of theaters in which you have seen movies. How do the size and shape of those theaters affect the size of the screen images you see and the effect that those images have on you as a viewer? What kind of theater do you like best? How does viewing a movie in a theater compare to watching it at home on TV?

4. Make a survey of television viewing patterns.

   a. Check the weekly television rankings in a newspaper or on TV. Ask students in your school to report which of the top 20 shows they watch regularly themselves. Then give each show a *show rank* from most to least popular. Compare your results to the national survey.

   b. Make a scatterplot of (*number of students*, *show rank*) data for your class.

   c. Are the variables *number of students* and *show rank* linearly related? Explain your reasoning.

5. Agricultural and medical scientists conduct experiments to uncover possible relationships between pairs of variables. They often speak about *explanatory variables* and *response variables*.

   a. Many scientists have studied how drug dosage relates to blood pressure. Which of these factors would you consider as the response variable? Which is the explanatory variable? Give reasons for your choice.

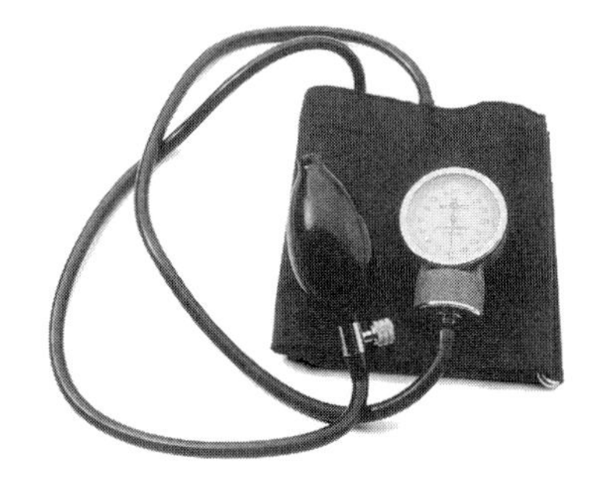

   b. How do you think the ideas of response and explanatory variables are related to the ideas, as described in Organizing Task 1, of independent and dependent variables?

Extending

1. Refer to your group's data on the bungee apparatus from Investigation 1 in Unit 2. See if you can find a linear model, including a rule, that predicts stretch length from weight.

a. Describe two ways in which you can use the rule and your graphing calculator or computer software to predict stretch lengths for given weights.

b. Apply the method of your choice to the weight data in your group's table. Compare your predictions with experimental results from Unit 2. Write a short report describing similarities and differences in the theoretical and experimental results. Be sure to provide explanations for any differences.

2. A common contest at carnivals or school game nights is to guess the number of beans or coins in a large jar. One way to arrive at good estimates for this amount, without actually counting every bean or coin, is to take some samples and look for a pattern. Test this strategy by performing the following experiment.

a. Weigh a large jar and then fill it with beans.

b. Take and then weigh some samples from the jar. Organize your data in a table like that below.

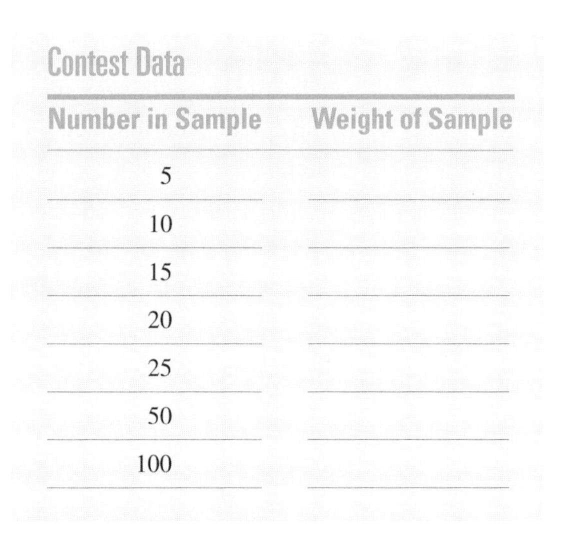

Contest Data

Number in Sample	Weight of Sample
5	
10	
15	
20	
25	
50	
100	

c. Draw a graph modeling the pattern in your data.

d. Find a rule relating *number in sample* and *weight of sample*.

e. Weigh the entire jar. Use the table, graph, and rule to calculate a good estimate of the number of beans in the jar.

3. Design and carry out an experiment to study the pattern relating the shadow length of an object as a function of the time of day. Make a table and a graph to display your data. Try to find an equation for *shadow length* as a function of *time of day*. Use your equation to predict the shadow lengths for various times of day. Then test your predictions with actual measurements. Would your equation model shadow length as a function of time of day for next month? Explain your response.

4. The ratio of height to width of many movies shown in a theater is typically about 1 to 2. For example, a picture 25 feet high is about 50 feet wide. Height and width of most television screens have a ratio of about 4 to 5.

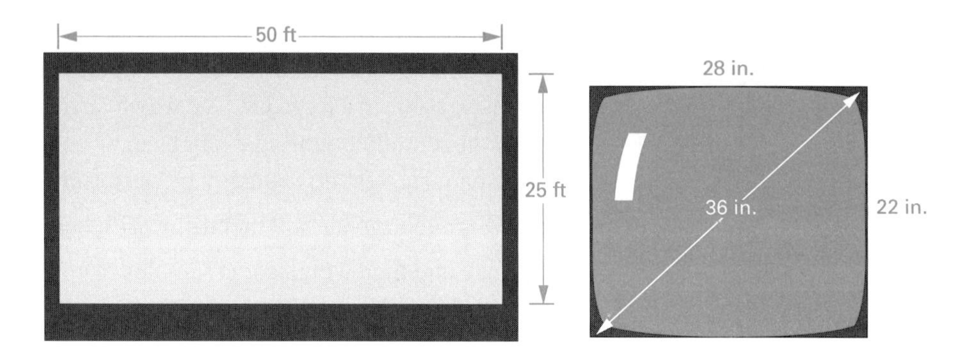

a. Usually, made-for-theater movies shown on TV are broadcast so that the image fills the entire height of the screen. This means that the whole width of the movie cannot be shown. What percentage of the movie image is cut off on a typical TV screen?

b. Some video cassettes and DVDs, and some television broadcasts use a "letterbox" format to show the full width. This means that the height of the image does not fill the full TV screen. What percentage of the TV screen is not used?

c. Which broadcast form do or would you prefer, and why?

Lesson 2

Linear Graphs, Tables, and Rules

When two variables are related in a pattern that is roughly linear, it is helpful to model the pattern with a straight-line graph. The scatterplot below shows data from a test of a single rubber band cord in a bungee-apparatus experiment. A linear graph has been drawn to model the pattern of change in length for increasing weight.

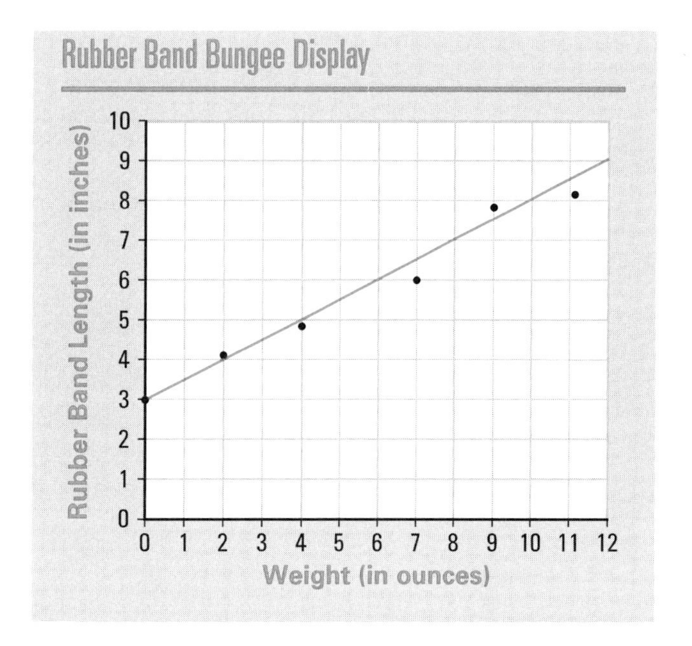

Think About This Situation

The graph shows the overall pattern relating *weight* and rubber band *length*.

a Based on the linear model (not the data points themselves), what pattern would you expect in a table of (*weight*, *length*) data pairs for weights that range from 0 to 10 ounces?

b How long is this rubber band with no weight attached, and how is that fact shown on the graph?

c How much does the rubber band stretch for each ounce of weight added, and how is that shown on the graph?

INVESTIGATION 1 Stretching Things Out

One of the most common examples of linear models is the way that forces stretch things such as rubber bands or coil springs. The more weight on the rubber band, the longer it stretches. The same is true about coil springs.

An equation for a linear model gives the simplest and most useful summary of the relation between the variables. In the following activities, you will explore the variety of patterns that can occur in linear equations as they model rubber bands being stretched by increasing forces. As you complete the activities in this investigation, look for clues that will help you answer this basic question:

How are patterns in linear graphs, tables, and equations related to each other?

1. One key to the connection between graphs, tables, and equations for linear models can be found by comparing patterns of change in the two related variables. In mathematics, the Greek letter *delta* (Δ) is used to represent "change." For example, the change in weight can be written **Δ weight** (read "delta weight"). The matching change in rubber band length would be **Δ length** (read "delta length"). The **rate of change** in length as a function of weight is given as a fraction:

$$\frac{\Delta \text{ length}}{\Delta \text{ weight}}$$

For example, look back at the graph model on page 181. The linear model passes through the points (0, 3) and (10, 8). The change in length between these points is (8 – 3) = 5; the change in weight is (10 – 0) = 10. The rate of change is $\frac{5}{10} = \frac{1}{2}$. Notice that the subtracted numbers, 3 and 0, came from the same point (0, 3).

a. Use the graph to estimate the changes in the length of the rubber band when the attached weight increases from:

0 ounces to 1 ounce;

1 ounce to 3 ounces;

5 ounces to 9 ounces.

In each case, find the rate of change in length as the attached weight changes. Explain the units that should be used to describe that rate of change.

b. How is the rate of change pattern in Part a illustrated by the shape of the graph?

c. The length of the rubber band with no weight attached is 3 inches. Use that fact and the rate of change pattern discovered in Part a to find the length of the stretched rubber band when the weight is:

2 ounces;

4 ounces;

10 ounces;

W ounces.

d. Write an equation using the words *NOW* and *NEXT* showing how the length of the stretched rubber band changes for each ounce of weight added.

e. Using the letters L (for *length in inches*) and W (for *weight in ounces*), write an equation that shows how the two variables are related: $L =$ __________.

f. How is the rate of change shown in the equations of Parts d and e?

g. How is the length of the rubber band with no weight attached shown in the equation for Part e?

h. Use your equations to predict stretch lengths for each of these weights:

4.5 ounces;

17 ounces;

0.25 ounces.

2. The diagram at the right shows linear models based on experiments with three coil springs. The models are all plotted on the same grid. What does the pattern of these graphs suggest about ways the springs are alike and different?

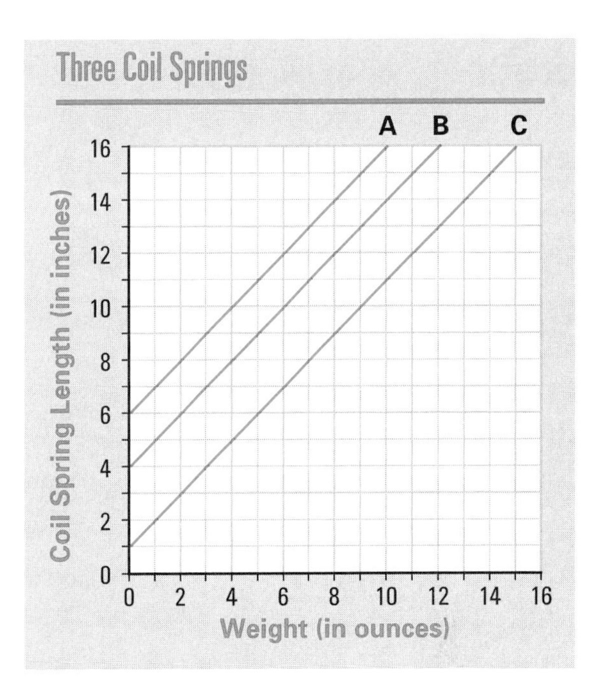

a. Splitting up the work among your group members, make three tables of (*weight*, *length*) pairs. Make one table for each linear model. Include weights from 0 ounces to 10 ounces.

b. According to the tables, how long were the different springs without any weights attached? How is that information shown on the graphs?

c. Looking at data in the tables, estimate the rates of change in length as weight is added to the three springs. How are those patterns shown on the graphs?

d. Write an equation using *NOW* and *NEXT* showing how the spring length changes with each added ounce of weight in each case. Indicate the starting value in each case.

e. Using the letters L (for *length in inches*) and W (for *weight in ounces*), write equations that show how the two variables are related in each case.

f. How can you use the equations from Part e to find the lengths of each spring with no weight attached?

g. How are the rates of change in length of the various springs related? How is this fact shown in the equations?

3. The diagram at the right shows graphs of linear models from another set of coil springs experiments. What do the patterns of these graphs suggest about ways the springs are alike and different?

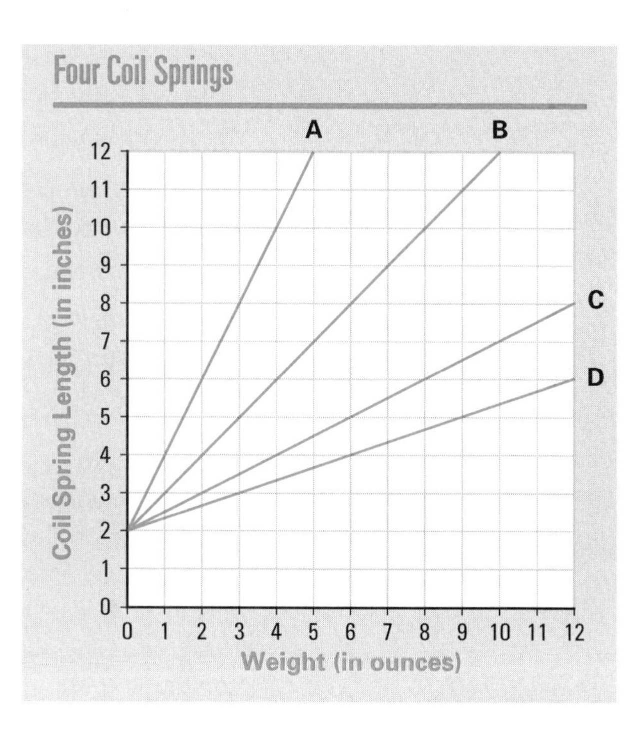

a. Sharing the work among your group members, make tables of (*weight*, *length*) pairs. Make one table for each linear model shown. Includc weights from 0 to 10 ounces.

b. According to the tables, how long were the springs without any weights attached? How is that information shown on the graphs?

c. Looking at the data in the tables or the graph, estimate the rates of change in length as weight is added. How can the rates be determined from the graphs? From the table?

d. For each linear model, write the starting value and an equation using *NOW* and *NEXT* showing how the length changes as each ounce of weight is added.

e. Using the letters L (for *length in inches*) and W (for *weight in ounces*), write equations that show how the two variables are related in each case.

f. What do the numbers in your equations tell you about each graph?

g. What differences in the springs could cause the differences in graph, tables, and equations that model the data from the experiments?

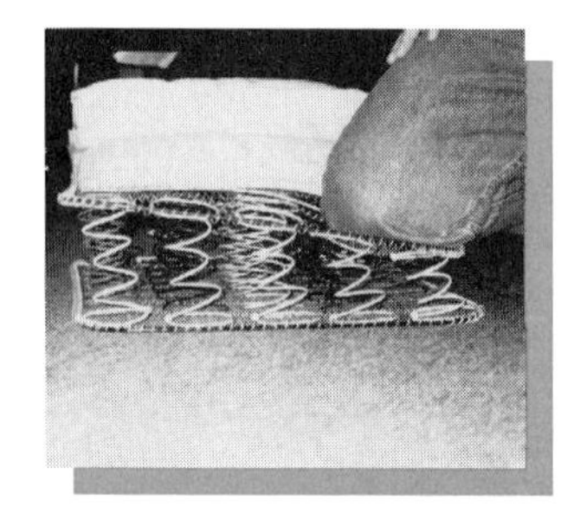

4. In some places where springs are used, the springs are made shorter by pressure from force or weights. For example, the springs in a bed mattress or chair are compressed when you sit on them. The springs in a scale are compressed when you stand on the scale.

The diagram at the top of page 185 shows graphs of four different linear relations. Those graphs model data from an experiment that tested bedsprings. What do the patterns of these graphs suggest about the similarities and differences in the springs?

a. Sharing the work with members of your group, make four tables of (*weight*, *length*) pairs. Make one table for each linear model. Include weights starting at 0 kg.

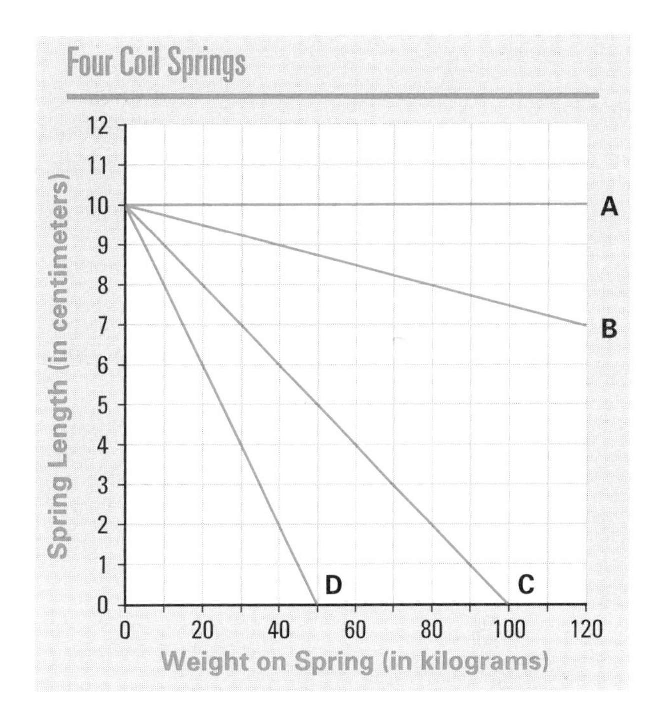

b. How long are the springs with no weight applied?

c. What are the rates of change in spring length as weight increases? How can you calculate these rates using points on the lines? Using pairs of values in the tables? (Remember, the subtracted numbers for both length and weight must come from the same point.)

d. Write the starting value and equations using *NOW* and *NEXT* showing how the springs change length as each kilogram of weight is added.

e. For each linear model, write an equation expressing L (*length in centimeters*) as a function of W (*weight in kilograms*).

f. Which linear model might correspond to springs in an extra-firm mattress? In a medium-firm mattress? Explain your choices.

g. For each of the tested springs, estimate the length of the spring with a weight of 30 kg. Which do you prefer to use in making estimates: a graph, table, or equation?

5. Each linear equation below models a stretching or compressing experiment with a spring. Identify the following:

- the initial length of the spring;
- the rate of change of the length;
- whether the experiment was designed to measure spring stretch or spring compression.

a. $L = 5 + 4W$

b. $L = 1 + 2.3W$

c. $L = 3 + (-1.5W)$

d. $L = 8 + (-0.1W)$

When studying a linear graph, it helps to think about weights and lengths of rubber bands or springs. But the connections between graphs, tables, and equations are the same for linear models relating *any* two variables x and y. By now you've probably noticed the following key features of linear models and their graphs.

- Linear models always have a **constant rate of change**. That is, $\frac{\Delta y}{\Delta x}$ is a constant.
- The constant rate of change can be seen in the **slope** of the linear graph. The slope is the direction and steepness of a walk along the graph from left to right.
- The ***y*-intercept** of the linear graph is the point where the graph intersects the y-axis.

How could the connections between tables, graphs, and equations of linear models be used to locate the y-intercept and to measure the slope of a graph?

6. Study the two linear models on this graph.

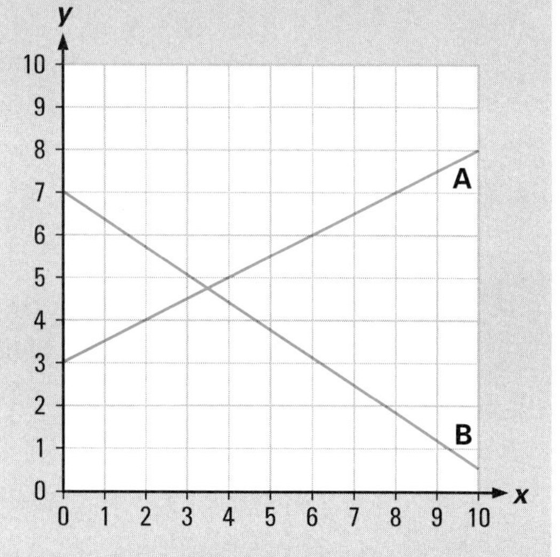

   a. Find the y-intercept of each graph. Then explain how those y-intercepts relate to the following:
   - tables of (x, y) values for each graph;
   - equations relating *NOW* and *NEXT* for each graph;
   - equations relating x and y for each graph.

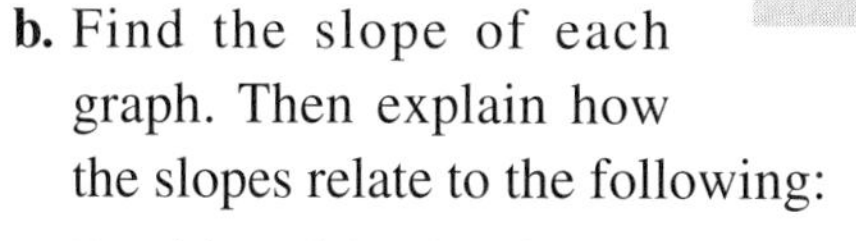

   b. Find the slope of each graph. Then explain how the slopes relate to the following:
   - tables of (x, y) values for each graph;
   - equations relating *NOW* and *NEXT* for each graph;
   - equations relating x and y for each graph.

7. Look back over the examples of linear models for (*weight, length*) data from experiments with rubber bands and springs.

   a. How are linear graphs related when they have the same slope?

   b. How are linear graphs with the same y-intercept, but different slopes, related to each other?

Checkpoint

Linear models relating any two variables x and y can be represented using tables, graphs, or equations. Important features of a linear model can be seen in each representation.

a How can the rate of change in two variables be seen:
- in a table of (x, y) values?
- in a linear graph?
- in an equation relating *NOW* and *NEXT* for the model?
- in an equation relating x and y?

b How can the y-intercept be seen:
- in a table of (x, y) values?
- in a linear graph?
- in an equation relating *NOW* and *NEXT* for the model?
- in an equation relating x and y?

Be prepared to share your group's descriptions with the whole class.

On Your Own

Every business has to deal with two very important patterns of change, called *depreciation* and *inflation*. When they buy new equipment, the value of that equipment declines as it is used. The cost of buying new replacement equipment increases from inflation as time passes.

The owners of *Laser Games Unlimited* operate a series of video game arcades in malls across the midwest. They keep a close eye on costs for new and used arcade games. One set of predictions is shown in the graph at the right.

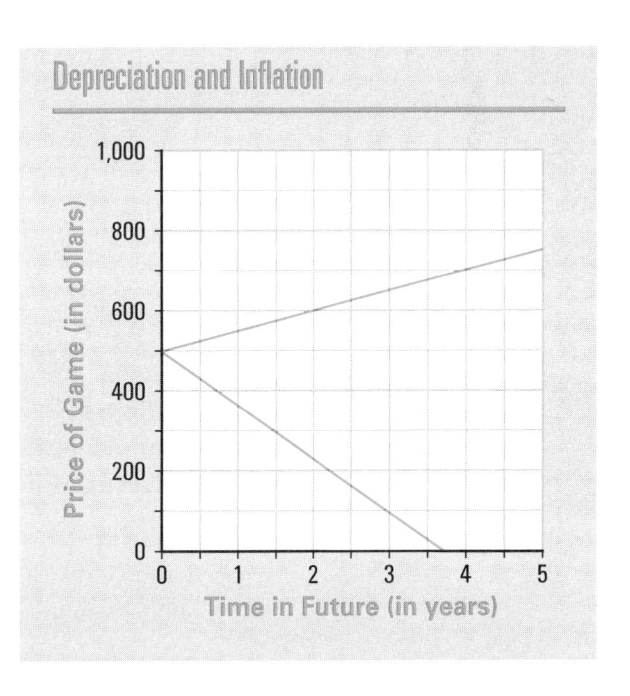

a. Which of the two linear models represents the price of a new game? The price of a used game?

b. For each linear graph, do the following:

- Find the slope and y-intercept.
- Make a table of (*time, game value*) data and find the rate of change.
- Write an equation relating *NOW* and *NEXT* showing how game price changes each year.
- Write an equation relating game price P and time T.

c. Compare:

- slopes and y-intercepts of the two graphs;
- rates of change in the tables;
- equations relating *NOW* and *NEXT*;
- equations relating P and T.

MORE

Modeling • Organizing • Reflecting • Extending

Modeling

1. Mary and Jeff Jordan both have jobs at their local baseball park selling programs. They get paid \$10 per game plus \$0.25 for each program they sell.

   a. Make a table showing the pay they can expect for any game as a function of the number of programs they sell. Include values for 0 to 100 programs in steps of 5.

   b. Write an equation relating *programs sold* S and *pay* P for a game.

   c. Graph the relation between *programs sold* and *pay* for 0 to 100 programs. Since pay depends on the number of programs sold, pay is the *dependent* (y-axis) *variable*. The number of programs sold is the *independent* (x-axis) *variable*.

   d. How do the numbers from the pay rule of \$10 plus \$0.25 per program sold relate to the patterns in the table, the graph, and the equation?

2. Jamal and Tanya Guinier work at a day camp for young children. Tanya is a coordinator with several years of experience, so she earns \$7.25 per hour. This is Jamal's first year on the job, so he earns only \$5.50 per hour.

a. Make a table showing how Tanya's and Jamal's earnings will grow over the summer as a function of hours worked.

b. Write equations relating *hours worked* H and pay earned for each worker. Use T for *Tanya's earnings* and J for *Jamal's earnings.*

c. Sketch graphs of the two pay rules on the same coordinate system.

d. How do the pay rates of \$5.50 per hour and \$7.25 per hour relate to the table, the graphs, and the equations?

3. Jose Alvarez got a job with a lawn service. He earned \$400 during the summer and saved quite a bit of that money. When school started again in the fall, he needed to use some of his savings. The following table shows his bank balance over a 9-week period in the fall.

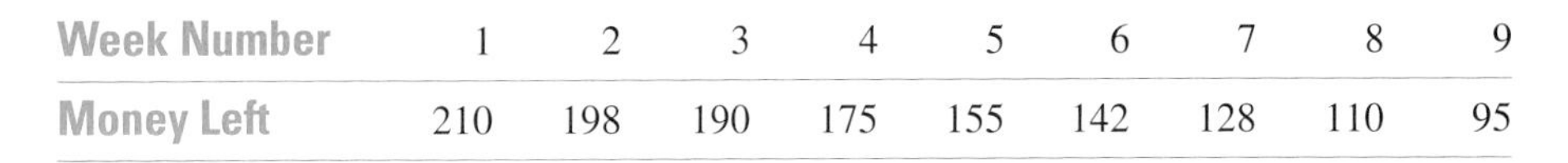

Week Number	1	2	3	4	5	6	7	8	9
Money Left	210	198	190	175	155	142	128	110	95

a. Make a scatterplot of Jose's bank balance data. Draw a line that models the trend in that plot.

b. Write an equation for the linear model. Use W for *weeks* and M for *money left*. Remember that money left depends on the week number.

c. Predict Jose's bank balance after 10, 15, and 20 weeks.

d. Explain how the patterns in the table, graph, and equation are related to each other. What do they say about Jose's spending habits?

4. Victoria DeStefano got a job at her school as scorekeeper for a summer basketball league. The job pays \$450 for the summer and the league plays on 25 nights. Some nights Victoria will have to get a substitute for her job and give her pay for that night to the substitute.

a. What should Victoria pay a substitute for one night?

b. Use the letters N for *nights a substitute works*, S for *pay to the substitute*, and E for *Victoria's total summer earnings*.

- Write an equation relating N and S, beginning $S = \ldots$.
- Write an equation relating N and E, beginning $E = \ldots$.

c. Produce graphs of the equations in Part b.

d. How do the equations and the patterns of the graphs show Victoria's earning prospects from two views?

Organizing

1. In the diagram shown here, there are five graphs of relations between variables. Following are five tables of sample (x, y) data. Match each graph with the table that best describes its pattern. In each case, explain the clues that you used to make the matches. Are there any other clues that could have been used?

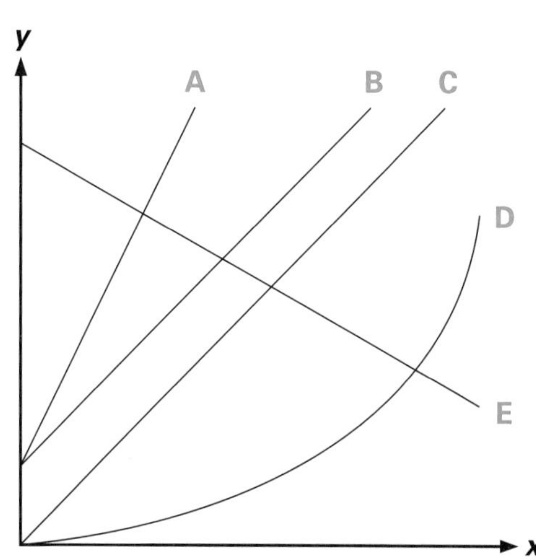

a.

x	y
1	1
2	2
3	3
4	4
5	5
6	6
7	7
8	8
9	9

b.

x	y
1	9.5
2	9.0
3	8.5
4	8.0
5	7.5
6	7.0
7	6.5
8	6.0
9	5.5

c.

x	y
1	3
2	4
3	5
4	6
5	7
6	8
7	9
8	10
9	11

d.

x	y
1	0.04
2	0.17
3	0.38
4	0.69
5	1.09
6	1.61
7	2.25
8	3.06
9	4.06

e.

x	y
1	4
2	6
3	8
4	10
5	12
6	14
7	16
8	18
9	20

2. The diagram below shows four linear graphs. For each graph, do the following:

   a. Find the rate at which y changes as x changes.

   b. Write an equation using *NOW* and *NEXT* that describes the pattern of change shown by the graph.

   c. Write an equation relating x and y beginning $y = \ldots$.

   d. How do your answers to Parts a through c relate to each other for each graph?

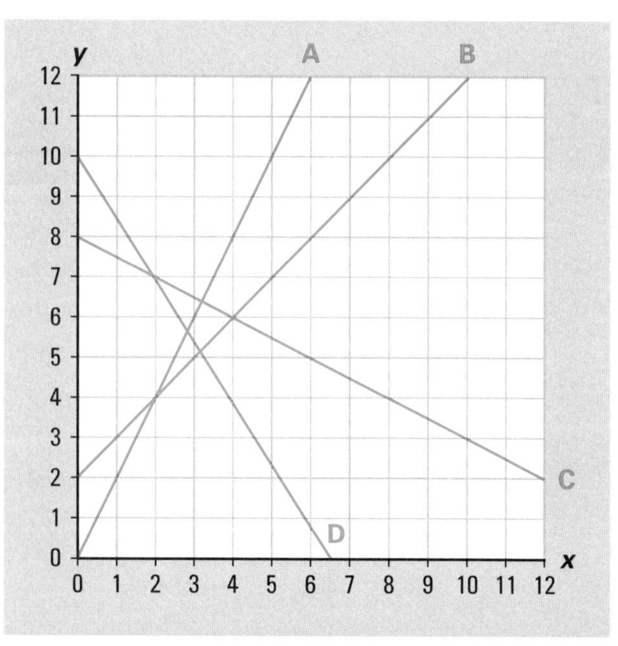

3. When the pattern relating two variables x and y can be described by a linear model, the rate of change in the variables tells the slope of the graph. The reverse is also true: the slope tells the rate of change.

a. Suppose (7, 2) and (9, 5) are two points on the graph of a linear model. What is the slope of the line containing the points?

b. Use the ordered pairs (r, s) and (t, u) to stand for coordinates of any two points on a graph or entries in a table of values for some linear model.

- Write an expression for calculating the slope of the linear graph.
- Write an expression for calculating the rate of change in y as x changes.

c. Explain the meaning of the numerator and denominator of the expressions you wrote for Part b.

d. What is the connection between slope or rate of change of a linear model and the equation for the model?

4. Linear models often are described by equations of the form $y = a + bx$. For each of the equations below, describe what the numbers corresponding to a and b tell you about the tables and graphs of that model.

a. $y = 4 + 2x$

b. $y = 12.5 + 7.3x$

c. $y = 200 - 25x$

5. Linear models also can be described using equations relating *NOW* and *NEXT*.

a. Consider the equation $y = 4 + 2x$. Use the equation to complete a table like the one shown here. Then use the table to help answer the following questions.

x	y
0	
1	
2	
3	
4	

- If $NOW = 4$, what is $NEXT$ when x increases by 1?
- Write an equation relating *NEXT* to *NOW* for increases of 1 in x.
- How does your equation relating *NOW* and *NEXT* show constant rate of change?

b. If $NOW - 6 = NEXT$, what is the slope of the linear model? Why?

Reflecting

1. Describe a situation from your experience in which one variable is:
   a. increasing at a constant rate;
   b. decreasing at a constant rate;
   c. increasing, but not at a constant rate.

2. Measuring the slope of a linear graph is very similar to measuring the pitch of the roof line on a house.

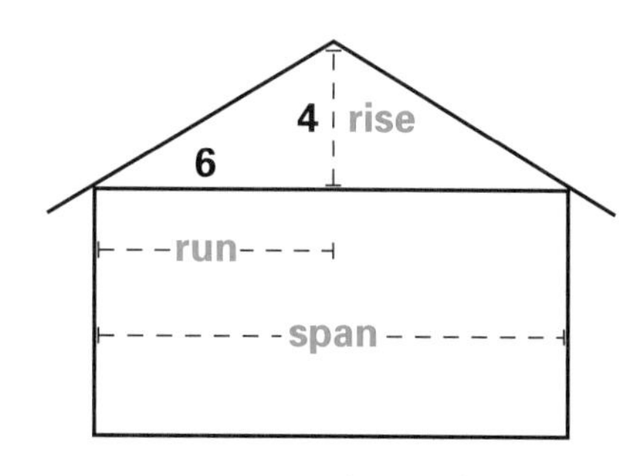

$$\text{slope} = \frac{\text{rise}}{\text{run}} = \frac{4}{6}$$

$$\text{pitch} = \frac{\text{rise}}{\text{span}} = \frac{4}{12}$$

   a. Describe five or more different roof slopes that you see around school, near home, or around your city or town. Estimate each slope using the $\frac{\text{rise}}{\text{run}}$ relation shown in the diagram. Estimate the pitch of the same roofs.
   b. Which roofs have the steepest slope? Which have the most gradual slope?
   c. What are some reasons for having steep or gradual slopes in buildings?

3. Sloping lines show up many other places in the world around us. Make sketches illustrating at least five examples other than roofs.

4. On hilly roads, you sometimes see signs warning of steep grades ahead. What do you think a sign like the one at the left tells you about the slope of the road ahead?

Extending

1. If a linear model is represented by an equation of the form $y = a + bx$, what do a and b tell you about the table of values and graph of that model?

2. The table at the right gives the amount of money spent on national health care for approximately every ten years from 1960 to 1998.

Annual U.S. Health Care Expenditures, 1960–1998 (in billions of dollars)

1960	1970	1980	1990	1998
26.9	73.2	247.3	699.4	1,149.1

Source: *The New York Times 2001 Almanac*. New York, NY: The New York Times, 2000.

   a. Make a scatterplot of these data.

b. What is the rate of change in health care expenditures from 1960 to 1970? From 1970 to 1980? From 1980 to 1990? From 1990 to 1998? From 1960 to 1998?

c. Do you think a line would be a good model for the trend in the scatterplot data? Explain.

d. How could you estimate health care costs for the year 2010?

3. The graph at the right illustrates the relationship between time in flight and height of a soccer ball kicked straight up in the air. The equation for that relation is $H = -5T^2 + 20T$, where T is in seconds and H is in meters.

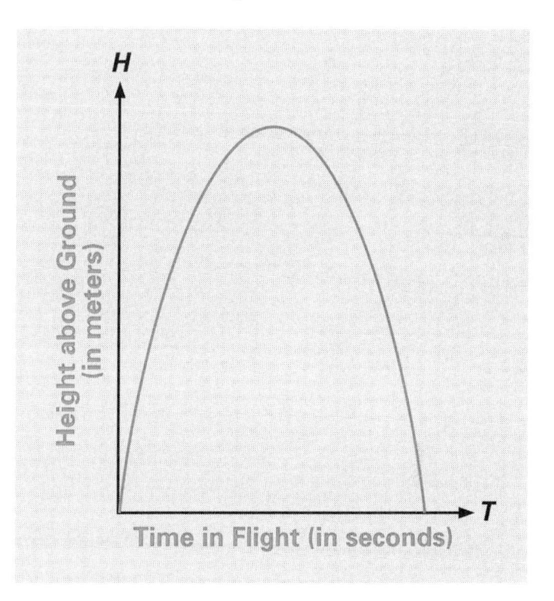

a. What could it mean to talk about the slope of this curved graph?

b. How would you measure the rate of change in the height of the ball?

c. How would rate of change and slope relate to each other?

d. What would rate of change and slope tell about the flight of the ball?

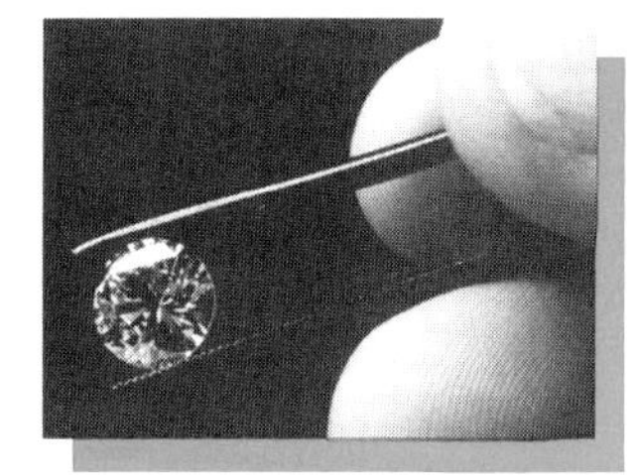

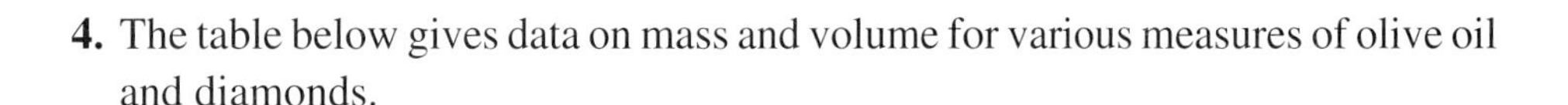

4. The table below gives data on mass and volume for various measures of olive oil and diamonds.

Volume (cm^3)	Oil (grams)	Diamonds (grams)
1	0.91	3.49
2	1.84	7.02
3	2.70	10.65
4	3.64	13.92
5	4.65	17.35
6	5.40	21.00
7	6.37	24.64
8	7.44	27.92

a. Make scatterplots of the relations between volume and mass of olive oil and between volume and mass of diamonds. On each plot, draw a graph that seems to be a good fit for the pattern of the data.

b. Write *NOW-NEXT* equations showing how the mass of oil or diamonds increases as the volume increases.

c. Use the models drawn in Part a to find rules for predicting mass M from volume V for the two substances.

- M_{oil} = __________
- $M_{diamonds}$ = __________

What connections do you see between the two types of equations for each relationship?

d. The ratio $\frac{\text{mass}}{\text{volume}}$ is called the *density* of a substance. It often is measured in grams per cubic centimeter.

- How can the density of oil and of diamonds be seen in the modeling equations?
- What is the density of each of these two substances?

e. What differences in the table data and the equations show differences in the densities of the two substances?

f. Mary Jo inherited a 5.5-carat gemstone from her great aunt. The volume of the stone is 0.4 cm^3.

- If one carat is equal to 0.2 grams, what is the mass of the stone?
- How could Mary Jo have figured out the volume of the stone?
- Is the stone a diamond?

INVESTIGATION 2 Finding Linear Equations

A symbolic rule showing how the variables x and y are related is a concise and simple way to record a linear model. The most common way to write rules for linear models is using equations of the form

$$y = a + bx$$

where a and b are specific numbers that set the relation between x and y. In earlier investigations, you've probably found several ways to determine the values of a and b in particular cases.

1. In many problems, you can discover the right equation simply by thinking about given facts relating the variables. For example:

The music department at Robert Kennedy High School operates a soft drink machine for students. Profits go to the band, orchestra, and chorus. The department pays \$100 per month to rent the machine and \$7.50 per case for soft drinks in cans.

a. What equation relates monthly sales S in cases of soda to monthly costs C paid to the distributor?

b. In your equation $C = a + bS$, what are the values of a and b? What do they tell about the situation?

c. If you made a graph of the (*sales*, *cost*) relation, what would the y-intercept and slope of the graph be?

d. Suppose you made a table of sample (*sales*, *cost*) values. What rate of change in cost, as a function of sales, would appear in that table?

e. What *NOW-NEXT* equation shows how monthly costs increase for each case of soft drinks sold? How does that equation relate to:

- the equation in $C = a + bS$ form?
- the slope of a graph for the model?
- the rate of change in cost as a function of sales?

2. In some cases, you can draw a straight-line model for the pattern in a scatterplot and find the equation relating x and y from the y-intercept and slope of that graph. For example:

The music department experimented with several different prices on soft drinks. The (*price*, *sales*) data are shown on the graph below. A linear model has been drawn to show the trend in those data.

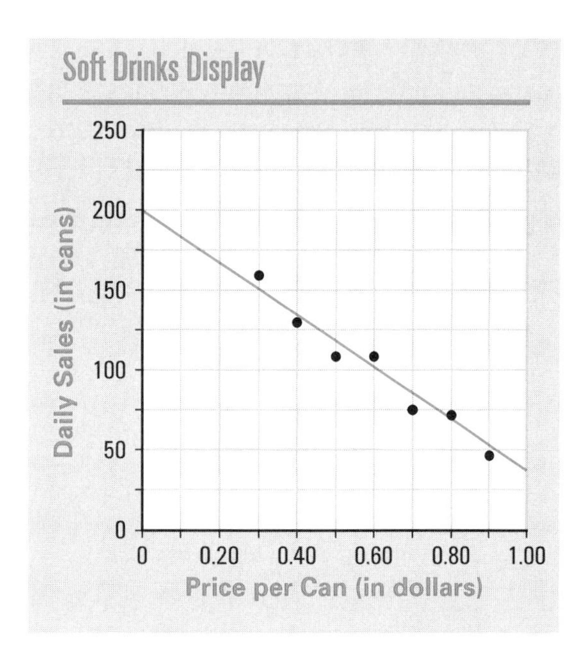

a. What overall pattern relates price and daily sales in this graph?

b. What are the slope and y-intercept of the linear model?

c. Using your answers to Part b, write an equation that fits the model.

3. In some cases, the modeling line does not show a clear choice for the slope or for the y-intercept. For example, the pattern of growth for one bean plant grown under special lighting is shown in the table and graph below.

Day	Height
3	4.2
4	4.7
5	5.1
6	6.3
7	7.4

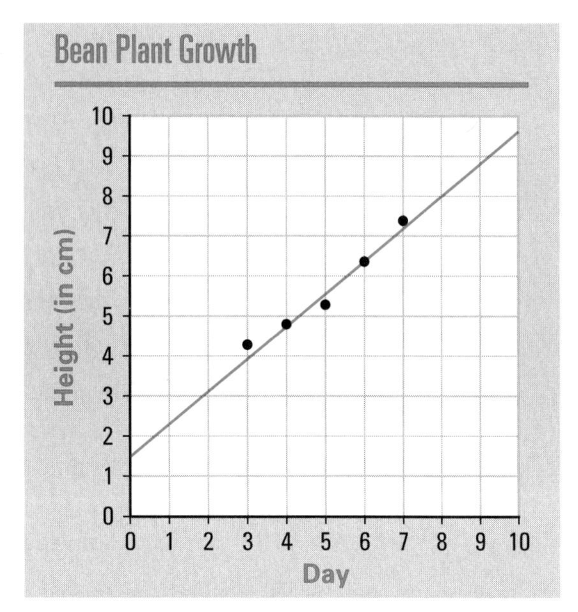

Unfortunately, neither the slope nor the y-intercept is easy to see accurately from the graph. In such a case, you can try the following steps to find a and b in an equation of the form $y = a + bx$.

- To find the slope b, pick any two points on the line for which you know coordinates. In this example, you could try (4, 4.7) and (6, 6.3). Using those two points to estimate the slope, $b = \frac{6.3-4.7}{6-4} = \frac{1.6}{2} = 0.8$. So the equation becomes $b = a + 0.8x$.
- To find the y-intercept a, pick the x and y values for any point on the line. For example, try (7, 7.1). These values of x and y must satisfy the model relation, $y = a + 0.8x$. So, it must be true that

$$7.1 = a + 0.8(7)$$
$$7.1 = a + 5.6$$

This means that, in this case, a must be 1.5. So the model equation is $y = 1.5 + 0.8x$.

a. Try the method just outlined using these three points on the line: (5, 5.5), (7, 7.1), and (3, 3.9).

- What slope do you get by using (5, 5.5) and (7, 7.1)?
- What value of a do you get by using the point (3, 3.9) and substituting 3 for x and 3.9 for y?

b. Try the method for finding a and b using a combination of just *two* points.

4. In many cases, you will need to fit a linear model to a scatterplot pattern that is more complicated. For example, it may include many data points, numbers that are not simple whole numbers, and a trend for which it is not clear what modeling line will be best. For example, the following table gives list prices for Toyota Celicas and Pontiac Firebirds.

Car Prices (in dollars)

Year	Base Price Toyota Celica	Base Price Pontiac Firebird	Year	Base Price Toyota Celica	Base Price Pontiac Firebird
1982	8,159	9,658	1992	15,708	18,105
1984	8,799	10,699	1994	18,428	19,895
1986	10,148	12,395	1996	19,468	21,414
1988	12,888	13,999	1998	20,111	25,975
1990	13,938	16,510	2000	21,195	26,630

Source: *N.A.D.A. Official Used Car Guide.* McLean, VA: N.A.D.A. Official Used Car Guide Company, 2001; *N.A.D.A. Official Older Used Car Guide*. Costa Mesa, CA: National Appraisal Guides, Inc., 2001.

a. Working in pairs, make a scatterplot of the Toyota Celica (*year*, *price*) data. Draw what you believe will be a good linear model for that pattern. It will simplify your work if you treat 1982 as year 1 on the time axis and scale the price axis in $1,000 units.

b. Find an equation for the linear model you've drawn.

c. Compare the graph and equation models you found to those of other members of your group. Try to explain any differences.

5. Most graphing calculators and computers offer efficient and systematic ways to find good equation models for linear data patterns. One way, called **linear regression**, finds the equation of the line that minimizes the mean of the *squares* of the differences between actual and predicted y values for the various x values. Like the linear models you sketched in Investigation 3, Lesson 1, the **linear regression model** passes through the point $(\bar{x}, \bar{y})$.

a. Use your calculator or computer software to enter the data for *year* and *Celica price* into the first and second data lists.

b. Produce a scatterplot of the (*year*, *Celica price*) data.

c. Compute the slope and y-intercept of the linear regression model.

d. Write the equation of the linear regression line.

e. What is the estimated rate at which Celicas increased in cost per year?

6. If you display the graph of the linear regression line on the scatterplot, you can use the equation model for estimating Celica prices in other years. Enter the appropriate equation in your calculator or software's functions list.

   a. What price for a new Celica is predicted for 1983? 1987? 1995?

   b. When is the new Celcia price predicted to first exceed $23,000?

   c. For how many years is the price of a new Celica predicted to stay under $28,000?

   d. What factors might cause actual prices to differ from predicted prices?

7. Enter the data for *Firebird price* into a new list. Using the (*year*, *Firebird price*) data, practice use of the calculator modeling procedure by finding the linear regression model for those data. Then use your model to answer the questions below.

   a. What new Firebird price is predicted for 1987? 1997?

   b. In what year is the new Firebird price predicted to reach $35,000?

Checkpoint

There are several different methods of finding an equation for a linear model.

a To find an equation in the form $y = a + bx$, how can you use information about:

- slope and y-intercept of the graph of that model?
- rate of change and other values in a table of (x, y) data?

b How can you find the equation if slope and y-intercept are not given?

c In what cases does it make sense to use a graphing calculator or computer software to find the modeling line and equation? What steps are involved?

Be prepared to share descriptions of your methods with the entire class.

On Your Own

The music department at Robert Kennedy High School also operates a bottled water machine near the gymnasium. The distributor regularly collects the money from the machine. The school is paid $50 per month plus $3 per case of drinks sold.

a. What equation gives the rule showing monthly profit P to the music department as a function of number S of cases sold?

b. Using the graph alone, estimate monthly profit figures for the following table entries. (Do not use the rule you wrote in Part a.) Then use your graphing calculator or computer software to find the *linear regression line* that fits those estimates. Compare the equation for that line to your equation in Part a.

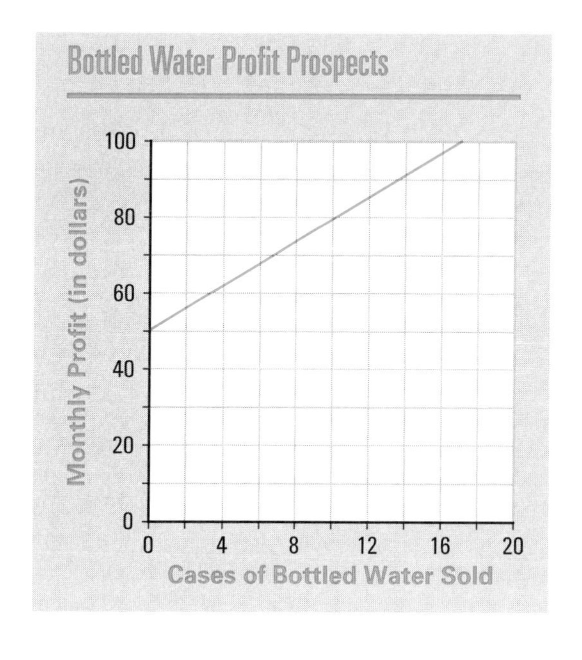

Cases Sold	1	2	4	6	8	10	12	14	16
Profit (in dollars)	53								

c. What might explain any difference between your two equations?

INVESTIGATION 3 Lines All Over the Plane

The investigations in this unit and in Unit 2, "Patterns of Change," have shown many places where linear models help you to describe and reason about relations between variables. But the range of linear models is even broader than what you have studied so far.

For example, you might recall that in Unit 2, the profits for operation of the Palace Theater were related to the daily ticket sales with this equation: $P = -450 + 2.50T$

If you graph this equation with a window like the one shown, some important parts of the Palace Theater business story will not appear.

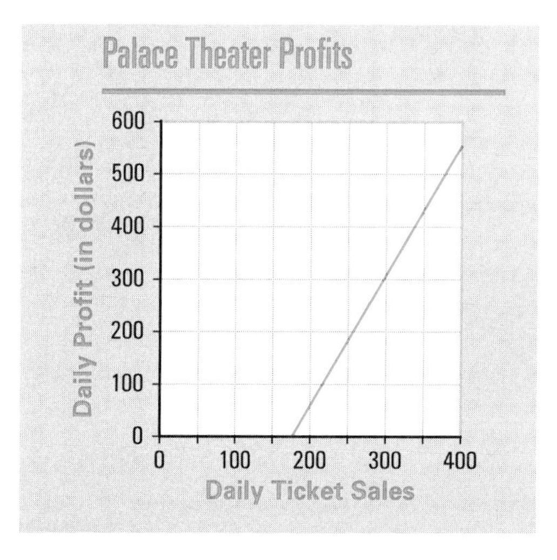

1. If you use your graphing calculator or computer to reproduce the graph of $P = -450 + 2.50T$ in the window shown on the previous page, what would the out-of-window points tell you about the business?

   a. Modify your window settings to show the entire linear graph from the case of no audience to a full house of 400 viewers. Use the trace function to find coordinates of some of the new points shown. Explain what they tell about theater business prospects.

   b. What values of P come from this model for negative values of T, such as -100? Where will those values appear on the graph of the model? What do they say about the theater profit situation?

The Palace Theater profit questions show again why it is important to know how to work with equations $y = a + bx$. To solve problems involving linear models, it is helpful to understand the connections between the equations and their tables and graphs, even for points outside the first quadrant. To sort out the possibilities, organize your group for the following exploration of cases.

2. Assign each member to explore two of the cases for the equation $y = a + bx$ which are listed below. Use the following guidelines for the exploration. Each explorer should do the following, using the graphing calculator or computer as a tool. Then each member should report and explain his or her conclusions to the group.

   - Choose 4 different sets of values for a and b.
   - Make tables and graphs for Xmin = –10 to Xmax = 10.
   - Identify properties common to all graphs and common to all tables.
   - Identify ways the graphs differ and ways the tables differ.
   - Identify ways that patterns in tables can be predicted from the values of a and b.
   - Identify ways that patterns in graphs can be predicted from the values of a and b.
   - Think about the reasons that the predictions work.

   Case 1: a and b are both positive numbers.

   Case 2: a is negative and b is positive.

   Case 3: a and b are both negative numbers.

   Case 4: a is positive and b is negative.

   Case 5: $a = 0$ and b is positive or negative.

   Case 6: a is positive or negative and $b = 0$.

3. As a result of your group work, each group member should be able to make a quick and accurate sketch of any linear equation. To test your skill, divide your group into pairs. Each pair should complete the following challenge.

The Brain-Machine Challenge: For each equation in list A, one partner should sketch a graph by hand while the other partner produces the graph using a calculator or a computer. Compare results. For equations in list B, partners should change roles and repeat the process.

List A	List B
$y = 5 + 3x$	$y = -3 - 5x$
$y = -5 + 3x$	$y = 8 + 2x$
$y = 5 - 3x$	$y = -10 + 0.2x$
$y = -2x + 8$	$y = -4 - 3x$
$y = 6 + 0x$	$y = 2 - 0x$

After completing the challenge, make a list of things you look for in an equation to help make a quick sketch.

Checkpoint

How can the equation of a linear model be used to predict:

a the slope and location of its graph?

b the rate of change in a table of the equation's values?

Be prepared to share your group's thinking with the entire class.

On Your Own

Examine the diagram at the right.

- Which line passes through the quadrants of the plane indicated below?
- Find values of a and b so that the graph of $y = a + bx$ passes through the indicated quadrants.

a. I, II, and III

b. I, II, and IV

c. I, III, and IV

d. II, III, and IV

e. I and III

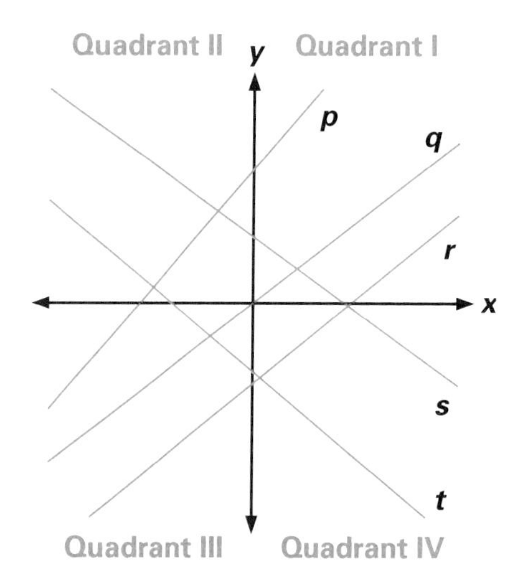

MORE
Modeling • Organizing • Reflecting • Extending

Modeling

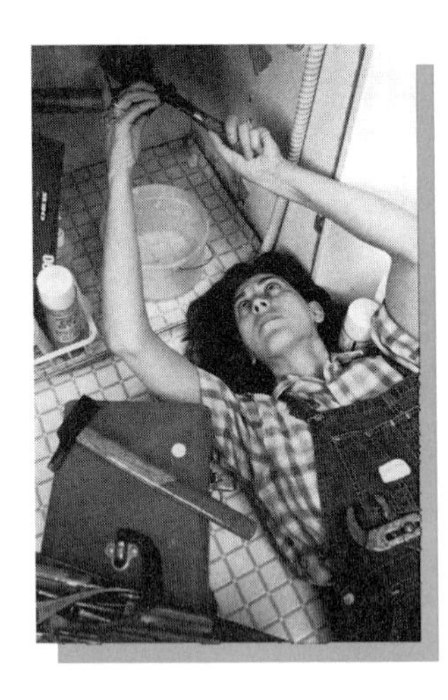

1. Phone calls to two plumbing repair shops gave the following quotations for repair work.

   Pride Plumbing: $35 for the service call, plus $32 per hour for repair time.

   Kovach and Co.: $25 for the service call, plus $40 per hour for repair time.

   a. For each shop, write an equation that shows the relation between total repair cost in dollars and the number of hours to complete the repair.

   b. If the repair took less than an eight-hour working day, what would be the possible values for the input variable and the output variable for your model of Pride Plumbing's cost? For your model of Kovach and Company's cost?

   c. Without actually graphing, describe how the graphs of your equations in Part a would be related in terms of location and slope.

   d. If the graphs of the equations in Part a intersect, what would the point of intersection tell you?

   e. Will the graphs of the equations in Part a intersect? Explain.

2. Many Americans love to eat fast food. But we also are concerned about weight gain and the cholesterol that is generated by eating saturated fat. Many fast-food restaurants now advertise special "lite" menus. They give information about the fat and calorie content of those foods, like the data in the following table.

Nutrition Data

Item	Grams Fat	Total Calories	Calories from Fat	Item	Grams Fat	Total Calories	Calories from Fat
McDonald's				**Wendy's**			
Chef Salad	8	150	70	Grilled Chicken Sandwich	8	300	70
Garden Salad	6	100	60	Deluxe Garden Salad	6	110	50
Chicken McGrill	18	450	160	Grilled Chicken Salad	8	190	80
Hardee's				Broccoli/Cheese Potato	14	470	130
Grilled Chicken Sandwich	16	350	150	**Taco Bell**			
Roast Beef Sandwich	16	310	140	Taco	10	170	90

Sources: www.macdonalds.com/countries/usa/food/nutrition_facts/index.html; www.wendys.com/the_menu/nutrition/; www.tacobell.com/2product/2menu/nutrition.htm; www.hardees.com/menu/

a. Make a scatterplot of data relating grams F of fat to total calories T in the menu items shown.

b. Draw a linear model joining the points (8, 190) and (18, 450). Find its equation in the form $T = a + bF$. Explain what the values of a and b tell about the model graph and about the relation between grams of fat and total calories in the food items.

c. Use your graphing calculator or computer to find the linear regression model for the (F, T) data in the table. Compare this result to what you found in Part b.

3. Use the table from Modeling Task 2 to analyze the relation between grams F of fat and calories C from fat in the menu items.

a. Make a scatterplot of data relating F and C.

b. Draw a linear model joining the points (10, 90) and (18, 160). Find its equation in the form $C = a + bF$. Explain what the values of a and b tell about the model graph and about the relation between grams of fat and calories from fat in the food items.

c. If a food item contains 13 grams of fat, how many calories of the item would you expect to be from the fat content?

d. Use your graphing calculator or computer to find the linear regression model for the (*grams of fat*, *calories from fat*) data. Compare that model with the model you found in Part b. Explain your findings.

4. Over the past 40 years, more and more women have taken full-time jobs outside the home. As women have entered the work force, there has been a great deal of controversy about whether they are being paid fairly. The table below shows patterns of change from 1970 to 1998 in median incomes for men and women employed full-time outside the home. These data do not show pay for comparable jobs, but average pay for all jobs.

Median Income (in dollars)

Year	Men	Women	Year	Men	Women
1970	9,184	5,440	1990	28,979	20,591
1975	12,934	7,719	1995	32,199	23,777
1980	19,173	11,591	1998	36,258	26,855
1985	25,999	16,252			

Source: *The World Almanac and Book of Facts 2001*. Mahwah, NJ: World Almanac Education Group, Inc., 2001.

a. What do you believe are the most interesting and important patterns in these data?

b. Did women's incomes improve in relation to men's incomes between 1970 and 1998?

c. The diagram below shows a scatterplot of the (*year*, *median income*) data for women, using 0 for the year 1970. A linear model for the pattern in those data is drawn on the coordinate grid. Use two points on the line to find the slope of the line. Estimate the y-intercept. Then write an equation for the line.

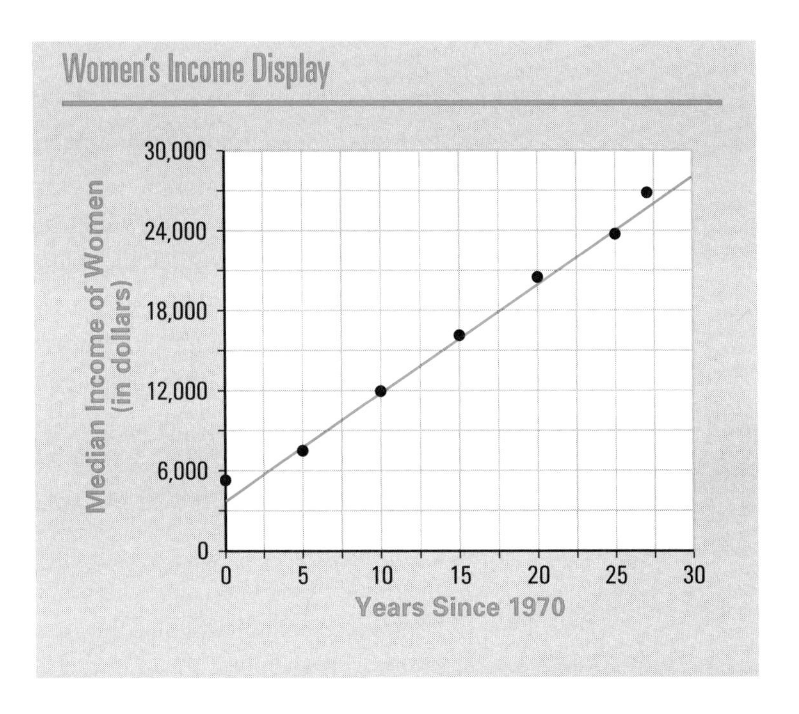

d. Using the linear model for median income, estimate the income of women in 1963, 1973, 1983, 1993, and 2003.

e. Use a linear regression procedure on your calculator or computer to find another equation for a linear model of the (*year*, *median income*) data. Compare the predictions of that model with your results from Part d.

5. Use a linear regression procedure to find an equation modeling the pattern of (*year*, *median income*) data for men.

a. Compare the slope and y-intercept of the model for men's income to those for women's income found in Modeling Task 4. Explain what each value tells about the situation.

b. Use the equation to estimate median income for men in 1963, 1973, 1983, 1993, and 2003. Compare the results to what you found for women.

Organizing

1. Find equations for each of the linear models graphed on the diagram at the right.

2. For each of the following ordered pairs of numbers:
   - plot the points on a coordinate grid and draw a line through the points;
   - find the equation of the line through the points.

   a. (0, 3) and (6, 6)

   b. (0, –4) and (5, 6)

   c. (–4, –3) and (2, 3)

   d. (–6, 4) and (3, –8)

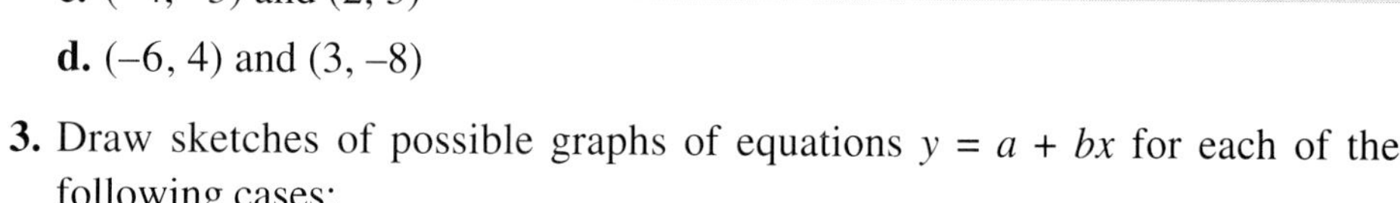

3. Draw sketches of possible graphs of equations $y = a + bx$ for each of the following cases:

   a. $a < 0$ and $b > 0$

   b. $a > 0$ and $b < 0$

   c. $a < 0$ and $b < 0$

   d. $a > 0$ and $b > 0$

4. Plot the two points $A(2, 4)$ and $B(8, 8)$ on graph paper. Label the points with the coordinates and the letters A and B. Then draw the line containing points A and B.

   a. What is the slope of the line containing points A and B?

   b. Sketch the line $y = \frac{2}{3}x$ on the same coordinate grid. Describe how this line compares to the line which contains points A and B.
      - Find the y-coordinates for the two points on the line $y = \frac{2}{3}x$ whose x-coordinates are 2 and 8. Label these points with their coordinates.
      - How far vertically (up or down) must you move each point of the line $y = \frac{2}{3}x$ to get corresponding points on the line containing the points (2, 4) and (8, 8)? Explain your answer.

   c. Explain how you can find the value of the y-intercept for the line containing points A and B.

   d. Write the equation for the line through the points (2, 4) and (8, 8).

   e. Use a similar method to find the equation of the line through the points (4, 5) and (14, 30).

Reflecting

1. What can you tell about the signs of the coordinates (positive or negative) for points:
   a. in quadrant I?
   b. in quadrant II?
   c. in quadrant III?
   d. in quadrant IV?

2. Think about how the values of a and b affect the graph of the equation $y = a + bx$.
   a. Describe the effect of increasing or decreasing the value of b on the graph of the equation $y = 4 + bx$.
   b. Describe the effect of increasing or decreasing the value of a on the graph of the equation $y = a + 2x$.

3. To use linear models wisely it helps to be in the habit of asking, "What sorts of numbers would make sense in this situation?" For example, in the relation between ticket sales T and profits P of the Palace Theater, it would not make much sense to substitute negative values for T in the equation $P = -450 + 2.50T$. In each of the following situations, decide what range of values for the variables would make sense.
   a. Temperature can be expressed in many different scales. You probably are familiar with two, Celsius and Fahrenheit. One linear model to convert from one scale to another is $F = \frac{9}{5}C + 32$, where F is degrees Fahrenheit and C is degrees Celsius. What range of values for F and C make sense for temperatures that you might encounter?
   b. Suppose a ball is tossed into the air with an upward velocity of 40 feet per second. Its upward velocity is a function of time in flight, modeled by the equation $V = 40 - 32T$. *Velocity* V is in feet per second and *time* T is in seconds. What range of values for T and V make sense in this context?
   c. The cost of buying a used Ford Mustang automobile can be modeled by $C = 10{,}000 - 560Y$, where C is the cost in dollars and Y is the age of the car in years. What range of values for Y and C make sense?
   d. In one apartment building, new renters are offered \$150 off their first month's rent, then they pay a normal rate of \$450 per month. The total rent paid for an apartment in that building is modeled by $R = 450T - 150$, where T is the number of months. What range of values for R and T make sense?

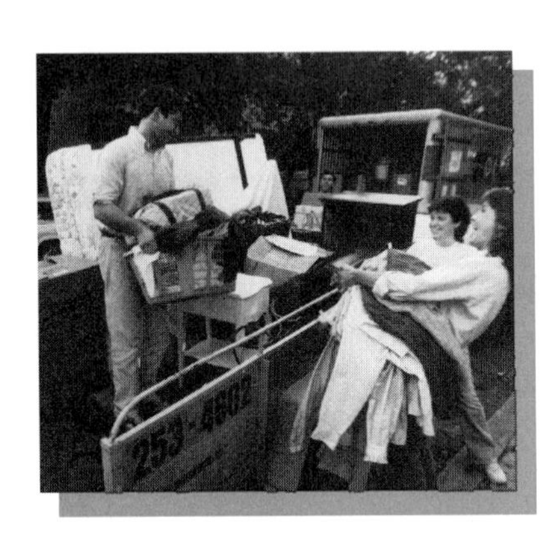

4. Discuss the pros and cons of each of the following methods for finding equation models of linear patterns in data.

a. Draw a line that seems to fit the data pattern. Estimate the slope of the line, estimate the y-intercept of the line, and write the equation.

b. Draw a line through two data points that seem typical of the pattern. Calculate the slope of that line from the data points. Estimate the y-intercept, then write the equation.

c. Draw a line that seems to fit the data pattern. Estimate coordinates of several points on that line. Use the coordinates to calculate the slope. Then substitute values from one of the points to get an equation for finding the y-intercept.

d. Enter the data in a calculator or computer. Then use the linear regression procedure for finding an equation to fit the data.

Extending

1. Investigate the linear regression procedure for finding a linear model to fit data patterns.

a. For each of the following data sets, use your graphing calculator or computer to make a scatterplot. Then find the modeling equation calculated by the linear regression procedure.

i.

x	0	1	2	3	4	5	6	7	8
y	3	5	7	9	11	13	15	17	19

ii.

x	0	1	2	3	4	5	6	7	8
y	1	5	7	9	11	13	15	17	29

iii.

x	0	1	2	3	4	5	6	7	8
y	4	6	8	10	12	14	16	18	20

iv.

x	0	1	2	3	4	5	6	7	8
y	1	2	5	10	17	26	37	50	65

v.

x	0	1	2	3	4	5	6	7	8
y	3	8	11	12	12	11	8	3	–4

b. Compare the graph produced by the linear regression equation to the actual data in each case of Part a. Comment on the fit in each case and explain places where the fit is not good.

c. What limitations of using the linear regression procedure are suggested by the results in Parts a and b?

2. Graph the line $y = 5 + 3x$.

a. Find the equations of all lines that cross the vertical axis 2 units from $y = 5 + 3x$ and that are parallel to it.

b. Imagine jumping from $y = 5 + 3x$ to any of the lines found in Part a. How long is that jump if you:

- jump vertically?
- jump horizontally?
- jump the direction that gives the shortest jump?

c. Describe the form of the equation of all the lines that cross the vertical axis exactly 2 units away from the given line.

3. The 100-meter run for men has been run in the Olympics since 1896. The winning times for each of the years through 2000 are given in the following table.

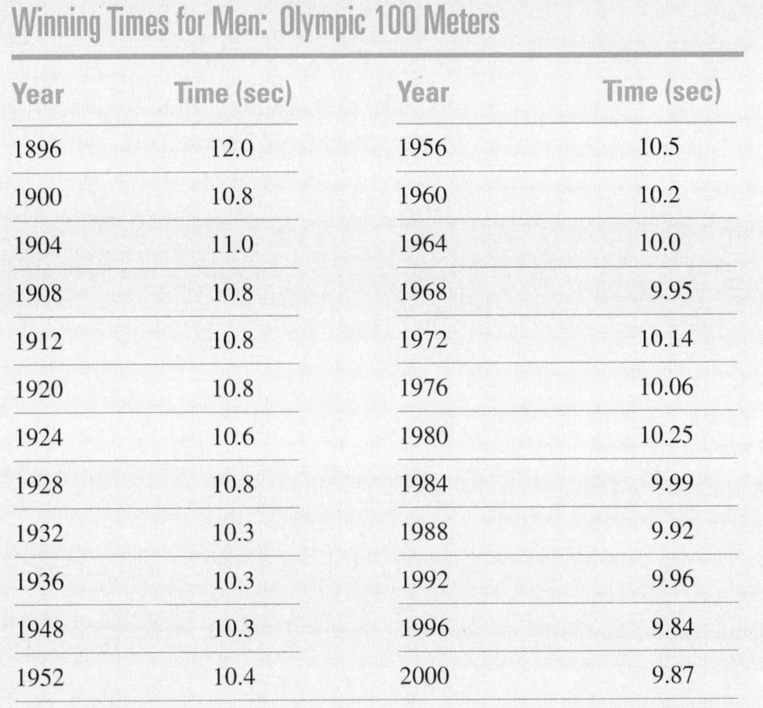

Winning Times for Men: Olympic 100 Meters

Year	Time (sec)	Year	Time (sec)
1896	12.0	1956	10.5
1900	10.8	1960	10.2
1904	11.0	1964	10.0
1908	10.8	1968	9.95
1912	10.8	1972	10.14
1920	10.8	1976	10.06
1924	10.6	1980	10.25
1928	10.8	1984	9.99
1932	10.3	1988	9.92
1936	10.3	1992	9.96
1948	10.3	1996	9.84
1952	10.4	2000	9.87

Source: *The World Almanac and Book of Facts 2001*, Mahwah, NJ: World Almanac Education Group, Inc. 2001.

a. Study the data. Notice the pattern in the years. Are there gaps that don't fit the pattern? Why were races not run those years?

b. Make a scatterplot of the (*year*, *time*) data using 1890 as year 0. Then decide whether you think a linear model is reasonable for the pattern in your plot.

c. Find the $(\bar{x}, \bar{y})$ point for the data. Locate it on your scatterplot. Then draw a linear model that goes through $(\bar{x}, \bar{y})$ and fits the trend in the data as well as you think is possible.

d. Compare your line with those of other students. Are they nearly the same? If not, how do they differ?

e. Find the linear regression equation for the data pattern using your calculator or computer. Compare its graph to the linear model you drew in Part c. If the hand-drawn and technological models differ, why do you think that happens?

f. Use the linear regression line model to predict the 100-meter winning time for the 1940, and 2004 Olympics.

g. Use your linear model to predict the approximate year in which the winning time for the 100-meter dash should be 9.80 seconds or less.

h. According to your model, in which Olympics should the race have been won in 10.4 seconds or less?

4. Women began running 100-meter Olympic races in 1928. The winning times for women are shown in the table below.

Winning Times for Women: Olympic 100 Meters

Year	Time (sec)	Year	Time (sec)
1928	12.2	1972	11.07
1932	11.9	1976	11.08
1936	11.5	1980	11.60
1948	11.9	1984	10.97
1952	11.5	1988	10.54
1956	11.5	1992	10.82
1960	11.0	1996	10.94
1964	11.4	2000	10.75
1968	11.0		

Source: *The World Almanac and Book of Facts 2001*, Mahwah, NJ: World Almanac, Education Group, Inc. 2001.

a. Study the data. What patterns do you see?

b. Make a scatterplot and then find a linear regression model. Use 1900 as year 0.

c. Use your linear model to answer each of the following questions.

- What winning time would you predict for 1944?
- What winning time does the model predict for 1996? Compare the actual winning time for 1996 to the prediction.
- In what Olympic year does the model suggest there will be a winning time of 10.7 seconds?
- According to the model, when should a winning time of 11.2 seconds have occurred?

d. According to the model, by about how much does the women's winning time change from one Olympic year to the next?

e. According to the model in Extending Task 3, by about how much does the men's winning time change from one Olympic year to the next? Compare this to your answer for Part d.

Women's Marathon at 1996 Olympics; Ethiopia's Fatuma Roba won the event.

Lesson 3

Linear Equations and Inequalities

For most of the 20th century, the vast majority of American medical doctors were men. However, during the past 40 years there has been a significant increase in the number of women graduating from medical schools. As a result, the percent of doctors who are women has grown steadily to nearly 25% in 2000. The graph at the right shows this trend.

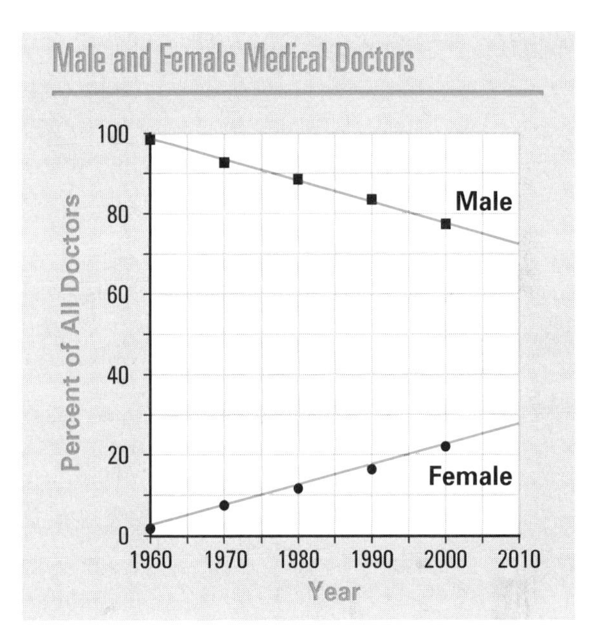

Source: www.ama-assn.org/ama/pub/article/171-195.html

Think About This Situation

The graph shows trends in the numbers of male and female medical doctors in the United States between 1960 and 2000.

a How would you describe the trends shown in the data points and the linear models that have been drawn to match patterns in those points?

b Why do you suppose the percent of women doctors has been increasing over the past 40 years?

c Would you expect the trend in the graph to continue 10 or 20 years into the future? This means 10 or 20 years beyond 2000.

d How would you go about finding equations for linear models of the data trends?

e If you were asked to make a report on future prospects for numbers of male and female doctors, what kinds of questions could you answer using the linear models?

INVESTIGATION 1 Using Tables and Graphs

There are several kinds of questions that occur naturally in thinking about the trends in numbers of male and female medical doctors. To plan for future educational programs and medical services, medical schools, hospitals, and clinics might wonder:

- When will the number of female doctors reach a 40% share?
- When will the numbers of male and female doctors be equal?
- How long will the number of male doctors remain above a 70% share?

The trends in percent of male and female medical doctors can be modeled by the following related linear equations.

Percent of Male Doctors: $Y_1 = 98 - 0.54X$
Percent of Female Doctors: $Y_2 = 2 + 0.54X$

Here, X stands for years since 1960. Y_1 and Y_2 stand for percent of all U.S. medical doctors. Using symbolic models, the three prediction questions above can be written as algebraic equations and inequalities:

$40 = 2 + 0.54X$
$98 - 0.54X = 2 + 0.54X$
$98 - 0.54X > 70$

The problem is finding values of X (years since 1960) when the various share conditions hold.

1. Write equations and inequalities that can be used to answer each of these questions about the relation between numbers of male and female medical doctors in the United States.

a. When might the share of male doctors fall to 40%?

b. How long will the share of female doctors remain below 60%?

c. When will the number of male doctors be double the number of female doctors?

2. Write questions matching each of these equations and inequalities.

a. $98 - 0.54X = 65$

b. $2 + 0.54X < 30$

c. $98 - 0.54X = 4(2 + 0.54X)$

Writing equations and inequalities to match important questions is only the first task in solving the problems they represent. The essential next step is to **solve the equations** or **solve the inequalities**. That is, find values of the variables that satisfy the conditions.

One way to solve equations and inequalities matching questions about numbers of male and female medical doctors is to make and study tables and graphs of the prediction models.

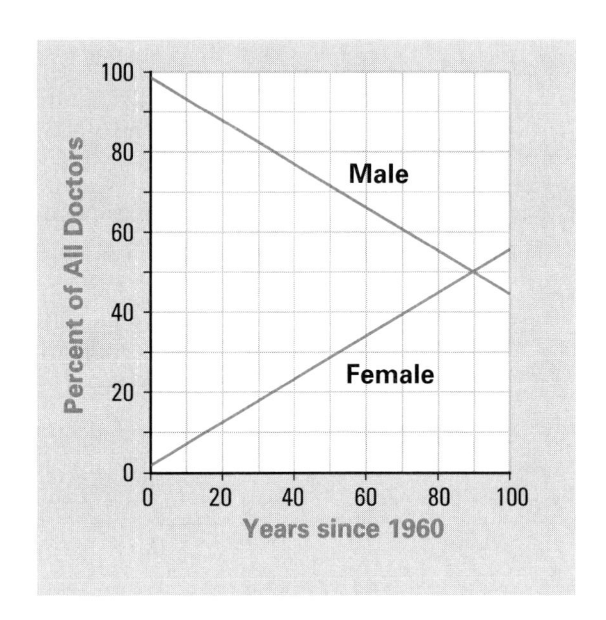

X	Y_1	Y_2
0	98	2
10	92.6	7.4
20	87.2	12.8
30	81.8	18.2
40	76.4	23.6
50	71	29

X	Y_1	Y_2
60	65.6	34.4
70	60.2	39.8
80	54.8	45.2
90	49.4	50.6
100	44	56

3. Solve each of these equations and inequalities by finding the value or range of values of X that satisfy the given conditions. Then explain what each solution tells about prospects for male and female percents of all U.S. medical doctors. Explain how the solutions can be found (or at least estimated) in tables and graphs of $Y_1 = 98 - 0.54X$ and $Y_2 = 2 + 0.54X$.

a. $70 = 2 + 0.54X$

b. $98 - 0.54X = 2 + 0.54X$

c. $98 - 0.54X > 80$

d. $98 - 0.54X = 65$

e. $2 + 0.54X < 40$

f. $98 - 0.54X = 4(2 + 0.54X)$ [Hint: Consider $Y_3 = 4Y_2$.]

g. $98 - 0.54X = 1.5(2 + 0.54X)$ [Hint: Consider $Y_4 = 1.5Y_2$.]

4. Write and solve (as accurately as possible) equations and inequalities to answer each of the following questions about male and female medical doctors in the United States, based on the linear models suggested by recent data. In each case, explain how you can use tables and graphs of linear models to find solutions.

   a. When will the percent of male doctors decline to only 55%?

   b. When will the percent of female doctors reach 35%?

   c. How long will the percent of male doctors be above 40%?

   d. When will the percent of male doctors be less than the percent of female doctors?

5. When you solve an equation or inequality, it is always a good idea to check the solution you find. If someone told you that the solution to $45 = 98 - 0.54X$ is $X = 10$, would you believe it? How could you check the suggestion, without using either the table or the graph of $Y_1 = 98 - 0.54X$?

6. If someone told you that the solution to $2 + 0.54X \leq 45$ is $X \leq 70$, how could you check the suggestion:

   a. In a table?

   b. In a graph?

   c. Without using either a table or a graph?

7. What percents of male and female medical doctors in the United States do the models predict for the year 2020 (60 years from 1960)? How confident would you be of such a prediction?

Checkpoint

Many important questions about linear models require solution of linear equations or inequalities, such as $50 = 23 + 5.2X$ or $45 - 3.5X < 25$.

a What does it mean to solve an equation or inequality?

b How do you check a solution?

c How could you use tables and graphs of linear models to solve the following equation and inequality:

- $50 = 23 + 5.2X$?
- $45 - 3.5X < 25$?

Be prepared to share your ideas with the class.

On Your Own

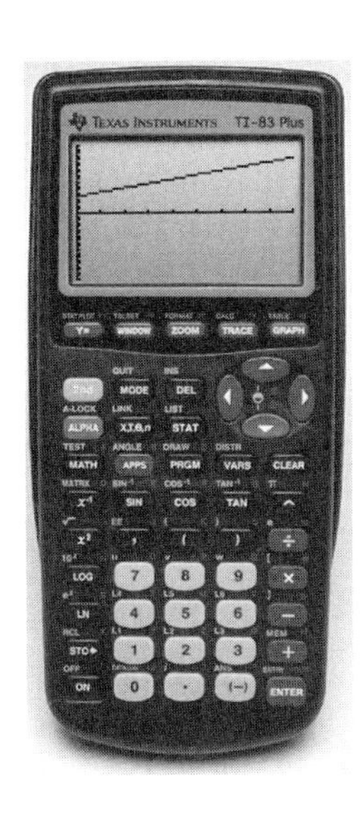

Bronco Electronics is a regional distributer of graphing calculators. When an order is received, a shipping company packs the calculators in a box. They place the box on a scale which automatically finds the shipping cost. The shipping cost C is a function of the number N of calculators in the box, with rule $C = 4.95 + 1.25N$.

Use your graphing calculator or computer to make a table and a graph showing the relation between number of calculators and shipping cost. Include information for 0 to 20 calculators. Use the table and graph to answer the following questions:

a. How much would it cost to ship an empty box? How is that information shown in the table, the graph, and the cost rule?

b. How much does a single calculator add to the cost of shipping a box? How is that information shown in the table, the graph, and the cost rule?

c. Write and solve equations and inequalities to answer the following questions about Bronco Electronics shipping costs.

- If the shipping cost is \$17.45, how many calculators are in the box?
- How many calculators can be shipped if the cost is to be held below \$25?
- What is the cost of shipping eight calculators?

d. What questions about shipping costs could be answered using the following equation and inequality?

$$27.45 = 4.95 + 1.25N$$

$$4.95 + 1.25N \leq 10$$

e. Bronco Electronics got an offer from a different shipping company. The new company would charge based on the rule $C = 7.45 + 1.00N$. Write and solve equations or inequalities to answer the following questions:

- For what number of calculators in a box will the two shippers make the same charge?
- For what number of calculators in a box will the new shipping company's offer be more economical for Bronco Electronics?

MORE

Modeling • Organizing • Reflecting • Extending

Modeling

1. Parents often weigh their child at regular intervals during the first several months after birth. The data usually can be modeled well with a line. For example, the rule $y = 96 + 2.1x$ gives the relationship between *weight in ounces* and *age in days* for Rachel.

a. How much did Rachel weigh at birth?

b. Make a table and a graph of this equation for Xmin = 0 to Xmax = 90.

c. For each equation or inequality below:

- Write a question about the infant's age and weight that the equation or inequality could help answer.
- Use the table or graph to solve the equation or inequality and then answer your question.

i. $y = 96 + 2.1(10)$

ii. $159 = 96 + 2.1x$

iii. $264 = 96 + 2.1x$

iv. $96 + 2.1x \leq 201$

2. A concession stand at the Ann Arbor Art Fair sells soft drinks in paper cups that are filled by a dispensing machine. There is a gauge on the machine that shows the amount of each drink left in the supply tank. Wendy collected the following data one day.

Number of Drinks Sold	0	50	100	150	200	250
Ounces of Soft Drink Left	2,500	2,030	1,525	1,000	540	20

a. Plot these data and draw a line that fits the pattern relating the *number N of drinks* and the *ounces L of soft drink left*.

b. Find the slope of the linear model. Explain what it tells about the soft drink business at the art fair.

c. Find the coordinates of the point where the linear model crosses the vertical axis. What do the coordinates tell about the soft drink business?

d. Find an equation for the linear model relating N and L.

e. Write calculations, equations, or inequalities that can be used to answer each of the following questions. Answer the questions. Show how the answers can be found on the graph.

- About how many drinks should have been sold when the machine had 1,200 ounces left in the tank?
- How many ounces were left in the tank when 125 drinks had been sold?
- How many drinks were sold before the amount left fell below 1,750 ounces?

3. Recall (from page 188) that Mary and Jeff Jordan each sell programs at the local baseball park. They are paid $10 per game and $0.25 per program sold.

a. Write a rule relating number X of programs sold and pay earned Y.

b. Write equations, inequalities, or calculations that can be used to answer each of the following questions:

- How many programs does Jeff need to sell to earn $25 per game?
- How much will Mary earn if she sells 75 programs?
- How many programs does Jeff need to sell to earn at least $35 per game?

c. Produce a table and a graph of the relation between sales and pay from which the questions in Part b can be answered. Show on the graph and in the table how the answers can be found. Find the answers.

4. Emily works as a waitress at Pietro's Restaurant. The restaurant owners have a policy of automatically adding a 15% tip on all customers' bills as a courtesy to their waitresses and waiters. Emily works the 4 P.M. to 10 P.M. shift. She is paid $15 per shift plus tips.

a. Write an equation to model Emily's evening wage W based on the total B of her customers' bills. Use your graphing calculator or computer software to produce a table and a graph of this relation.

b. If the customers' bills total $110, what calculation will give Emily's wage for the evening?

c. If Emily's wage last night was $47, write an equation showing the total for her customers' bills. Solve the equation.

d. What is the smallest amount that Emily could make in an evening? Which point on the graph represents that amount? How is her smallest amount reflected in the equation?

e. After six months at Pietro's, Emily will receive a raise to $17 per shift. She will continue to receive 15% tips.

- Write an equation to model her new wage. Graph it along with her previous wage equation in the same viewing window.
- Compare the smallest amounts Emily could earn for each wage scale and compare the rates of change in the wages. Describe the similarities and differences in the equations and graphs.

Organizing

1. Suppose two variables x and y are related by the rule $y = 4 - 0.5x$. Make a table and a graph of this relation for Xmin = –20 to Xmax = 20. Use the table and graph to solve each equation or inequality below.

   a. $y = 4 - 0.5(12)$ **b.** $-1 = 4 - 0.5x$

   c. $-5 = 4 - 0.5x$ **d.** $4 - 0.5x \geq 0$

2. Graph the two equations $y = 2 + 0.25x$ and $y = -8 + 1.5x$ for Xmin = –5 to Xmax = 10. Use those graphs to solve these equations and inequalities.

   a. $2 + 0.25x = -8 + 1.5x$ **b.** $2 + 0.25x \geq -8 + 1.5x$

   c. $2 + 0.25x \leq -8 + 1.5x$ **d.** $-8 + 1.5x \geq 0$

3. For any linear relation with rule of the form $y = a + bx$:

   a. Explain, with sketches, how to solve equations of the form $c = a + bx$ using the graph of the linear model.

   b. Explain, with sketches, how to solve inequalities of the form $c \leq a + bx$ using the graph of the linear model.

   c. Explain, with sketches, how to solve equations of the form $a + bx = c + dx$ using graphs of linear models.

   d. Explain, with sketches, how to solve inequalities of the form $a + bx \leq c + dx$ using graphs of linear models.

4. The linear equations you have solved in this investigation have been of two forms.

   a. How many different values of x can be found that satisfy any specific equation of the form $c = a + bx$? Explain your answer with sketches of linear models.

   b. How many different values of x can be found that satisfy any specific equation of the form $a + bx = c + dx$? Explain your answer with sketches of linear models.

5. The linear inequalities you have solved in this investigation have been of two forms.

   a. How many different values of x can be found that satisfy any specific inequality of the form $c \leq a + bx$? Explain your answer with sketches of linear models.

   b. How many different values of x can be found that satisfy any specific inequality of the form $a + bx \leq c + dx$? Explain your answer with sketches of linear models.

Reflecting

1. To study a linear model like $y = 125 + 35x$, would you prefer to use a graph or a table of this relation? Give reasons for your preference.

2. Describe a problem situation which could be modeled by the equation $y = 10 + 4.35x$.

   a. What would solving $109 \geq 10 + 4.35x$ mean in your situation?

   b. Solve $109 \geq 10 + 4.35x$.

3. Describe a strategy for solving inequalities of the form $c \leq a + bx$ using the table-building capability of your graphing calculator or computer software. How would you modify your strategy in order to solve inequalities of the form $c > a + bx$?

4. When solving equations or inequalities that model real situations, why is it important not only to check the solution, but also to check if the solution makes sense? Show your reasoning with an example.

Extending

1. The diagram at the right shows graphs of two relations between variables:

   $y = x + 3$ and $y = x^2 - 3$

   Reproduce that diagram on your graphing calculator or computer. Use the trace function to solve the equation $x + 3 = x^2 - 3$.

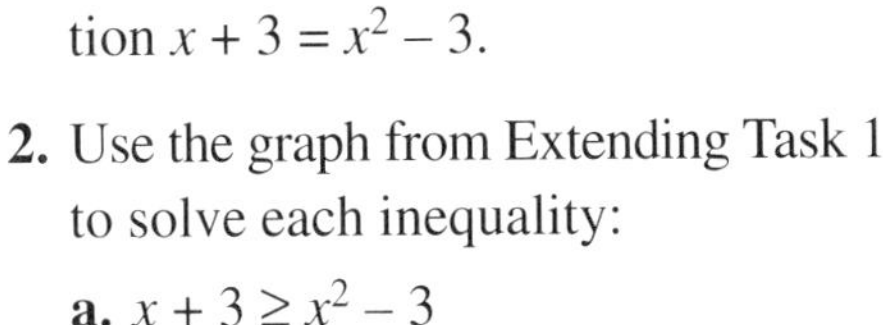

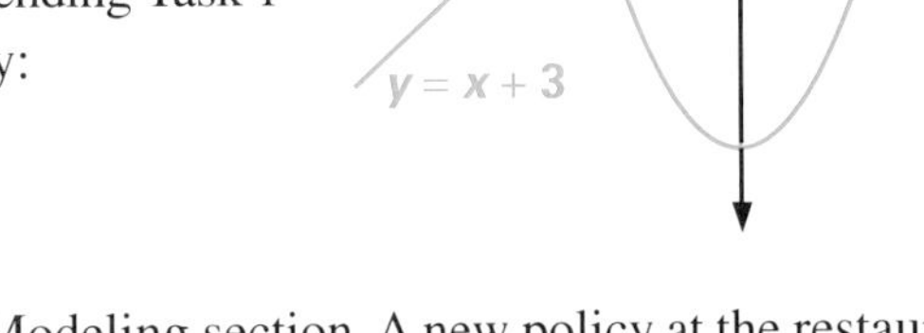

2. Use the graph from Extending Task 1 to solve each inequality:

   a. $x + 3 \geq x^2 - 3$

   b. $x + 3 < x^2 - 3$

3. Refer to Task 4 of the Modeling section. A new policy at the restaurant requires wait staff to share their tips with busers. Wait staff will receive \$20 per shift plus 10% in tips. Busers will receive \$25 per shift plus 5% in tips.

   a. Write one equation to model the busers' wages, and write another equation to model the wait staff's wages. Graph these two equations on the same coordinate axes.

   b. Write and answer three questions about the wages for busers and wait staff. Based on your analysis of the graphs, write equations or inequalities corresponding to your questions.

INVESTIGATION 2 Quick Solutions

Linear models are common. They are easy to find with or without a calculator or computer. They are also easy to use. In fact, it is often possible to solve problems that involve linear equations without the use of tables or graphs. For example, to solve $3x + 12 = 45$ you might reason like one of these students:

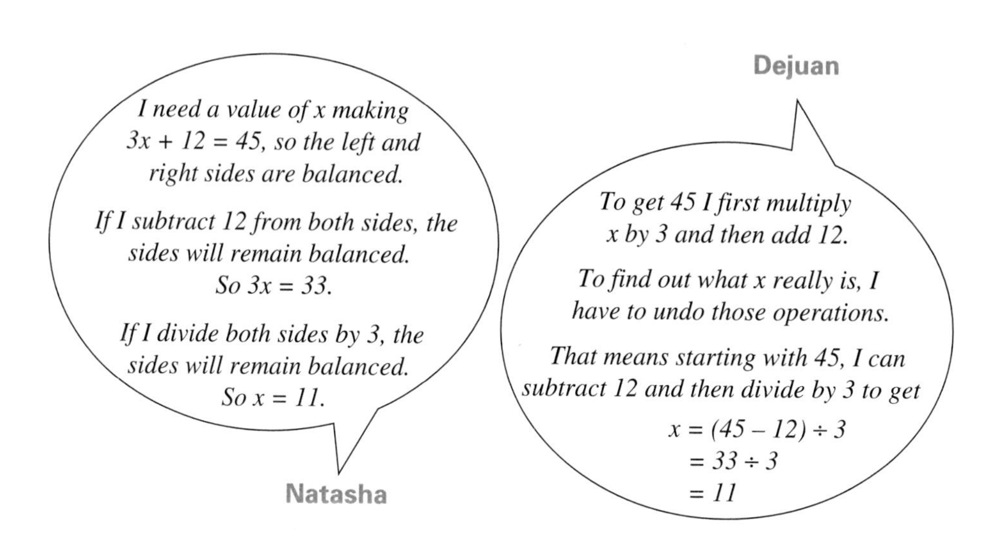

1. In your group, try to figure out both students' reasoning.

a. Natasha and Dejuan both found that $x = 11$. How can you be sure the answer is correct?

b. Analyze Natasha's thinking.

- Why did she subtract 12 from both sides? Why didn't she add 12 to both sides? What if she subtracted 10 from both sides?
- Why did she divide both sides by 3?

c. Analyze Dejuan's thinking.

- What did he mean by *undoing* the operations?
- Why did he subtract 12 and then divide by 3? Why not divide by 3 and then subtract 12?

2. Solve the equation $8x + 20 = 116$ without the use of tables or graphs in a way that makes sense to you. Check your answer.

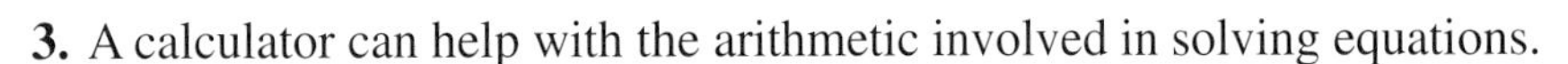

3. A calculator can help with the arithmetic involved in solving equations.

a. When one student used her calculator to solve an equation by undoing, her screen showed this:

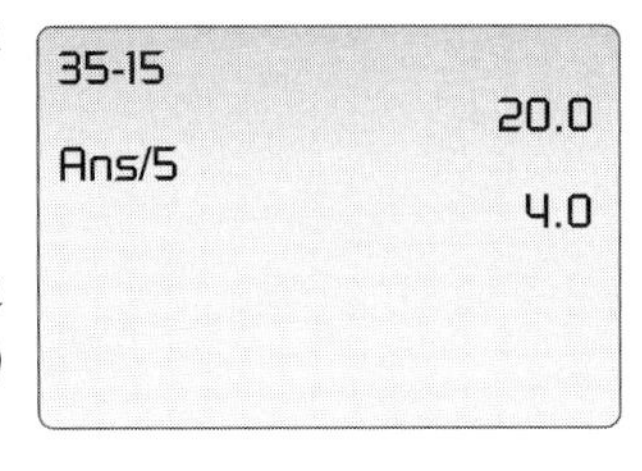

What equation could she have been solving?

b. What would appear on your screen if you used a calculator to solve the equation $30x + 50 = 120$ by the "undoing" method?

4. When people shop for cars or trucks, they usually look closely at data on fuel economy. The data are given as miles per gallon in city and in highway driving. Let C stand for *miles per gallon in city driving* and H stand for *miles per gallon in highway driving*. Data from tests of the 20 most popular American cars and trucks show that the equation $H = 1.4 + 1.25C$ is a good model for the relation between the variables. Solve the following equations without making a calculator table or graph. Explain what the results tell about the relation between city and highway mileage. Be prepared to explain the reasoning you used to find each solution.

a. $35 = 1.4 + 1.25C$ **b.** $10 = 1.4 + 1.25C$ **c.** $H = 1.4 + (1.25)(20)$

5. Profit at the Palace Theater is a function of number of tickets sold according to the rule $P = -450 + 2.5T$. Without making a table or graph, solve the following equations and explain what the results tell about ticket sales and theater profit. Again, be prepared to explain your reasoning in each case.

a. $-200 = -450 + 2.5T$ **b.** $0 = -450 + 2.5T$ **c.** $500 = -450 + 2.5T$

6. Martina and Ann experimented with the strengths of different springs. They found that the length of one spring was a function of the weight attached with rule $L = 4.8 + 1.2W$. The length was measured in inches and the weight in ounces. To find the weight that could be attached to produce a stretched spring length of 15 inches, they reasoned as follows:

- We need to solve $15 = 4.8 + 1.2W$.
- This is the same as $10.2 = 1.2W$.
- This is the same as $10.2 \div 1.2 = W$.
- This means $8.5 = W$.

a. Is each step of their reasoning correct? If so, how would you justify each step? If not, which steps contain errors and what are those errors?

b. What does the answer tell about the spring?

7. The Yogurt Shop makes several different flavors of frozen yogurt. Each new batch is 650 ounces, and a typical cone uses 8 ounces. As sales take place, the amount A of each flavor remaining from a fresh batch is a function of the number N of cones of that flavor that have been sold. The rule relating number of cones sold to amount of yogurt left is $A = 650 - 8N$.

   a. Solve each equation. Show all your work.

   $570 = 650 - 8N$

   $250 = 650 - 8N$

   $A = 650 - 8(42)$

   b. Use the rule $A = 650 - 8N$ to write and solve equations to answer the following questions:

   - How many cones have been sold if 390 ounces remain?
   - How much yogurt will be left when 75 cones have been sold?
   - If the machine shows 370 ounces left, how many cones have been sold?

8. Bronco Electronics received bids from two shipping companies that wanted the business of shipping all calculator orders. For shipping N calculators, Speedy Package Express would charge $3 + 2.25N$. The Fly-By-Night Express would charge $4 + 2N$. Solve the equation $3 + 2.25N = 4 + 2N$. Explain what the solution tells about the shipping bids.

Checkpoint

It is often relatively easy to solve problems involving linear equations without the use of tables or graphs.

a Suppose you are going to tell someone how to solve an equation like $43 = 7 - 4x$ without the use of a table or graph. What steps would you recommend? Why?

b When would you recommend solving an equation like the ones you've seen so far without a table or graph? When would you advise use of the calculator methods?

Be prepared to explain and defend your procedures.

On Your Own

When a soccer ball, a volleyball, or a tennis ball is hit into the air, its upward velocity changes as time passes. The ball slows down as it reaches its maximum height and then speeds up in its return flight toward the ground. Suppose the upward velocity of a high volleyball serve is given by the rule $V = 64 - 32T$ where T is *time* in seconds and V is *velocity* in feet per second.

a. Solve each of the following equations and explain what each tells about the flight of the ball. In your report, show your reasoning in finding the solutions.

$16 = 64 - 32T$
$-24 = 64 - 32T$
$0 = 64 - 32T$
$96 = 64 - 32T$

b. Use your graphing calculator or computer software to make a table and a graph of the relation between time and velocity of the ball for $0 \leq T \leq 5$. Show how to find the solution for each equation in Part a in the table and on the graph.

MORE Modeling • Organizing • Reflecting • Extending

Modeling

1. Recall from page 189 that Victoria DeStefano can earn as much as \$450 as scorekeeper for a summer basketball league. She learned that she must pay \$18 per game for substitutes when she misses a day of work, so her *summer earnings E* will depend on the *number of days D* she misses according to the rule $E = 450 - 18D$. Solve each of the following equations and explain what it tells you about Victoria's summer earnings.

a. $306 = 450 - 18D$
b. $360 = 450 - 18D$
c. $0 = 450 - 18D$
d. $E = 450 - 18(2)$
e. $315 = 450 - 18D$
f. $486 = 450 - 18D$

2. For Mary and Jeff Jordan, who work at a baseball park selling programs, their pay P per game is given by $P = 10 + 0.25S$ where S is number of programs sold. Solve each of the following equations. Explain what the answers tell about Mary's and Jeff's summer earnings.

a. $10 + 0.25S = 17.50$
b. $10 + 0.25S = 20$
c. $10 + 0.25S = 7.50$
d. $10 + 0.25(38) = P$

3. The ball park manager offered Mary and Jeff a different pay plan, $P_2 = 5 + 0.4S$.

a. What do the 5 and 0.4 each tell you about this new pay plan?

b. Solve the equation below. Explain the meaning of the solution.

$$10 + 0.25S = 5 + 0.4S$$

c. Graph the two possible pay plans using your graphing calculator or computer. Explain how the graph shows which is better for different numbers of sales.

d. Suppose the manager offered yet another pay plan P_3: \$0.75 per program sold and no minimum pay. What equation relating S and P_3 would fit this rule? When would this rule give the same earnings as each of the other two plans?

4. Garrett and Luis conducted a bouncing ball experiment. They dropped a tennis ball from various heights x (in cm) and measured the height y (in cm) of the rebound. They modeled their data with this rule:

$$y = 5 + 0.4x$$

Write three questions involving this model that can be answered by solving corresponding equations. Write and then solve the equations without using tables or graphs.

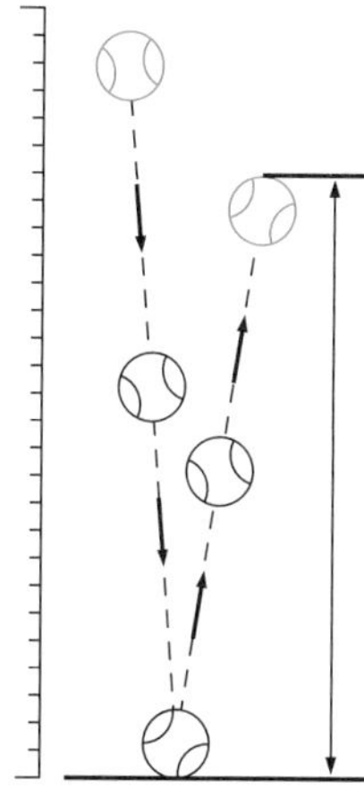

Organizing

1. Solve each of the following equations and check your answers. Show the steps in your solutions and in your checks.

a. $25 = 13 + 3x$

b. $74 = 8.5x - 62$

c. $34 + 12x = 76$

d. $76 = 34 - 12x$

e. $3{,}141 = 2{,}718 + 42x$

2. If a, b, and c are any numbers, what operations are required to solve these equations?

a. $c = a + bx$

b. $c = a - bx$

c. $c = -a + bx$

3. If a, b, c, and d are any numbers, describe the operations that are required to solve equations of the type $a + bx = c + dx$.

4. Below are two properties of arithmetic operations. The first relates addition and subtraction. The second relates multiplication and division.

- *For any numbers a, b, and c, then $a + b = c$ is true if and only if $a = c - b$.*
- *For any numbers a, b, and c, then $a \times b = c$ is true if and only if $a = c \div b$ and $b \neq 0$.*

Erik solved the equation $3x + 12 = 45$ given at the beginning of this investigation as follows:

If $3x + 12 = 45$, then $3x = 45 - 12$ or 33.

If $3x = 33$, then $x = 33 \div 3$ or 11.

So $x = 11$.

a. Explain how the above properties of operations can be used to support each step in Erik's reasoning.

b. Solve the equation $130 + 30x = 250$ using the properties of operations. Check your answer.

Reflecting

1. Why is it a good idea to check solutions to equations?

2. If a question about two related variables involves solving an equation of the form $c = a + bx$, how do you decide which of the following methods to use to search for a solution?

a. Guess and test the guesses.

b. Look at a graph of the equation $y = a + bx$.

c. Look at a table of values for $y = a + bx$.

d. Reason with the symbolic form itself as follows:

$$c = a + bx$$
$$c - a = bx$$
$$(c - a) \div b = x$$

3. Consider Mary and Jeff Jordan's pay possibilities for selling programs at the ball park. Let S represent the number of programs sold and P represent their pay.

a. Using the rule stated as $P = 10 + 0.25S$, what operations are needed to find the pay for selling 36 programs in a single night?

b. What operations are needed to solve the equation $19 = 10 + 0.25S$?

c. In what sense do the operations in Part b "undo" the operations in Part a?

d. How does the order in which you do the operations in Part a compare with the order in Part b? Why does this make sense?

4. What are two different, but reasonable, first steps in solving $2x + 8 = 5x - 4$?

a. What does the solution to the equation tell you about the graphs of $y = 2x + 8$ and $y = 5x - 4$?

b. What does the solution to the equation tell you about tables of values for $y = 2x + 8$ and $y = 5x - 4$?

Extending

1. How is $y = a + bx$ related to $c = a + bx$ where x and y are variables and a, b, and c are fixed numbers?

2. One linear model relating *grams of fat F* and *calories C* in popular "lite" menu items of fast food restaurants is given by the equation $C = 300 + 13(F - 10)$. Solve each equation below in the following three different ways:
   - Use a graph of the equation.
   - Use a table for the equation.
   - Use symbolic reasoning as in the examples of this investigation.

   a. $430 = 300 + 13(F - 10)$

   b. $685 = 300 + 13(F - 10)$

   c. $170 = 300 + 13(F - 10)$

   d. $685 \leq 300 + 13(F - 10)$

INVESTIGATION 3 Making Comparisons

In many problems where linear equations occur, the key question asks for comparison of two or more different models. For example, in Investigation 1 of Lesson 3 you used linear models to compare the patterns of change in percentage of female and male doctors in the United States. In this investigation, you will examine methods for making sense of situations modeled by *systems of linear equations*.

Regional Exchange

You can have reliable and efficient telephone service for only \$8.00 per month plus \$0.15 per call.

General Telephone

We offer top quality phone service for only \$14.00 per month plus \$0.10 per call.

You probably have noticed from ads on television and in newspapers that the market for long-distance telephone service has become very competitive. Deregulation of the communication industry in 1996 promised that similar competition would come to local telephone services as well. Suppose that two telephone companies, Regional Exchange and General Telephone, advertise their services for rural areas, as shown. Installation charges are the same for both companies. Which telephone company offers a more economical deal for local service?

1. For both companies, the monthly charge is a function of the number of calls made.

a. Write linear equations giving the relations between number of calls and monthly charge for each company.

b. Compare the monthly charges by each company for 80 calls.

c. How many calls could you make in a month for $30 under the pricing plans of the two companies?

d. For what number of calls is Regional Exchange more economical? For what number of calls is General Telephone more economical?

e. Which plan would cost less for the way your family uses the telephone?

2. To compare the price of service from the two telephone companies, the key problem is finding the number of calls for which these two rules give the same monthly charge. That means finding a value of x for which each rule gives the same y.

a. How could you use tables and graphs to find the number of calls for which the two telephone companies have the same monthly charge?

b. When one class discussed their methods for comparing the price of service from the two companies, they concluded, "All we have to do is solve the equation $8 + 0.15x = 14 + 0.10x$." Is this correct?

c. How could you solve the equation in Part b by using reasoning like that applied by Natasha or Dejuan on page 220?

The questions about monthly telephone charges from two different companies involve comparisons of two linear models. The models can be expressed with equations.

Regional Exchange: $y = 8 + 0.15x$

General Telephone: $y = 14 + 0.10x$

The pair of linear models is called a **system of linear equations**. Finding the single pair of numbers x and y that satisfy both equations is called **solving the system**. In this case, the solution was $x = 120$ and $y = 26$ because

$$26 = 8 + 0.15(120) \quad \text{and} \quad 26 = 14 + 0.10(120)$$

3. How is the solution to a system of equations seen in the graphs of the equations? In tables of values?

4. Each of the following systems of linear equations could represent the cost y of a service that is used x times. Use tables, graphs, or reasoning with the symbolic forms themselves to solve each system. Check each answer by substituting the solution values of x and y back in the original equations. If a system does not have a solution, explain why.

a. $y = 2x + 5$
$y = 3x + 1$

b. $y = 0.5x + 8$
$y = x + 1$

c. $y = 1.5x + 2$
$y = 1.5x + 5$

d. $y = -1.6x + 10$
$y = 0.4x + 2$

Checkpoint

In solving a system of linear equations like $y = 5x + 8$ and $y = -3x + 14$, be able to answer the following questions:

a What is the objective?

b How could the solution be found on a graph of the two equations?

c How could the solution be found in a table of (x, y) values for both equations?

d How could the solution be found using reasoning with the symbolic forms themselves?

e What patterns in the tables, graphs, and equations of a system will indicate that there is no pair of values for x and y that satisfies both equations?

Be prepared to explain your solution methods and reasoning.

On Your Own

Charter boat fishing for walleyes is popular on Lake Erie. The charge for an eight-hour charter trip is as follows:

Charter Company	Boat Rental	Charge per Person
Wally's	$200	$29
Pike's	$50	$60

Each boat can carry a maximum of ten people in addition to the crew.

a. Model the cost for charter service by Wally's and by Pike's with equations.

b. Determine which service is more economical for a party of four and for a party of eight.

c. Assuming you want to minimize your costs, under what circumstances would you choose Wally's charter service?

MORE
Modeling • Organizing • Reflecting • Extending

Modeling

1. Competition between telephone companies is intense. General Telephone Company wanted to become more competitive for customers who make fewer calls per month. They decided to change the monthly base charge from $14 to $12, but maintained the $0.10 per call charge.

a. Write an equation that models the monthly charges under the new program.

b. How are the graphs of the new and old service charges related?

c. What would be the monthly bill for 30 calls using the new program?

d. How many calls would you need to make in order for the new program to be more economical than Regional Exchange?

e. If your bill is $16.30 under the new program, how many calls did you make? What is Regional Exchange's bill for the same number of calls?

2. General Telephone did not notice any large increase in subscriptions when they changed their base monthly charge from $14 to $12. They decided to change it back to $14 and reduce the per-call charge from $0.10 to $0.08.

a. Write an equation that models their new service charge.

b. How are the graphs of the new and original service charges related?

c. What is the cost of 30 calls under this new plan?

d. How many calls would need to be made for the General Telephone monthly bill to be competitive with Regional Exchange?

e. Compare the cost of service under this plan with that proposed in Modeling Task 1. Which plan do you think will attract more customers? Explain your reasoning.

3. Suppose General Telephone decides to lower its base monthly charge to \$10 but is unsure what to charge per call. They want to advertise monthly bills that are lower than Regional Exchange if one makes more than 40 calls per month. Regional Exchange has an \$8 base monthly charge plus \$0.15 per call.

   a. To meet their goal, at what point will the General Telephone graph need to cross the Regional Exchange graph?

   b. What charge per call by General Telephone will meet that condition?

   c. Suppose a customer makes 60 calls per month. By how much is the new General Telephone plan lower than the Regional Exchange plan for this many calls?

   d. If a family makes only 20 calls per month, how much less will they spend by using Regional Exchange rather than the new General Telephone plan?

4. From the situations described below, choose two situations that most interest you. Identify the variables involved and write equations showing how those variables are related. Graph the equations. Then explain how the graphs show the costs of different decisions in each case.

   a. A school club decides to have customized T-shirts made. For their design, the Clothing Shack will charge \$15 each for the first ten shirts and \$12 for each additional shirt. The cost of having them made at Clever Creations is a \$50 initial fee for the setup and \$8 for each T-shirt.

   b. The *Evening News* charges \$4 for the first 3 lines and \$1.75 for each additional line of a listing placed in the Classified Section. The *Morning Journal* charges \$8 for the first five lines and \$1.25 for every additional line.

   c. Speedy telegram service charges a \$30 base fee and \$0.75 for each letter or symbol. Quick Delivery charges a \$25 base fee and \$0.90 for each character.

   d. Cheezy's Pies charges \$5 for a 12-inch sausage pizza and \$5 for delivery. The Pizza Palace delivers for free, but charges \$7 for a 12-inch sausage pizza.

Organizing

1. The table on the following page shows the winning times for women and men in the Olympic 100-meter freestyle swim.

Olympic 100-meter Freestyle Swim Times

Year	Women (Time in seconds)	Men (Time in seconds)	Year	Women (Time in seconds)	Men (Time in seconds)
1912	82.2	63.4	1964	59.5	53.4
1920	73.6	61.4	1968	60.0	52.2
1924	72.4	59.0	1972	58.59	51.22
1928	71.0	58.6	1976	55.65	49.99
1932	66.8	58.2	1980	54.79	50.40
1936	65.9	57.6	1984	55.92	49.80
1948	66.3	57.3	1988	54.93	48.63
1952	66.8	57.4	1992	54.64	49.02
1956	62.0	55.4	1996	54.50	48.74
1960	61.2	55.2	2000	53.83	48.30

Source: *The World Almanac and Book of Facts 2001*, Mahwah, NJ: World Almanac Education Group, Inc., 2001.

a. Make scatterplots for the (*year*, *winning time*) data for men and for women. Use 0 for the year 1900.

b. Find the linear regression model for each scatterplot. Write the equations.

c. Which group of athletes has shown a greater improvement in time? Explain.

d. Use your calculator or computer to graph the regression lines. Where do they intersect?

e. What is the significance of the point of intersection of the two lines in Part d? How much confidence do you have that the lines accurately predict the future? Explain.

2. Solve each of the following systems of linear equations using at least two different methods (tables, graphs, or reasoning with the symbolic forms themselves). Check each solution.

a. $y = x + 4$
$y = 2x - 9$

b. $y = -2x + 18$
$y = -x + 10$

c. $y = 3x - 12$
$y = 1.5x + 3$

d. $y = x$
$y = -0.4x + 7$

3. Write a system of equations for which there is no pair of values for x and y that satisfies both equations. Explain how one can predict this result from the equations, without using tables or graphs.

Reflecting

1. Describe a situation for which a graph of a system of linear equations would help you make a good decision.

2. Comparing various consumer service plans often involves analyzing and solving a system of equations of the form:

$$y = a + bx$$
$$y = c + dx$$

How do you decide which method to use in solving the system of equations?

3. What seems to be the best way to use the information from tables and graphs to solve a system of linear equations?

4. When asked to solve the system of linear equations $y = 2x + 9$ and $y = 5x - 18$, Sabrina reasoned as follows:
   - I want x so that $2x + 9 = 5x - 18$.
   - Adding 18 to each side of that equation must give an equivalent equation $2x + 27 = 5x$.
   - Subtracting $2x$ from each side of the new equation must give an equivalent equation $27 = 3x$.
   - Dividing each side of that equation by 3 must give an equivalent equation, so $x = 9$.
   - If $x = 9$, then one equation is $y = 2(9) + 9$ and the other equation is $y = 5(9) - 18$. Both equations give $y = 27$.
   - The solution of the system must be $x = 9$ and $y = 27$.

   Do you believe each step of her reasoning? Why or why not?

Extending

1. Refer back to the "On Your Own" on page 228. Suppose it was noticed that most fishing parties coming to the dock were four or fewer persons.

   a. How should Wally revise his boat rental fee so that his rates are lower than the competition's (Pike's) for parties of three or more? Write an equation that models the new rate system.

   b. How much less would a party of four pay by hiring Wally's charter service instead of Pike's?

   c. Which service should you hire for a party of two? How much more would you spend on the other service?

   d. Suppose Pike's charter service lowers the per-person rate from \$60 to \$40. For what size parties would Pike's be less expensive?

e. If Wally wants to change his per-person rate so that both services charge the same for parties of four, what per-person rate should Wally charge? Write an equation that models the new rate structure.

2. A discount card at a local second-run movie theater costs \$10.00. It is valid for 3 months. With this card it costs only \$3.00 to attend a movie, instead of the usual \$5.00.

a. How many movies would you need to attend in order to spend the same amount with and without the discount card?

b. Movie theaters view discount card plans as a way to increase attendance and profits. Devise a data-gathering plan that would help a theater set the price of a discount plan in your community. Write a summary outlining your recommendation.

3. Make up a linear system relating cost to number of uses of a service for which Company A's rate per service is 1.5 times that of Company B's, but Company B's is not more economical until 15 services have been performed.

INVESTIGATION 4 Equivalent Rules and Equations

From network television, movies, and concert tours to local school plays and musical shows, entertainment is a big business in the United States. Each live or recorded performance is prepared with weeks, months, or even years of creative work and business planning.

For example, a reasonable cost for production of a CD by a popular recording artist is \$100,000. Depending on various things, each copy of the CD could cost about \$1.50 for materials and reproduction. Royalties to the composers, producers, and performers could be about \$2.25 for each CD. The record label might charge about \$5 per copy to the stores that will sell the CD. Using these numbers, how does the *label's net profit P* relate to the *number of copies N* that are made and sold?

1. Here are four possible rules relating P and N.

$P = 5.00N - 1.50N - 2.25N - 100{,}000$

$P = 5.00N - 3.75N - 100{,}000$

$P = 1.25N - 100{,}000$

$P = -100{,}000 + 1.25N$

a. How was each rule constructed from the information given?

b. Which of the four equations seems the best way to express the relation between copies sold and profits? Explain your reasoning in making a choice.

c. Compare tables of (*sales*, *profit*) data to see the pattern defined by each of the four equations. (Consider CD sales of 0 to 500,000 in steps of 10,000.)

d. Create graphs of the four rules with your calculator or computer. Compare the graphs to see the pattern each gives relating sales and profits. Use the trace function to examine each graph.

Your tables, graphs, and thinking about possible profit rules for a new CD release show an important fact about linear models:

Several different equations can each model the same linear pattern.

Since tables and graphs require some effort to construct (even with a graphing calculator), it is helpful to be able to tell quickly when two equations are **equivalent**—that is, when they describe the *same* linear model. As you explore the following situations, look for patterns in linear expressions that help in predicting equivalence of different forms.

2. Studios that make motion pictures deal with many of the same kinds of cost and income questions as music producers. Contracts sometimes designate parts of the income from a movie to the writers, directors, and actors. Suppose that for one film those payments are:

4% to the writer of the screenplay;

6% to the director;

15% to the leading actors.

a. Suppose the studio receives income of $50 million from the film. What payments will go to the writer, the director, the actors, and to all these people combined?

b. Let I represent film income and E represent expenses for the writer, director, and actors. Write two equivalent equations showing how E is a function of I. Make one of those equations in a form that shows the breakdown to each person or group. Make the other the shortest form that shows the combined payments.

c. A movie studio might have other expenses, before any income occurs. For example, there will be costs for shooting and editing the film. Suppose those pre-release expenses are $20 million. Assuming there are no other expenses, what will the studio's *profit* be if income is $50 million? (Remember, the profit is income minus expenses.)

d. Here is one expression for profit P as a function of income I for that movie.

$$P = I - (20 + 0.25I)$$

- In your group, discuss how this rule was constructed from the given information.
- Write two other equivalent expressions for profit as a function of income. Make one of those equations as short as possible. Make the other a longer expression that shows how the separate payments to the writer, the director, and the actors affect profit.

3. For theaters, there are two main sources of income. Money is collected from ticket sales and from concession stand sales. Suppose that for a major new film a theater is charging \$8 for each admission ticket and that concession stand income averages \$3 per person.

a. Write two equivalent equations showing how theater income I depends on number N of people who come to the show. Make one equation as short as possible. Make the other a longer expression that shows the two components of income.

b. Suppose that the theater has to send 35% of its income from ticket sales to the producer of the film. The cost of stocking the concession stand averages about 15% of its income from those items. Suppose also that the theater has to pay rent, electricity, and staff salaries of about \$15,000 per month. Combine these facts with the information from Part a to write two different but equivalent equations showing how monthly theater profit P depends on the number N of ticket buyers.

4. The movie theater described in Activity 3 charges \$8 per admission ticket sold and receives an average of \$3 per person from the concession stand. The theater has to pay taxes on its receipts, which it includes in the admission and concession prices. Suppose the theater has to pay taxes equal to 9% of its receipts.

a. Calculate the tax due if 1,000 tickets are sold. Report the results in two ways—first showing the tax on ticket sales and concession stand sales separately, and second showing the combination of those two tax sources.

b. Write two different, but equivalent, equations giving tax due T as a function of the number N of tickets sold. First, show how taxes can be calculated on each income source separately and then combined to give total tax due. Second, show how these sources can be combined before calculating tax due.

c. After paying the 35% rental fee for the film, the theater's income from ticket sales is \$5.20 per ticket sold. The concession stand income is about \$2.55 per person, after paying its stocking costs. In addition to rental and stocking costs, the theater has operating expenses of \$15,000 per month. A new proposal will tax profits, which are receipts minus expenses, at 5%. The tax due can be given by

$$T = 0.05(7.75N - 15{,}000)$$

- Explain how this rule was constructed from the given information.
- Write an equivalent equation without use of parentheses.

In Activities 1–4, you translated information about variables and relations into equations and then into different, but equivalent, equations. You used facts about the numbers and variables involved to guide and check the writing of new equivalent symbolic expressions.

For example, if it costs \$1.50 per copy to make a music CD and \$2.25 per copy to pay royalties for the performers, it makes sense to combine $1.50x$ and $2.25x$ to get $3.75x$ expressing the cost for x copies. If the record label gets income of \$5.00 per copy from retail stores, it makes sense to simplify the expression for the profit on x copies from $(5.00x - 3.75x)$ to $1.25x$.

Do the examples in Activities 1–4 suggest some ways to rewrite symbolic expressions that will produce equivalent forms regardless of the situation being modeled? Can you rewrite expressions and equations involving variables x and y in equivalent forms even if you do not know what the variables represent?

5. For each of the following symbolic expressions, write three different forms that you believe are equivalent to the original. In each case:

- Write one equivalent form that is as short as possible.
- Write one equivalent form that is longer than the original.
- Test the equivalence of each new expression by comparing its table and graph to those of the original.

a. $7x + 11x$

b. $7x - 11x$

c. $8x + 5 - 3x$

d. $5 + 3x + 12 + 7x$

e. $2 + 3x - 5 - 7x$

f. $10 - (5x + 3)$

g. $10 - (5x - 3)$

h. $5(x + 3)$

i. $7(2x - 12)$

j. $2 - 3x + 5x^2 + 7x$

Symbolic expressions like $2 - 3x + 5x^2 + 7x$ are built by combining numbers and variables through the operations $+$, $-$, $\times$, and $\div$, and by using exponents. Each part of the expression that involves only multiplication, division, or exponents is called a **term**. So the terms of this expression are 2, $3x$, $5x^2$, and $7x$.

6. Writing symbolic expressions in equivalent forms involves rearranging, combining, and expanding terms of the original. Based on your experience in Activities 1–5, answer the following questions. Give specific examples to illustrate your ideas. Be prepared to explain why you think the rewriting procedures you recommend seem likely to work consistently.

   a. In what ways can terms in an expression be rearranged to produce equivalent forms?

   b. In what ways can terms of an expression be combined into shorter or longer equivalent forms?

   c. How can expressions be rewritten in equivalent forms without parentheses?

Checkpoint

In many situations, two people can suggest rules for linear models that are equivalent, but look quite different. For example, these two symbolic expressions for linear models are equivalent:

$$y = 15x - (12 + 7x) \quad \text{and} \quad y = 8x - 12$$

a How could you test the equivalence using tables of (x, y) values?

b How could you test the equivalence using graphs of the relations?

c What reasoning with the symbolic forms alone would confirm the equivalence of the expressions?

Be prepared to explain your responses to the entire class.

On Your Own

Many college basketball teams play in winter tournaments that are sponsored by businesses who want the advertising opportunity. For one such tournament, the projected income and expenses are as follows:

- Income is \$60 per ticket sold, \$75,000 from television and radio broadcast rights, and \$5 per person from concession stand sales.
- Expenses are \$200,000 for the teams, \$50,000 for rent of the arena and its staff, and a tax of \$2.50 per ticket sold.

a. Find the projected income, expenses, and profit if 15,000 tickets are sold for the tournament.

b. Write two equivalent expressions giving tournament income I as a function of the number N of tickets sold. In one expression, use terms that show each source of income. In the other, combine and rearrange terms to give the shortest possible expression.

c. Write two equivalent expressions giving tournament expenses E as a function of the number N of tickets sold. In one expression, use terms that show each source of expense. In the other, combine and rearrange terms to give the shortest possible expression.

d. Write two equivalent expressions giving tournament profit P as a function of the number N of tickets sold. In one expression, use terms that show each source of income. In the other, combine and rearrange terms to give the shortest possible expression.

MORE Modeling • Organizing • Reflecting • Extending

Modeling

1. To advertise a concert tour, the concert promoter paid an artist \$2,500 to design a special poster. The posters cost \$2.50 apiece to print and package in a cardboard cylinder. They are to be sold for \$7.95 apiece.

a. Write equations giving *production cost* C, *income* I, and *net profit* P as functions of the *number* N *of posters* printed and sold.

b. Write two equations that are different from, but equivalent to, the profit equation you wrote in Part a. Explain how you are sure they are equivalent.

2. One equation modeling the growth of median salary for working women since 1970 is $S = 4{,}000 + 750(Y - 1970)$.

a. Write an equivalent equation in the form $S = A + By$. Explain how you know the new equation is equivalent to the original.

b. What do the numbers 4,000, 750, and 1970 tell about the salary pattern?

c. What do the numbers A and B you found in the simpler rule of Part a tell about the salary pattern?

3. The video game industry is a big business around the world. Development of a new game might cost millions of dollars. Then to make and package each game disc will cost several more dollars per copy. Suppose the development cost for one game is \$5,000,000; each disc costs \$4.75 to make and package; and the wholesale price is set at \$53.50 per disc.

   a. Write equations giving the *total cost C* of designing and making *N* discs and the *income I* that would be earned from selling those *N* discs.

   b. Write two different equivalent equations relating *number N of copies sold* and *net profit P* from those sales.

   c. Use evidence in tables, graphs, and properties of the operations involved in the equations to prove the equivalence of the two equations from Part b.

4. Pick any number. Multiply it by 2 and subtract 10. Multiply the result by 3 and add 30. Finally, divide by your original number. Repeat the process several times with different starting numbers.

   a. What are your answers in each case?

   b. Let x represent the starting number and y represent the ending number. Write an equation showing how y is calculated from any value of x.

   c. Write the equation from Part b in simplest equivalent form. Explain how it makes the results in Part a reasonable.

Organizing

1. The reordering and combining of terms to produce equivalent expressions illustrates some basic properties of numbers and operations. Here are two of the most useful properties.

   For any numbers a, b, and c,

   Commutative Property of Addition: $a + b = b + a$

   Distributive Property of Multiplication over Addition: $a(b + c) = ab + ac$

   Show how the commutative and distributive properties are involved in writing simple equivalent forms for the following expressions.

   a. $3x + 5 + 8x$

   b. $7 + 3x + 12 + 9x$

   c. $8(5 + 2x)$

   d. $2(5x + 6) + 3 + 4x$

2. Invent and test a distributive property that would apply to expressions involving multiplication and subtraction. Use your ideas to find simple equivalent forms for the following expressions, recording each step of your reasoning.

a. $5(2x - 8)$

b. $5x + 7 - 3x + 12$

c. $3x - 7 + 4(3x - 6)$

d. $-7x + 13 + 12x - 10$

3. In Unit 2, you revisited the formula $A = \frac{1}{2}bh$ for the area of a triangle where b is the length of the base and h is the height of the triangle. Shown below is a *trapezoid* with bases of lengths b_1 and b_2 and height h.

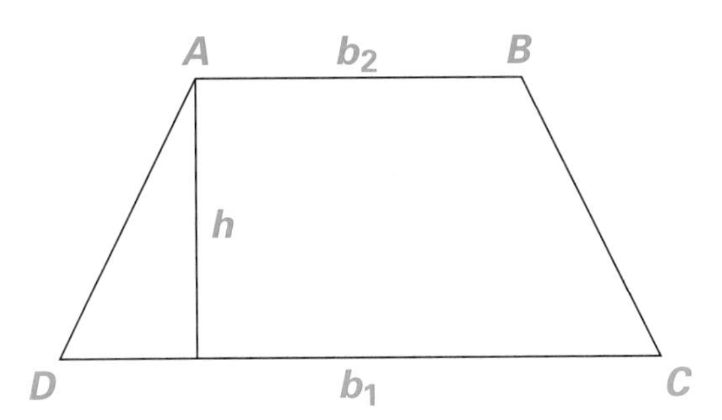

a. Make a copy of the diagram.

b. Draw $\triangle ACD$ and write an expression for its area.

c. Write an expression for the area of $\triangle ABC$.

d. Write an expression for the area of trapezoid $ABCD$.

e. Write an equivalent expression for the area of the trapezoid.

4. In transforming algebraic expressions to equivalent forms, it's easy to make some slips and use "illegal" moves. Given below are six pairs of algebraic expressions. Some are equivalent and some are not.

- Use tables, graphs, and reasoning with algebraic properties to decide which pairs are actually equivalent and which involve slips in reasoning.
- In each case of equivalent expressions, identify the use of commutative and distributive properties that justify the equivalence.
- In each case of an algebra "slip," spot the reasoning error. Write an explanation that you would use to help clear up the problem for the student who made the error.

a. Is $3(2x + 8)$ equivalent to $6x + 8$?

b. Is $4x - 6x$ equivalent to $2x$?

c. Is $8x(2 - 6x)$ equivalent to $16x - 48x^2$?

d. Is $10 + 3x - 12$ equivalent to $3x + 2$?

e. Is $10x + 5$ equivalent to $10(x + 5)$?

f. Is $-4(x - 3)$ equivalent to $-4x - 12$?

Reflecting

1. In Modeling Task 3 about a video game disc, any one of the following equations would show the relation of net profit to number sold:

$$P = 53.50N - 4.75N - 5{,}000{,}000$$

$$P = 48.75N - 5{,}000{,}000$$

$$P = 53.50N - (5{,}000{,}000 + 4.75N)$$

   a. Which equation do you believe shows the factors involved in the best way? Explain the reasons for your choice.

   b. Explain how you can be sure that all three equations are equivalent.

2. Describe a real situation which could be modeled by at least two different, but equivalent, symbolic rules.

3. How do you prefer to check whether one form of an equation is equivalent to another form: tables of values, graphs, or rewriting one of the equations using commutative and distributive properties? Why?

4. In Unit 1, "Patterns in Data," you encountered the rule

$$C = \tfrac{5}{9}(F - 32)$$

   for transforming temperature in degrees Fahrenheit F to temperature in degrees Celsius C.

   a. Use the number properties to rewrite this equation in the standard form of a linear model $y = a + bx$.

   b. Write a question that is more easily answered using the original form of the rule. Explain why you think the original form is better.

   c. Write a question that is more easily answered using the standard form in Part a. Explain why you think the standard form is better.

Extending

1. Are the following pairs of equations equivalent?

   a. $y = 7 - 5(x + 4 - 3x)$ and $y = 7 - 5x + 20 - 15x$

   b. $y = 7x - 12 + 3x - 8 + 9x - 5$ and $y = 7x - 4 + 2x + 8 + 10x - 13$

2. If you graph the equations in Extending Task 1, all will be lines, though no equation looks exactly like $y = a + bx$, the familiar form of an equation for a linear model.

   a. What features of the equations like those in Task 1 give a clue that the graph will be a line?

   b. What might appear in an equation that would give a clue that the graph would not be a line? Give some examples and sketch the graphs of those examples.

3. The real value of linear models for data patterns—relating two variables—occurs when you can recognize one of those patterns in a problem. Then you can build a good model and use it to solve the problem. Using the examples of this unit as a guide:

a. Identify a situation from your experience in which two related variables change in a pattern that seems linear.

b. Collect data on the two variables. Make a scatterplot of those data and fit a linear model to the data. Find an equation that matches that model.

c. Write five different questions about the relation between the variables that you are studying. Use your model to answer your questions.

d. Prepare a brief report describing the situation you investigated and your findings.

Lesson 4 Looking Back

The lessons of this unit have shown that variables in many situations are related in patterns which can be modeled by linear graphs. You have learned how to find equations that describe those graphs and the corresponding tables of values. You also have learned to use the equations to answer questions about the variables and relations. This final lesson of the unit has problems which will help you review, pull together, and apply your new knowledge.

1. Athletic shoes are made by dozens of international companies and worn by young people and adults all over the world. To help buyers in different countries pick shoes that will fit, most shoes show sizes on little tabs sewn inside the shoe.

The sizes are given in three systems—US, UK (for United Kingdom), and EUR (for European). Do you know the relation between shoe size and shoe length? Do you know the relation between US, UK, and EUR sizes? You can figure out the pattern relating shoe size and shoe length and the patterns relating shoe sizes in different systems by collecting and analyzing data from your classmates.

Scottie Pippen's shoes

a. To get started, ask everyone in your class who is wearing some sort of athletic shoe to look for the size tab inside his or her shoe. Then measure the lengths of the shoes using an agreed-upon unit. Record the data in a table with headings like those below. Make separate tables for data from females and males.

US	11
UK	10
EUR	46

Shoe Length	US Size	UK Size	EUR Size

b. Next, have various groups in the class make different scatterplots of these combinations of data for females and for males.

(*Length of shoe*, *US Size*)

(*US Size*, *UK Size*)

(*US Size*, *EUR Size*)

- Find linear models and equations that fit the patterns of data points.
- When you have an equation of the form $y = a + bx$ relating the shoe length and US size, use the information from that model to write an equation using *NOW* and *NEXT* to show how US size increases as shoe length increases.

c. When groups report the models they have found, do the following:

- Identify the slope of each linear graph from its modeling equation. Explain what the slope tells about patterns of change in shoe size and length.
- Compare the patterns for women's and men's sizes in each case.

d. Use the appropriate linear model of size and length to answer these questions. In each case, write an appropriate equation or inequality.

- For women, what EUR size matches a US size 6?
- For women, what is the approximate length of a US size 10 shoe?
- If a man's athletic shoe is 14 inches long, what US size do you think it would be?
- Shaquille O'Neal wears a US size 22 basketball shoe. About how long is his shoe?

2. Most Americans can afford nutritious and varied diets, even if we do not always eat what is best for us. In many countries of the world, life is a constant struggle to find enough food. This struggle causes health problems such as reduced life expectancy and infant mortality.

a. The following data show how daily food supply (in calories) is related to life expectancy (in years) and infant mortality rates (in deaths per 1,000 births) in a sample of countries in the western hemisphere. Working in pairs, make scatterplots of the (*calories*, *life expectancy*) and (*calories*, *infant mortality*) data.

Health and Nutrition

Country	Daily Calories	Life Expectancy	Infant Mortality
Argentina	3,113	71	29
Bolivia	1,916	60	82
Canada	3,482	78	7
Dominican Republic	2,359	68	53
Haiti	2,013	55	93
Mexico	3,052	70	35
United States	3,671	76	9
Venezuela	2,582	70	33

Source: *The Universal Almanac 1996*. Kansas City, MO: Andrews and McMeel, 1995.

Study the patterns in the table and the scatterplots. Then answer these questions as a group.

- What seems to be the general relation between average daily calorie supply and length of life in the sample countries?
- What seems to be the general relation between average daily calorie supply and infant mortality in the sample countries?
- What factors other than daily calorie supply might affect the two variables of life expectancy and infant mortality?

b. Economists try to predict the likely increase of life expectancy or decrease of infant mortality for various increases in food supply. Working in pairs, find equations for linear models of the (*calories*, *life expectancy*) and of the (*calories*, *infant mortality*) data patterns.

c. What do the slopes of your linear models say about the pattern relating daily calorie supply and life expectancy in the sample countries? How about the relation between calorie supply and infant mortality?

d. Average daily calorie supply in Chile is 2,581. What life expectancy and infant mortality would you predict from the calorie data?

e. Brazil has a life expectancy of 66 years.

- What daily calorie supply would you predict in Brazil?
- The actual data for Brazil is 2,751 calories. What does the difference between your predicted value and the actual value tell about the usefulness of the model you have found?

f. What life expectancy does your model predict for a daily calorie supply of 5,000? How close to that prediction would you expect the life expectancy to be in a country with a daily calorie supply of 5,000?

3. Many people who go to movies like to have popcorn to munch on during the show. But movie theater popcorn is often very expensive. The manager of a local theater wondered how much more she might sell if the price was lower. She also wondered whether such a reduced price would actually bring in more popcorn profit.

One week she set the price for popcorn at $1.00 and sold an average of 120 cups per night. The next week she set the price at $1.50 and sold an average of 90 cups per night. She graphed a linear model to predict number of cups sold at other possible prices.

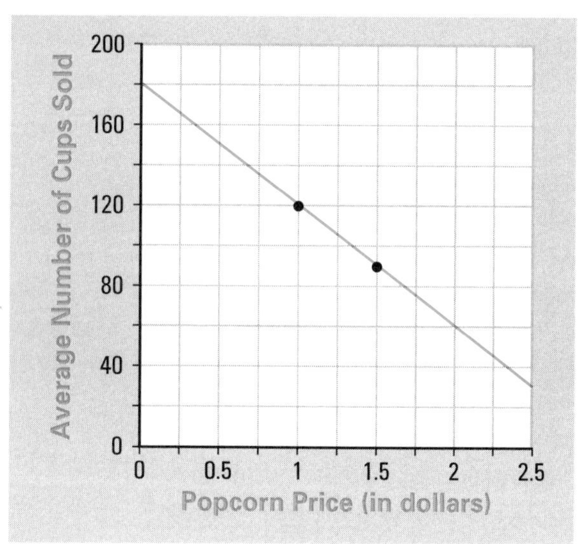

a. Find the equation of the linear model. Explain what the slope and intercept of the model tell about the prospective number of popcorn cups sold at various prices.

b. Write and solve equations or inequalities to answer the following questions:

- What price results in average daily number sold of about 150 cups?
- What price results in average daily number sold of fewer than 60 cups?
- What number sold is predicted for a price of $1.80?

c. Use the graph or rule relating price to average daily number of cups sold to make another table relating price to revenue from popcorn. Explain what the pattern in that table tells about the relation between price, number of cups sold, and revenue.

d. Solve this equation and inequality, which are similar to those about popcorn price and number of cups sold. Use at least three different methods to solve them: table, graph, and reasoning with the symbolic form itself. Show how each answer could be checked.

$9 + 6x = 24$

$1.5x + 8 \leq 3 + 2x$

4. The ninth grade class at Frederick Douglass High School has a tradition of closing the school year with a big party. The class officers researched costs for the dance they have in mind and came up with these items to consider:

Item	Cost
DJ for the dance	\$250
Food	\$3.75 per student
Drinks	\$1.50 per student
Custodians, Security	\$175

The question is whether the class treasury has enough money to pay for the dance or whether a special charge will be needed.

a. Which of the following equations correctly models *dance cost C* as a function of the *number N of students* who plan to come to the dance?

$C = 250 + 3.75N + 1.50N + 175$

$C = 425 + 5.25N$

$C = 5.25N + 425$

$C = 430.25N$

b. Write equations or inequalities whose solutions would answer these questions:

- How many students could come to the dance without extra charge if the class treasury has \$950?
- How many students could come to the dance with a charge of only \$2 if the class treasury has \$950?

5. Using algebraic expressions to help make sense out of problem situations is an important part of mathematics. Writing modeling expressions and equations is often a first step. Developing an understanding of equivalent algebraic expressions and how to obtain them is another important aspect of mathematics.

a. Find equations for the linear models passing through these pairs of points.

(0, 0) and (5, 8) (0, –3) and (4, 1)

(0, 3) and (6, 0) (2, –6) and (8, 12)

b. Compare the following pairs of linear models. Are they equivalent?

$y = 4.2x + 6$ and $y = (1 - 0.7x)6$

$N = 4C - 3(C + 2)$ and $C = N + 6$

$P = 0.3S - 0.4S + 2$ and $S = 10(P - 2)$

c. Solve each equation.

$34 = 6 - 4x$ $286 = 7p + 69$

$-25 = 8 + 1.1x$ $17y - 34 = 8y - 16$

$14 + 3k = 27 - 10k$

d. Solve the following system of equations using calculator- or computer-based methods and by reasoning with the symbolic forms.

$$y = 35 + 0.2x$$
$$y = 85 + 0.7x$$

Checkpoint

Linear patterns in data and linear relationships between quantities can be recognized in graphs, tables, and symbolic rules, or in conditions stated as applications or problems.

a Describe how you can tell whether a situation can be (or is) represented by a linear model by looking at:

- a scatterplot;
- a table of values;
- the form of the modeling equation;
- a description of the problem.

b Linear models often describe relationships between an input variable x and an output variable y.

- Write a general form for the rule of a linear model. What do the parts of the equation tell you about the relation being modeled?
- Explain how to find a value of y corresponding to a given value of x, using:

  i. a graph; **ii.** a table; **iii.** a symbolic rule.
- Explain how you can solve a linear equation using:

  i. a graph; **ii.** a table; **iii.** symbolic reasoning.
- Explain how you can solve a system of linear equations using:

  i. a graph; **ii.** a table; **iii.** symbolic reasoning.

Be prepared to share your descriptions and explanations with the whole class.

On Your Own

Write, in outline form, a summary of the important mathematical concepts and methods developed in this unit. Organize your summary so that it can be used as a quick reference in future units and courses.

Graph Models

Unit 4

Lesson 1

Careful Planning

Planning is something that people do every day. You might be planning a party or a vacation. You and your classmates might be planning a school play or the latest issue of the school newspaper. Businesses plan everything from advertising campaigns to new construction projects. City planners plan for the growth of their cities. Concert promoters must plan ahead to avoid scheduling conflicts, arrange for ticket sales, and ensure crowd control. With almost any project, careful planning is essential for success.

But it's not always easy to plan carefully. That's where mathematics can be helpful. Many mathematical models have been developed to help make planning more systematic and efficient. In this unit, you will use graph models to plan efficient routes, to schedule events and projects, and to plan ahead to avoid conflicts. This unit will give you another example of the power of mathematics. You can plan on it!

Think About This Situation

Suppose you have a job with a local school district for the summer. Your first assignment is to paint all the lockers in the high school.

a What are some tasks that would need to be completed before you could begin the actual painting?

b Would some tasks need to be completed before others? If so, in what order should the tasks be completed?

c Sketch a diagram of the arrangement of the lockers on one floor of your school.

d In what order would you paint the lockers? Do you think your plan involves the most efficient procedure? Why or why not?

INVESTIGATION 1 Planning Efficient Routes

You can save time, energy, and expense by studying a complex project before you begin your work. The order in which to complete the parts of the project may be important. There may be many ways to do this. However, one way may be judged to be the "best" or *optimal*, in some sense.

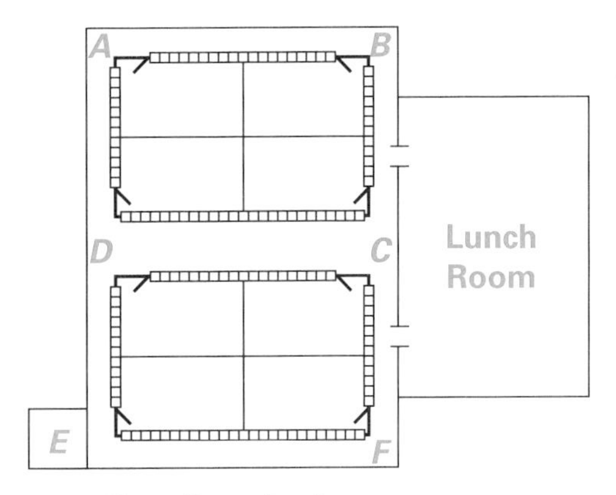

First-floor Lockers

Suppose you are expected to paint all the lockers around eight classrooms on the first floor of a high school. The lockers are located along the walls of the halls as shown above. Letters are placed at points where you would stop painting one row of lockers and start painting another. Five-gallon buckets of paint, a spray paint compressor, and other equipment are located in the first-floor equipment room *E*. You must move this bulky equipment with you as you paint the lockers. You also must return it to the equipment room when you are finished painting. (The lockers in the center hall must be painted one row at a time.)

1. Since you are being paid for the job, not by the hour, you would like to paint the lockers as quickly and efficiently as possible.
   a. Which row would you paint first? Is there more than one choice for the first row to paint?
   b. Which row would you paint last? Why?
2. Here are three plans that have been suggested for painting the first-floor lockers.

   Plan I: Paint from *E* to *F*, *F* to *C*, *C* to *D* (one side), *D* to *E*, *D* to *A*, *A* to *B*, *B* to *C*, *C* to *D* (the other side).

   Plan II: Paint from *A* to *B*, *B* to *C*, *C* to *D* (one side), *D* to *A*, *D* to *C* (other side), *C* to *F*, *F* to *E*, *E* to *D*.

   Plan III: Paint from *E* to *D*, *D* to *A*, *A* to *B*, *B* to *C*, *C* to *F*, *F* to *E*, *D* to *C* (one side), *C* to *D* (other side).

   a. Do you think any one of these plans is optimal; that is, the "best" way to do the painting? Explain your reasoning.

b. Without help from others in your group, find a plan you think is optimal for painting the lockers.

c. Compare your plan with those of others in your group. How are they alike? How are they different?

d. Write a list of criteria to decide whether a plan is optimal.

3. Suppose the lockers on the second floor of a high school are located as shown here. Suppose the equipment room located at G is at the bottom of a stairway leading to the second floor. Find two optimal plans for painting the lockers that satisfy the criteria you listed in Part d of the previous activity.

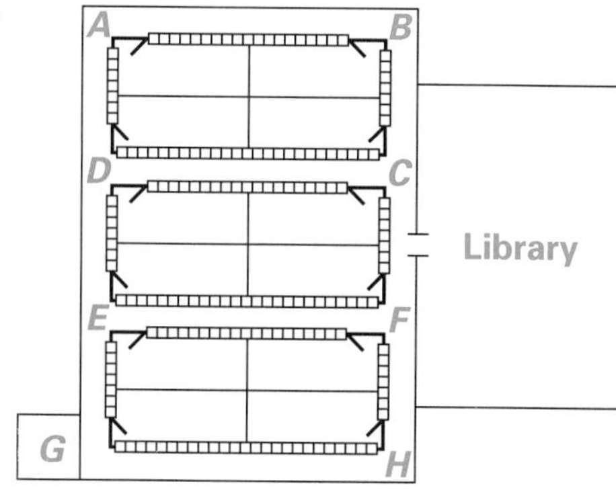

4. A **mathematical model** is a symbolic or pictorial representation including only the essential features of a problem situation. The floor-plan maps of the first and second floors of the school show the rows of lockers, classrooms, equipment room, hallways, outer walls, and other rooms. There are some features of these maps that you do not need in order to solve the locker-painting problems.

a. Which of the features of the maps did you use as you tried to solve the locker-painting problems? Which features were not needed?

b. Refer to the first-floor map of the school on page 251. Draw a simplified diagram (a mathematical model) that includes only the essential features of the locker-painting problem. For example, the lettered points on the map are important because E is the beginning and ending point. The other letters mark where one row of lockers ends and another begins. Complete the diagram.

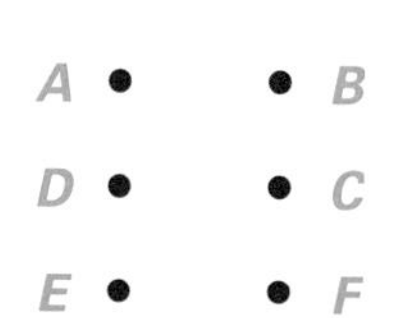

5. Now examine mathematical models drawn by some other students.

a. Michael drew the diagram at the right. Does his diagram show all the essential features of the locker-painting problem? If so, explain. If not, describe what is needed.

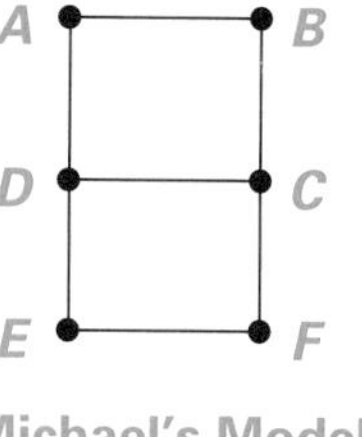

Michael's Model

b. Deonna drew the diagram at the right. Is it an appropriate model for the locker-painting problem? Explain.

c. Why do you think Deonna joined points C and D with two segments or arcs?

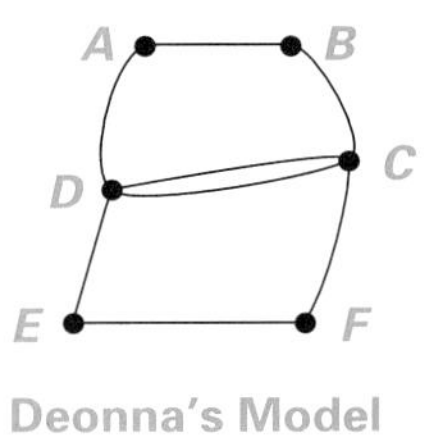

Deonna's Model

6. In Activity 2, you were asked to find an optimal plan for painting the first-floor lockers.

a. Use that plan to trace an optimal painting route on the diagram you drew in Activity 4. If you cannot trace your optimal route on your diagram, carefully check both your optimal plan and your diagram.

b. Trace the same painting route on Deonna's model. Does it matter if the points are connected by straight line segments or curved arcs? Does it matter how long the segments or arcs are?

7. To the right is a diagram that models one arrangement of lockers.

a. Draw a school floor-plan map that corresponds to this diagram. Assume that the equipment room is at V.

b. Find, if possible, an optimal route for painting these lockers.

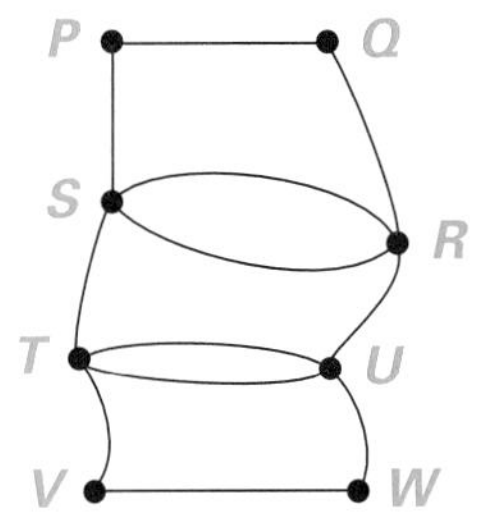

Checkpoint

In this investigation, you saw how special diagrams consisting of points and connecting segments and arcs can be used to model situations in which an efficient route is to be found.

a What is the difference between a floor-plan map of a school showing the lockers to be painted and a mathematical model of the locker-painting problem?

b Refer back to Deonna's model. What do the points and the connecting segments and arcs represent in terms of the locker-painting problem?

c Can two diagrams that have different shapes and sizes represent the same problem situation? Explain.

d In Activity 2, you wrote a list of criteria for an optimal locker-painting plan. Restate those criteria in terms of tracing around a diagram that models the situation.

Be prepared to share your thinking with the entire class.

A diagram consisting of a set of points along with segments or arcs joining some of the points is called a **graph**. The points are called **vertices**, and each point is called a **vertex**. The segments or arcs joining the vertices are called **edges**. A key step in modeling a problem situation with a graph is to decide what the vertices and edges will represent.

The word "graph" is used to mean different things at different times. In this unit, the word "graph" typically refers to a diagram consisting of vertices and edges. To clarify, the phrase *vertex-edge graph* will sometimes be used.

On Your Own

Suppose the lockers and an equipment room on the west wing of a high school are located as shown below.

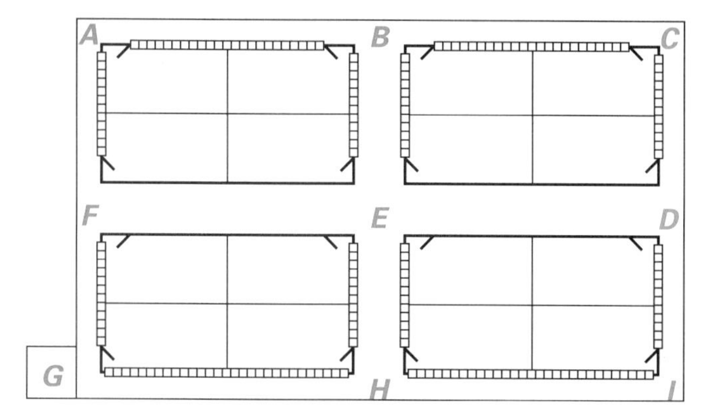

a. If you were to model the problem of painting these lockers with a vertex-edge graph, what would the vertices represent? The edges?

b. Draw a graph model for this problem.

c. Determine an optimal plan for painting the lockers. Use the criteria for tracing the edges and vertices of a graph that you arrived at in the Checkpoint on page 253.

INVESTIGATION 2 Making the Circuit

Your criteria for the optimal sequence for painting the lockers are the defining characteristics of an important property of a graph. An **Euler** (pronounced *oy'lur*) **circuit** is a route through a connected graph such that (1) each edge of the graph is traced exactly once, and (2) the route starts and ends at the same vertex. Given a connected graph, it often is helpful to know if it has an Euler circuit. (The name "Euler" is in recognition of the eighteenth-century Swiss mathematician Leonhard Euler. He was the first to study and write about these circuits.)

1. Shown below are graph models of the sidewalks in two sections of a town. Parking meters are placed along these sidewalks.

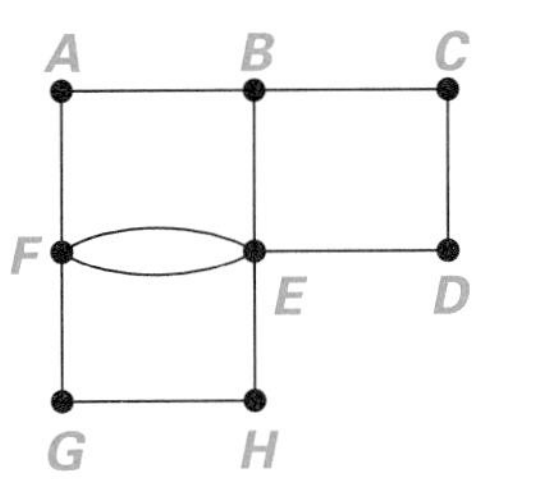

East Town Model

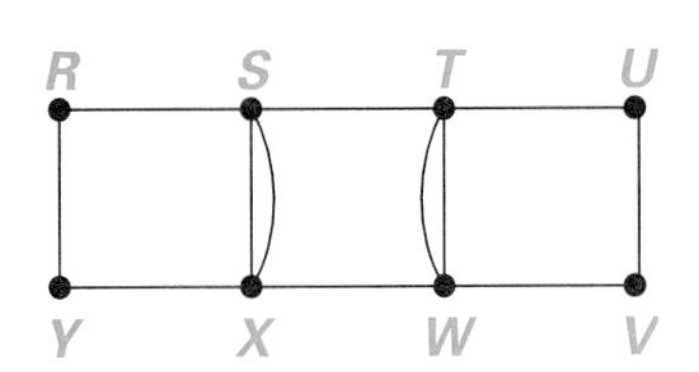

West Town Model

   a. Why would it be helpful for a parking control officer to know if these graphs had Euler circuits?

   b. Does the graph model of the east section of town have an Euler circuit? Explain your reasoning.

   c. Does the graph model of the west section of town have an Euler circuit? Explain.

2. Graphs a, b, and c below are similar to puzzles enjoyed by people all over the world. In each case, the challenge is to trace the figure. You must trace every edge exactly once without lifting your pencil and return to where you started. That is, the challenge is to trace an Euler circuit through the figure or graph. Place a sheet of paper over each graph and try to trace an Euler circuit. If the graph has an Euler circuit, write down the vertices in order as you trace the circuit.

a.

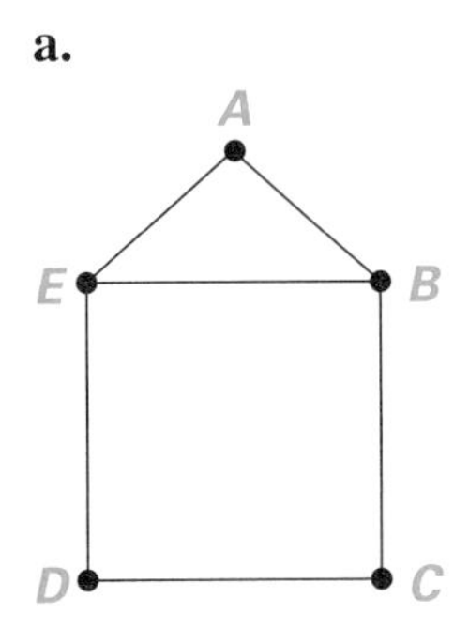

b.

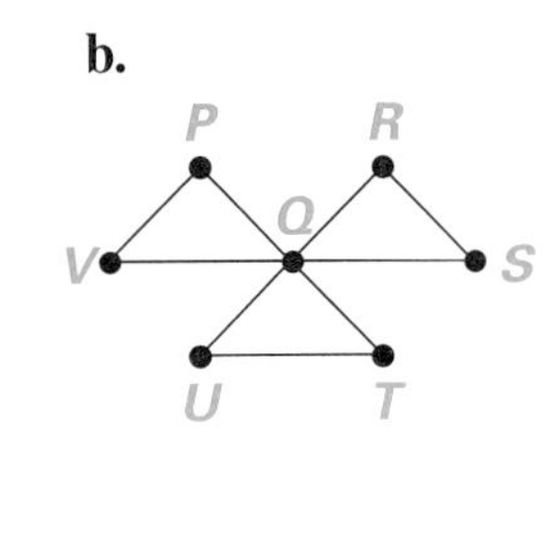

c.

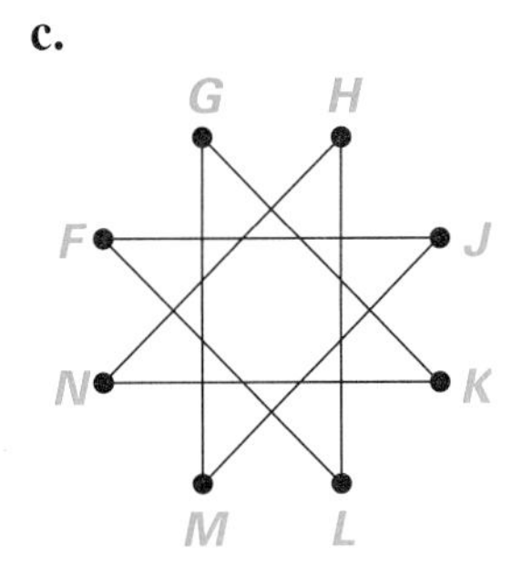

3. By looking at the form of an equation you often can predict the shape of the graph of the equation without plotting any points. Similarly, it would be helpful to be able to examine a vertex-edge graph and predict if it has an Euler circuit without trying to trace it.

a. Have each member of your group draw a graph with five or more edges that has an Euler circuit. On a separate sheet of paper, draw a graph with five or more edges that does *not* have an Euler circuit.

b. Sort your group's graphs into two piles, those that have an Euler circuit and those that do not.

c. Carefully examine the graphs in the two piles. Describe key ways that graphs with Euler circuits differ from those with no Euler circuit.

d. Try to figure out a way to predict if a graph has an Euler circuit simply by examining its vertices. Check your method of prediction using the graphs on page 255.

e. Make a conjecture about the properties of a graph that has an Euler circuit. Explain why you think your conjecture is true for *any* graph with an Euler circuit.

4. Once you can predict whether a graph has an Euler circuit, it is often still necessary to find the circuit. Consider the graphs below.

i.

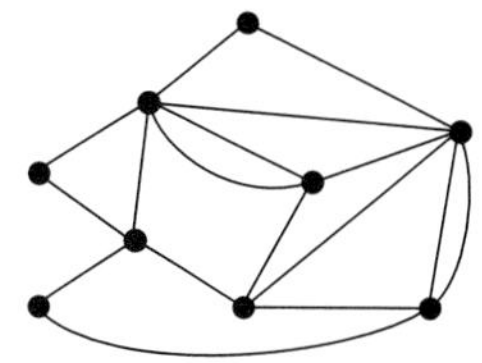

ii.

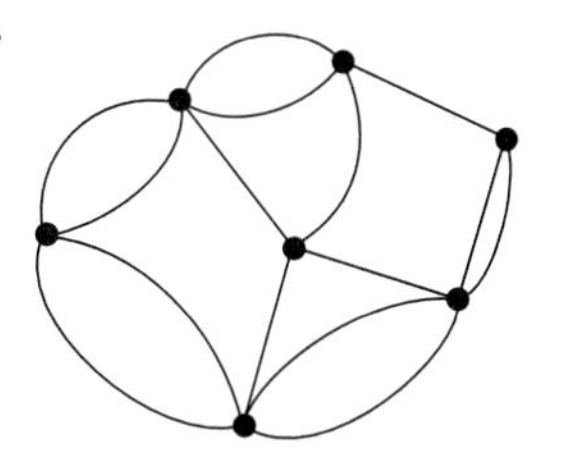

a. For each graph, predict whether it has an Euler circuit.

b. Find an Euler circuit in the graph that has one.

c. Describe the method you used to find your Euler circuit. Describe other possible methods for finding Euler circuits.

5. One systematic method for finding an Euler circuit is to trace the circuit in stages. For example, suppose you and your classmates want to find an Euler circuit that begins and ends at A in the graph below. You can trace the circuit in three stages.

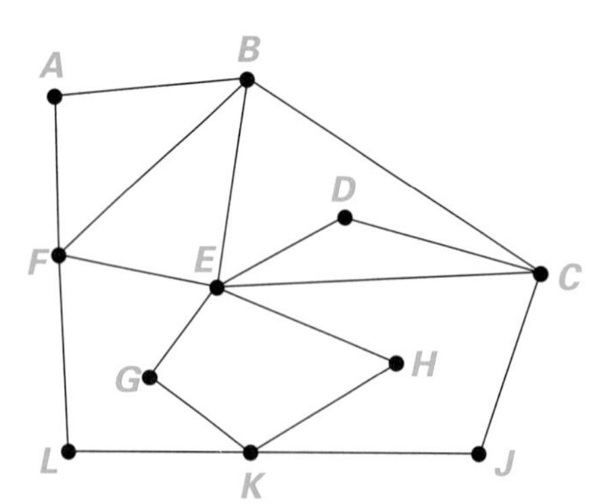

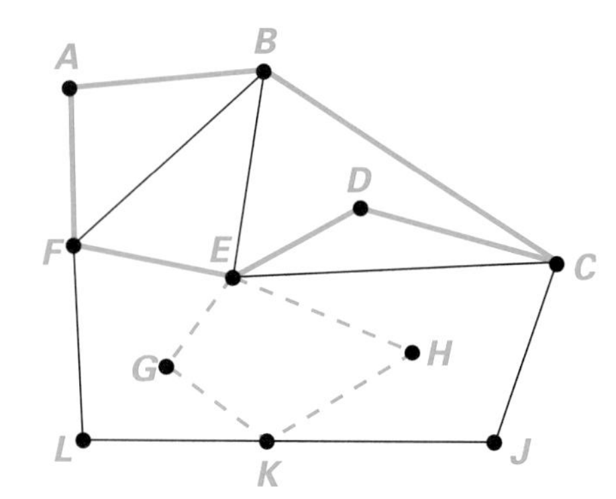

Stage I: Alicia began by drawing a circuit that begins and ends at A. The circuit she drew, shown in the diagram by the heavy edges, was A-B-C-D-E-F-A. But this does not trace all edges.

Stage II: George added another circuit shown by the dashed edges starting at E: E-G-K-H-E.

a. Alicia's and George's circuits can be combined to form a single circuit beginning and ending at A. List the order of vertices for that combined circuit.

Stage III: Since this circuit still does not trace each edge, a third stage is required.

b. Trace a third circuit which covers the rest of the edges.

c. Combine all the circuits to form an Euler circuit that begins and ends at A. List the vertices of your Euler circuit in order.

6. Choose your preferred method for finding Euler circuits from Activities 4 and 5. Write specific step-by-step instructions that describe the method you chose. Your instructions should be written so that they apply to *any* graph, not just the one that you may be working on at the moment. Such a list of step-by-step instructions is called an **algorithm**.

Checkpoint

It is possible to examine a graph to tell if it has an Euler circuit. If it does, there are algorithms to find such a circuit.

a How can you tell if a graph like the one at the right has an Euler circuit without actually trying to trace the graph?

b Use your algorithm from Activity 6 to find an Euler circuit in the graph.

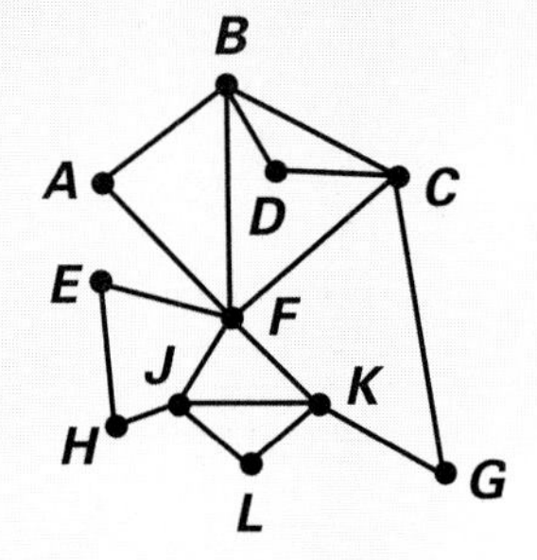

Be prepared to compare your method for determining if a graph has an Euler circuit and your algorithm for finding it with those of other groups.

Creating algorithms is an important aspect of mathematics. Two questions you should ask about any algorithm are *Does it work?* and *Is it efficient?* You will consider these questions in more detail later.

In devising a way to predict whether a graph has an Euler circuit, you probably counted the number of edges at each vertex of the graph. The number of edges touching a vertex is called the **degree of the vertex**. (If an edge loops back to the same vertex, that counts as two edge touchings. For an example, see Extending Task 4 on page 264.)

On Your Own

For each of the graphs below, check the degree of each vertex and then decide if an Euler circuit exists. If an Euler circuit exists, use your algorithm to find one.

a.

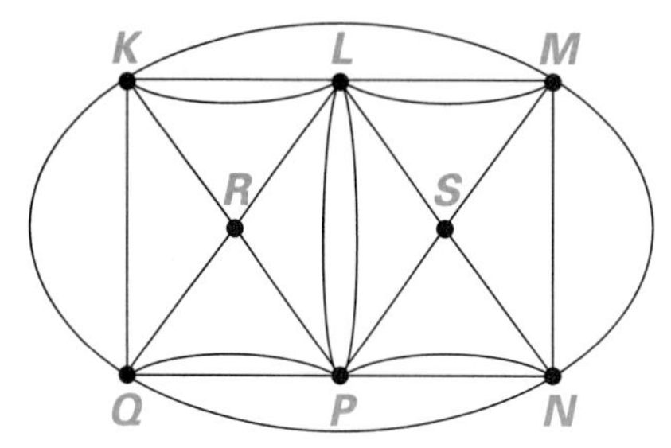

b.

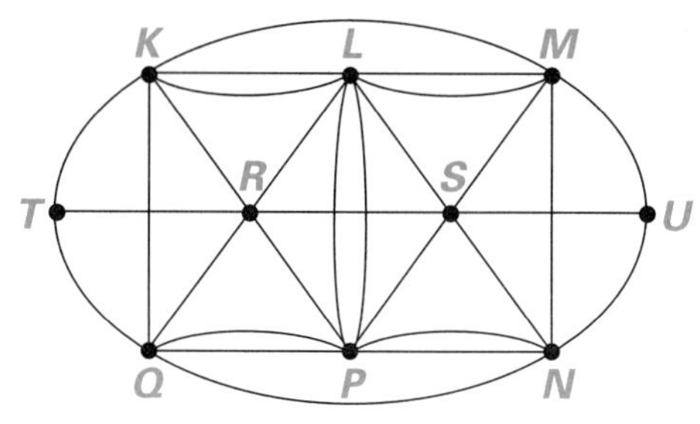

MORE

Modeling • Organizing • Reflecting • Extending

Modeling

1. Some popular puzzles involve trying to trace a figure without lifting your pencil or tracing an edge more than once, and starting and ending at the same vertex. That is, you try to find an Euler circuit.

a. Identify those graphs below and on the next page which you believe do not have an Euler circuit. Explain why you believe that.

b. For each of the graphs which has an Euler circuit, use the algorithm you developed to find a circuit.

i.

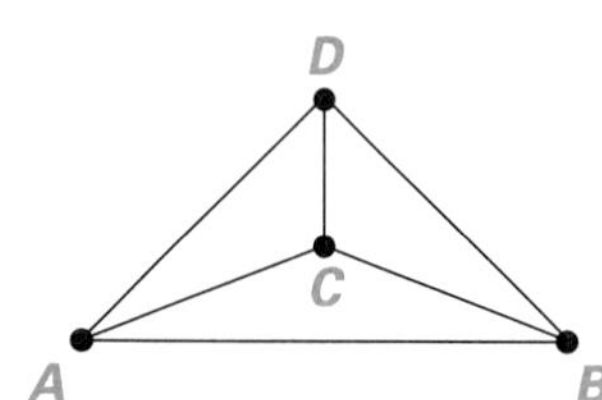

ii.

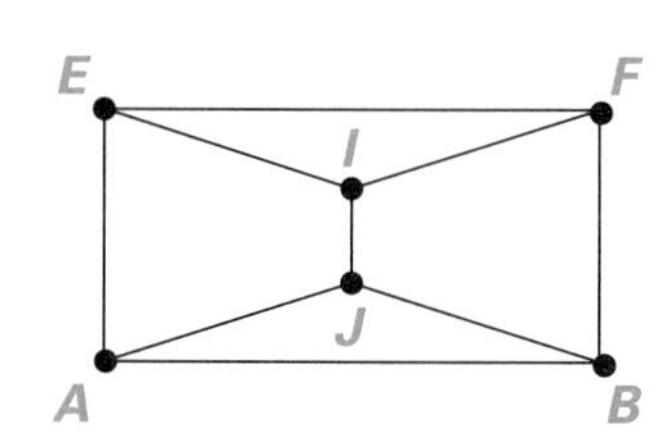

iii.

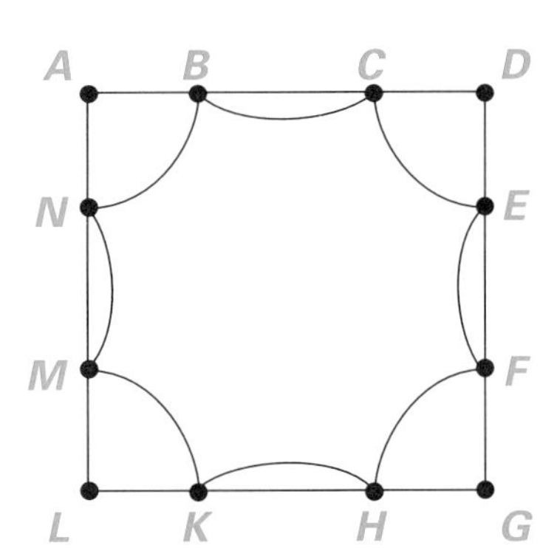

iv.

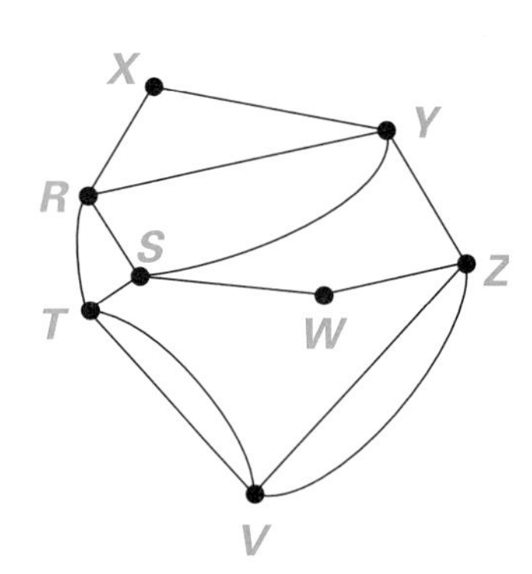

c. Draw two graphs that would be difficult to trace without lifting your pencil from the page or tracing an edge more than once. Draw one so that it has an Euler circuit and the other so that it does not. Challenge someone outside of class to trace your graphs, starting and ending at the same point. Then ask them to challenge you in the same way with any graph they draw. See if you can amaze them with how quickly you can tell whether or not it is possible to trace the graph without lifting your pencil or tracing any edge more than once.

2. The city of Kaliningrad in Russia is located on the banks and on two islands of the Pregel River. In the eighteenth century, the city was named Königsberg. Various parts of the city were connected by seven bridges as illustrated here. Citizens often would take walking tours of the city by crossing over the bridges. Some people wondered whether it was possible to tour the city by beginning at a point on land, walking across each bridge exactly once, and returning to the same point. As mentioned earlier, the problem intrigued the mathematician Leonhard Euler, who lived at that time. Euler used a graph model to solve the problem.

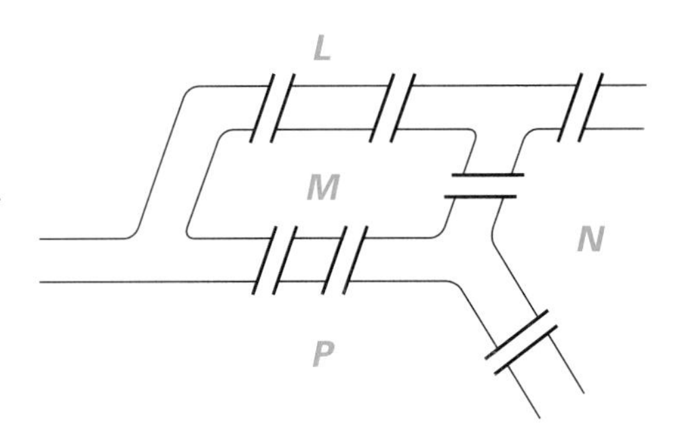

a. Draw a graph in which the vertices represent the four land areas (lettered in the figure) and the edges represent bridges.

b. What do you think Euler's solution was? Explain your response.

c. In the time since Euler solved the problem, two more bridges were built. One bridge was added at the left to connect areas labeled L and P. A second bridge was added to connect areas labeled N and P.

- Draw a graph model for this new situation of land areas and bridges.
- Use your graph to determine if it is possible to take a tour of the city that crosses each of the nine bridges exactly once and allows you to return to the point where you started.

3. Examine the school floor-plan showing locker placements and the graph model below.

a. If E represents the equipment room and C and D are the endpoints of the center hallway, identify the row(s) of lockers represented by each edge of the graph.

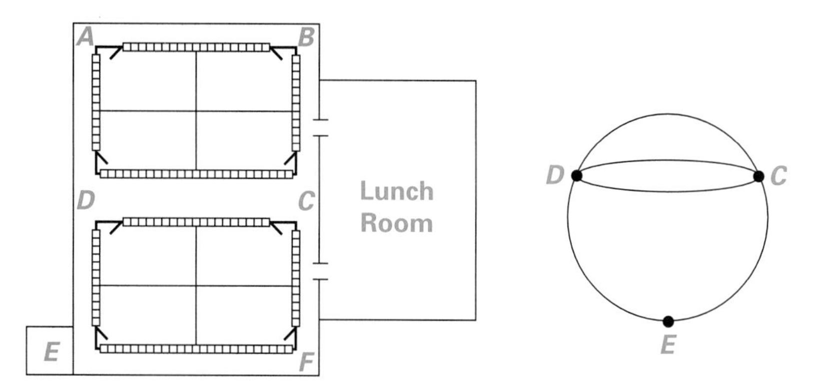

b. Trace a path around the graph that begins and ends at E and that traces each edge exactly once. Describe the order of painting the lockers that your path suggests. Is that order optimal?

c. Why is the graph still an accurate representation of the locker-painting problem when locations A, B, and F are not represented by vertices?

d. Are there any other points which could be deleted while maintaining an accurate graph model? Explain why or why not.

4. Suppose the lockers on the third floor of the high school in the locker-painting problem are located as shown here.

a. Draw a graph that represents this situation. Be sure to describe what the vertices and edges of your graph represent.

b. Is there a way to paint the lockers by starting and ending at the equipment room and never moving equipment down a hall without painting lockers on one side?

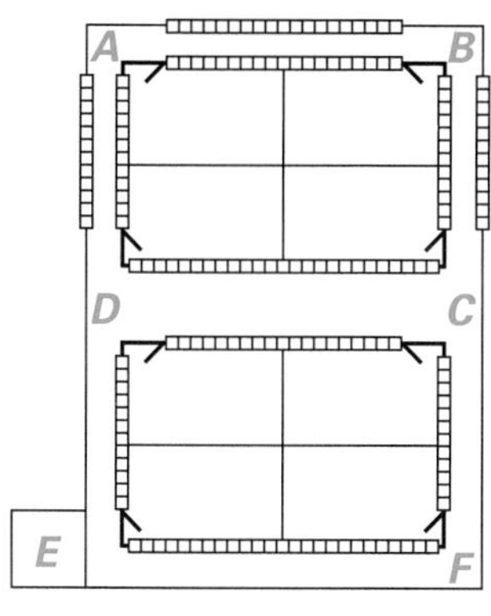

Third-Floor Lockers

5. A newspaper carrier wants to complete a delivery route without retracing steps. Some streets on the route have houses facing each other. Whenever there are houses on both sides of a street, papers must be delivered to both sides by going along one side and then along the other side.

a. Suppose the paper carrier only delivers to the houses on blocks 1, 2, and 3. Construct a vertex-edge graph model for this situation. What do the edges and vertices represent? Find an optimal delivery route.

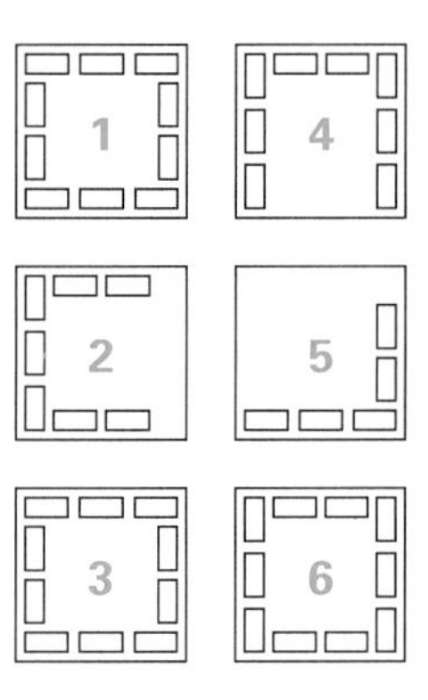

b. Suppose the paper carrier delivers to the houses on all six blocks. Construct a vertex-edge graph model for this situation. Find an optimal delivery route.

c. Now assume that *all* blocks have houses on all four sides and all streets continue in both directions. Add three more blocks that are adjacent to the given blocks on the street map. Find an optimal delivery route.

d. Can you find an Euler circuit no matter where the three new blocks are placed on the route? Explain your response.

e. Is it possible to place any number of new blocks on the route and still have an Euler circuit? Explain your reasoning.

Organizing

1. Graphs have interesting properties that can be discovered by collecting data and looking for patterns.

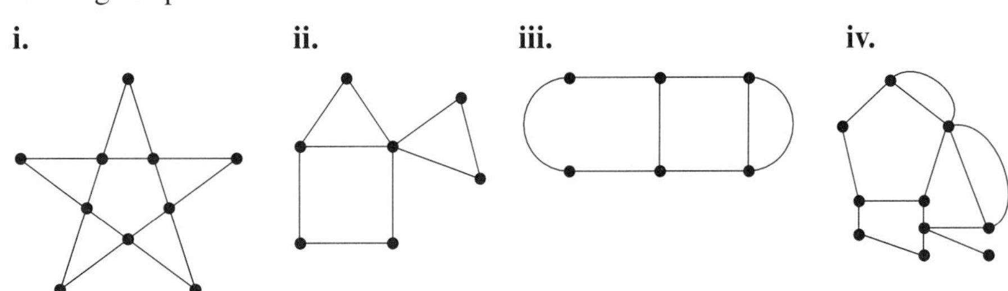

a. Complete a table like the one on the following page using the graphs above.

Graph	Sum of the Degrees of All Vertices	Number of Vertices of Odd Degree
i	30	
ii		2
iii		
iv		

b. Write down any patterns you see in the table.

c. Explain why the sum of the degrees of all the vertices in *any* graph is an even number.

d. Explain why *every* graph has an even number of vertices with odd degree.

2. The following graphs divide the plane into several *regions*. The exterior of the graph is an infinite region. The interior regions are enclosed by the edges. For example, graph i divides the plane into four regions.

i.

ii.

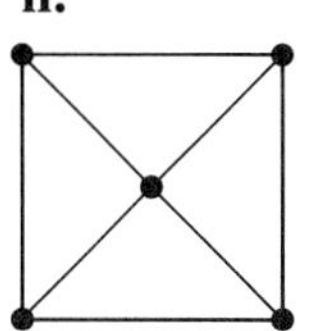

iii.

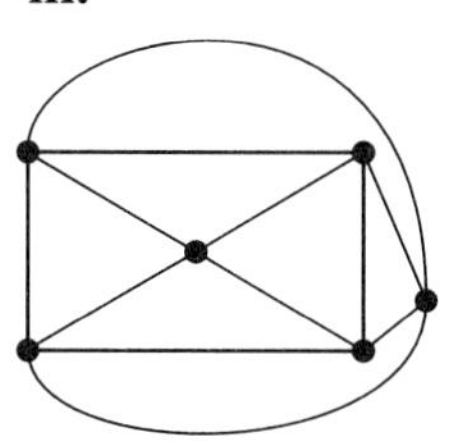

iv.

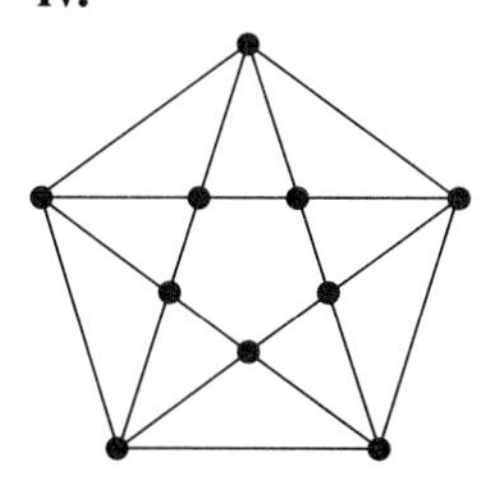

a. Complete the following table for each graph above. Be sure to count the exterior of the graph as one region.

Graph	Number of Vertices (V)	Number of Regions (R)	Number of Edges (E)
i			
ii			
iii			
iv			

b. Find a rule relating the numbers of vertices V, regions R, and edges E by using addition and subtraction to combine V, R, and E.

c. Draw several more graphs, and count V, R, and E. Does your rule also work for these graphs?

d. How many regions would be formed by a graph with 5 vertices and 12 edges? Draw such a graph to verify your answer.

3. Decide whether each of the following statements is true or false. If a statement is true, explain why it is true. If a statement is false, draw a graph that illustrates why it is false. (An example showing that a statement is false is called a **counterexample**.)

   a. Every vertex of a graph with an Euler circuit has degree greater than 1.

   b. If every vertex of a graph has the same degree, the graph has an Euler circuit.

4. The word "graph" in this unit means a diagram consisting of vertices and edges. List and draw sketches of the other types of graphs that you have used in this course.

Reflecting

1. What did you find most challenging in the first two investigations in this lesson? Why was it difficult for you?

2. Make a list of businesses or professions in which knowledge of Euler circuits might play an important role in lowering operating costs. Explain specifically how Euler circuits might be used in each case.

3. Write a question that you think would test whether your classmates understood the difference between graphs that have an Euler circuit and those that do not.

4. Explain Euler circuits to a friend or family member. Ask them to work a few problems like the ones you have worked. Write a summary of how well that person understood the ideas.

5. An Euler circuit is described as a certain kind of route through a connected graph. A **connected graph** is a graph that is all in one piece. That is, from each vertex there is at least one path to every other vertex. Draw a graph that is connected and one that is not. Why do you think Euler circuits are only considered for connected graphs?

Extending

1. Find information in the library or on the Internet about the life of Leonhard Euler. Write a report of what you find with particular attention to the contributions he made to graph theory.

2. Write an argument to support each of the following statements.

   a. If a graph has an Euler circuit, then all of its vertices are of even degree.

   b. If the vertices of a connected graph are all of even degree, then the graph has an Euler circuit.

3. Decide whether you agree with the following statement, and then write an argument to support your position: If a graph has an Euler circuit that begins and ends at a particular vertex, then it will have an Euler circuit that begins and ends at any vertex of the graph.

4. Many new housing developments have houses built on a street that is a "cul-de-sac" so that traffic past the houses is minimized.

a. Suppose a cul-de-sac is located at the end of the street between blocks 5 and 6 as shown here. Draw a vertex-edge graph that represents this housing development.

b. Find an optimal path for delivering papers to houses in this development.

c. You know from this lesson that the degree of a vertex is the number of edges that touch it, except that loops count as two edge touchings. Find the degree of each vertex in your graph model.

d. Repeat Parts a, b, and c with a second cul-de-sac constructed at the end of blocks 1 and 4.

e. How does adding a cul-de-sac affect the graph model?

f. Does the condition about degrees of vertices for graphs with Euler circuits still hold for graphs with loops?

5. Euler circuits are also useful in manufacturing processes where a piece of metal is cut with a mechanical torch. To reduce the number of times the torch is turned on and off, it is desirable to make the cut continuous. The metal piece must be clamped in air so that the torch does not burn the surface of the workbench. This leads to another condition; namely, any piece that falls off should not require additional cutting. Otherwise, it would have to be picked up and re-clamped, a time-consuming process. Find a way to make all the cuts indicated on the pictured piece of metal, so that you begin and end at S and the above conditions are satisfied.

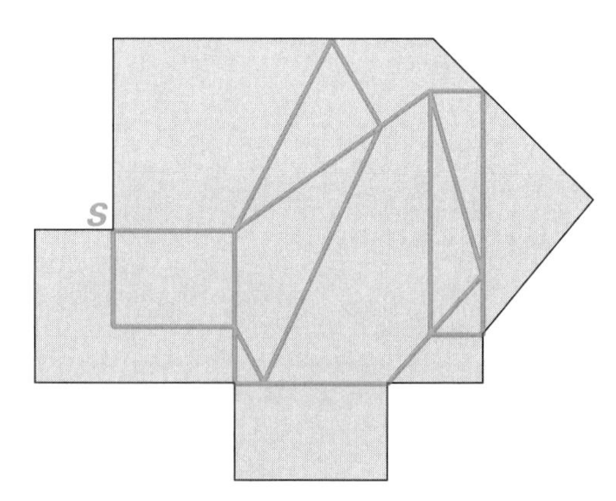

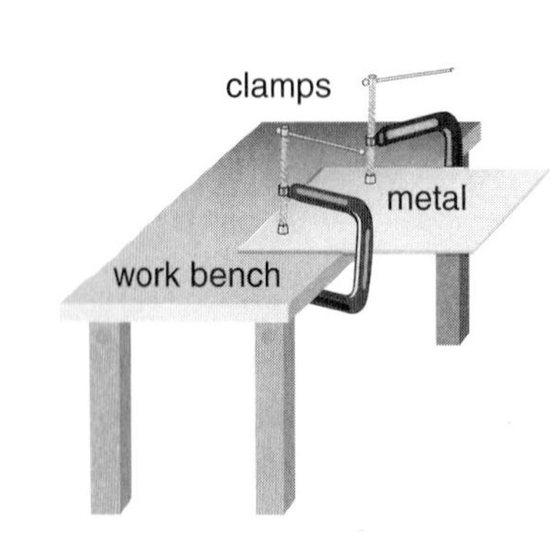

INVESTIGATION 3 Tracing Figures from One Point to Another

Graph-like figures have a rich and long history in many cultures, as illustrated in the following activities.

1. The Bushoong are a subgroup of the Kuba chiefdom in the Democratic Republic of Congo (changed from Zaire in 1997). Bushoong children have a long tradition of playing games that involve tracing figures in the sand using a stick. The challenge is to trace each line once and only once without lifting the stick from the sand. Two such figures are given below.

   Place a sheet of paper over the figures. Try to trace each figure without lifting your pencil and without any retracing.

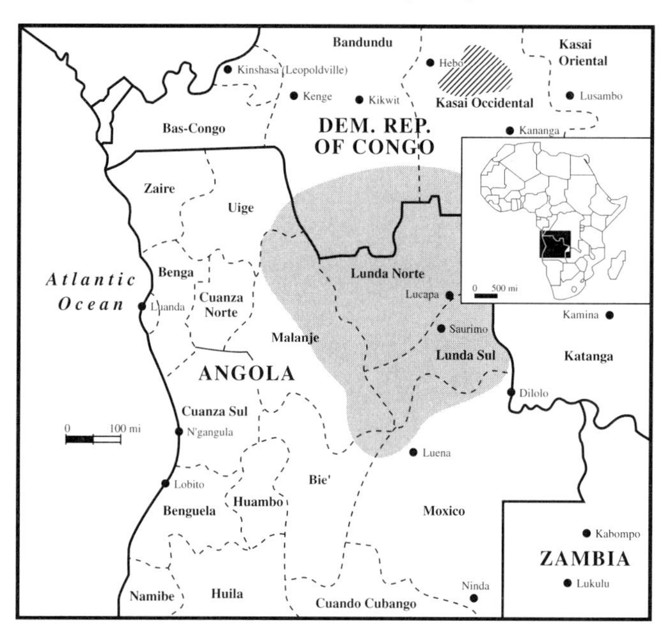

a.

b.

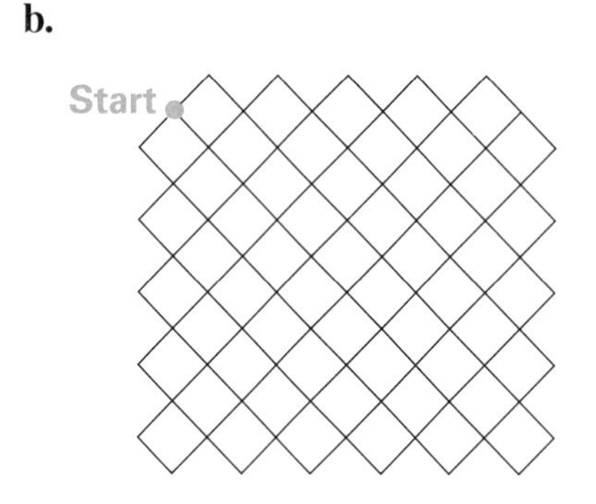

2. The Tshokwe (pronounced *shō - kwā′*) people, located near the Bushoong in Africa, made games of trying to trace pictures of objects. The two figures below represent intertwined rushes (tall flexible plants). Starting at S, is it possible to trace each figure without lifting your pencil? If so, where do you end your tracing?

a.

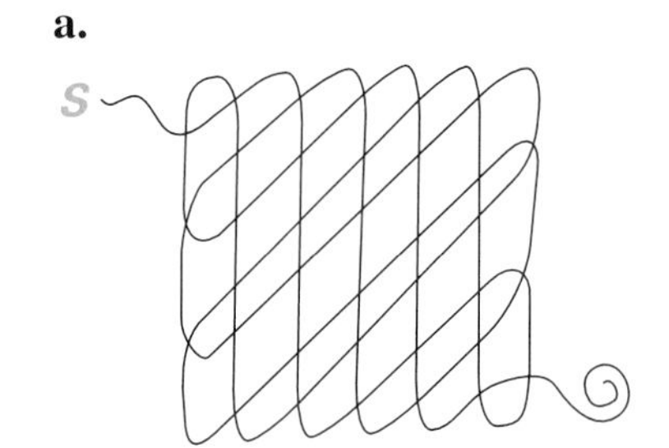

b.

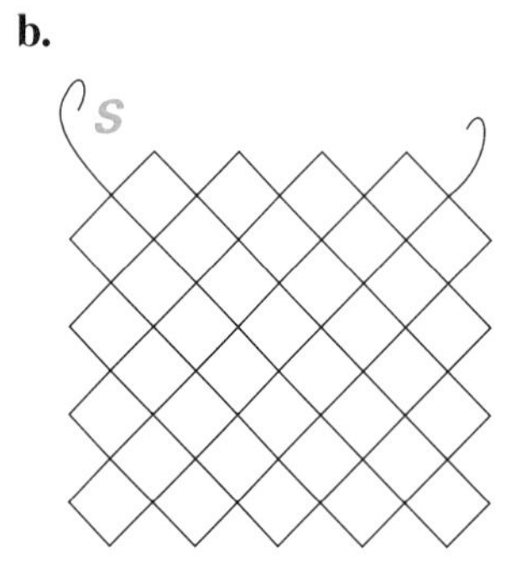

3. In Investigation 2, you discovered that some graphs do not have an Euler circuit.

   a. If you do not have to start and end at the same vertex, do you think the edges of every graph can be traced exactly once without lifting your pencil? Why or why not?

   b. Place a sheet of paper over the graphs below. Try to copy the graphs by tracing each edge exactly once.

   c. For those graphs which can be traced in this manner, how do the starting and ending vertices differ from the other vertices?

i.

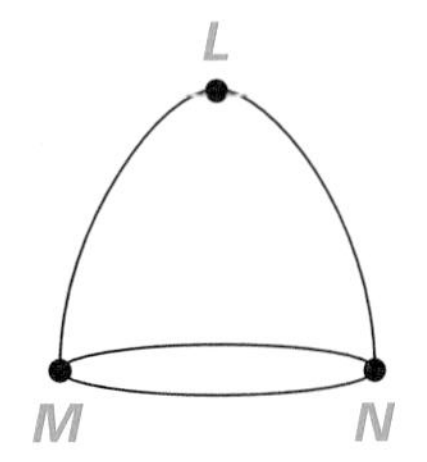

ii.

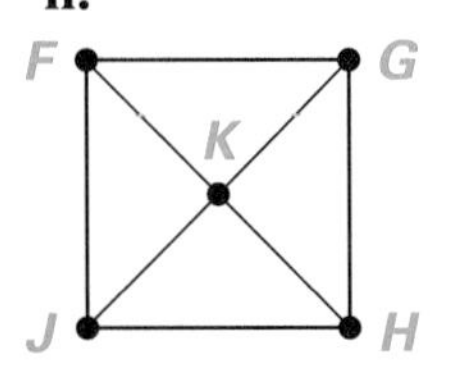

iii.

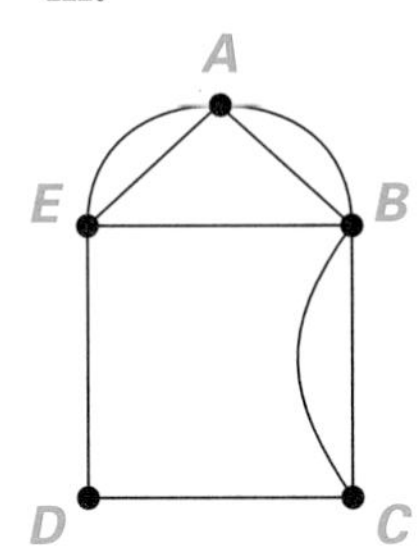

Checkpoint

You know that a graph has an Euler circuit whenever all vertices of the graph have even degree.

a Find a similar criterion for determining whether a graph *has* a traceable path when it does *not* have an Euler circuit.

b Test your criterion on the graphs in Activity 3 and on a graph that you create.

c Write an explanation for why your criterion is correct.

Be prepared to share your criterion and explanation with the entire class.

An **Euler path** is a path which traces each edge of the graph exactly once. Thus, an Euler circuit is a special type of Euler path—one which starts and ends at the same vertex.

On Your Own

Draw three graphs, each with at least five vertices, that meet the following criteria:

- One has an Euler circuit.
- One has an Euler path but no Euler circuit.
- One has neither an Euler circuit nor an Euler path.

Explain to a classmate or to someone at home which is which and why.

For the remainder of this investigation, work with your group to explore how to revise a graph so that it has an Euler circuit.

4. The graph shown here is a model of the arrangement of lockers along hallways of the second floor of a high school.

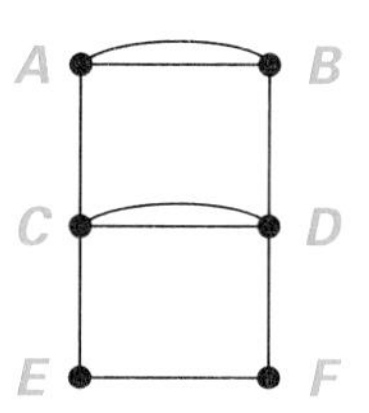

a. Explain why these lockers *cannot* be painted by starting and ending at the equipment room E and never moving down a hall without painting lockers on one side.

b. Find a way to paint the lockers so that the process starts and ends at E, and the number of already-painted rows along which the equipment must be moved is as small as possible. Write down the route that you would walk.

c. Revise the graph to represent your route. What do the vertices and edges of the revised graph represent?

d. How many additional edges did you add? Can you use fewer additional edges?

The process of revising a graph by adding edges so that the revised graph has an Euler circuit is called **Eulerizing** the graph. When Eulerizing a graph, you should only add edges that are duplicates of existing edges.

5. The graph at the right represents a network of streets.

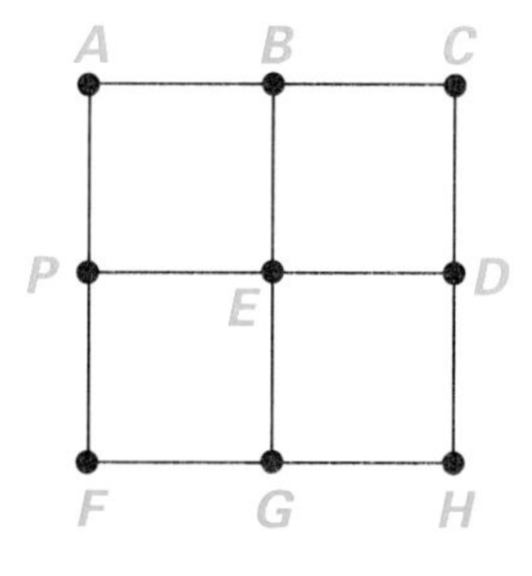

a. Is it possible for a police car to patrol the streets beginning and ending at P (a police station) and traveling each street exactly once? Explain your reasoning.

b. Albert proposed the solution route at the right in response to the problem. Is this an acceptable solution? Why or why not?

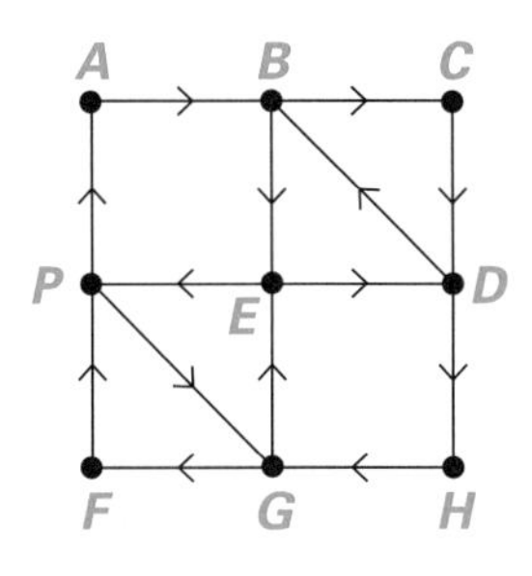

c. Design a route for a police car so that (i) it begins and ends at P; (ii) it only uses the existing streets shown in the graph in Part a; (iii) it travels each street at least once; and (iv) the number of streets it travels more than once is as small as possible.

d. Provide an argument for why the number of streets your route travels more than once is as small as possible.

Checkpoint

A graph that does not have an Euler circuit can be Eulerized by adding appropriate edges.

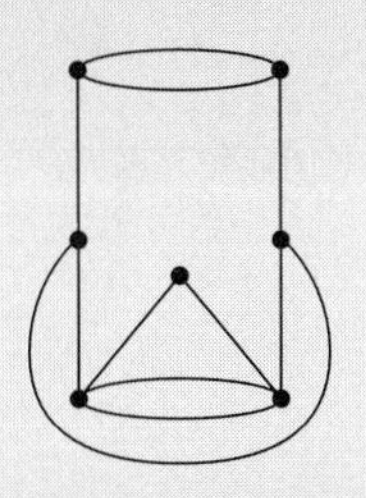

a As a group, write an algorithm to Eulerize a graph.

b Test your algorithm by Eulerizing the graph shown here.

Be prepared to compare your algorithm with those of other groups.

On Your Own

Eulerize each of the following graphs.

a.

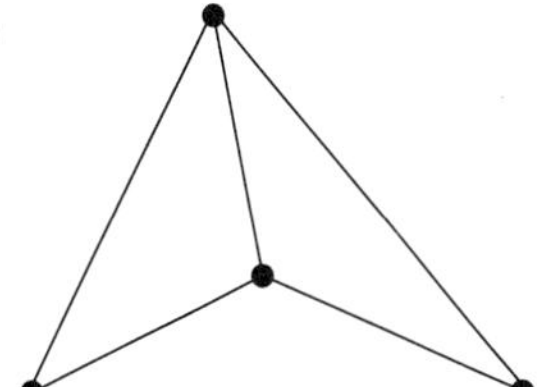

b.

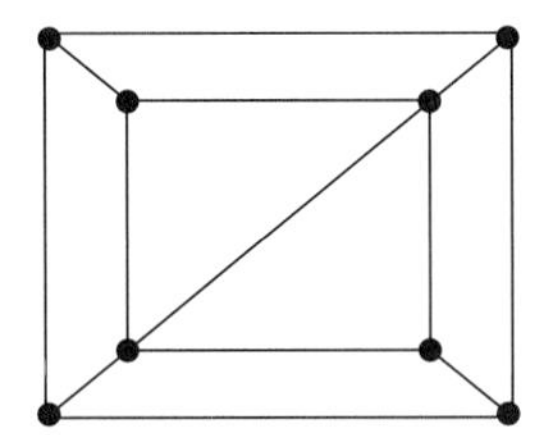

INVESTIGATION 4 Graphs and Matrices

Information often is organized and displayed in tables. The use of tables to summarize information can be seen in almost every section of most newspapers. In this investigation, you will explore how table-like arrays also can be used to represent information contained in vertex-edge graphs.

1. As a group, examine this information on medals awarded at the 1998 Winter Olympics.

Medal Count

Country	G	S	B	Total
Germany	12	9	8	29
Norway	10	10	5	25
Russia	9	6	3	18
Austria	3	5	9	17
Canada	6	5	4	15
United States	6	3	4	13

Source: *The World Almanac and Book of Facts.* Mahwah, NJ: World Almanac Books, 2000.

a. What do each of the numbers in the first row represent?

b. What is the meaning of the number in the fifth row and second column? (Don't count the row and column headings.) In the third row and third column?

c. China did not win any gold medals. However, the Chinese team did take home six silver medals and two bronze medals. How could you modify this chart to include this additional information?

2. Arrays of numbers, like the one above, are sometimes called **matrices**. Matrices can be used to represent graphs.

One way in which a graph can be represented by a **matrix** is shown here.

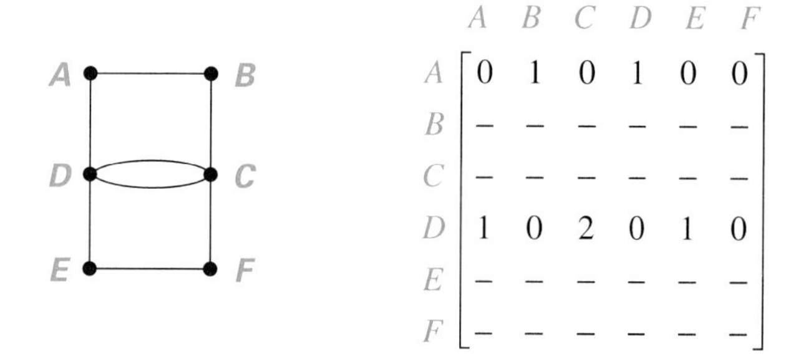

a. Study the first and fourth rows of the matrix. Explain what each entry means in terms of the graph.

b. Copy the matrix and then fill in the missing entries.

c. Construct a similar matrix for each of the three graphs below.

i.

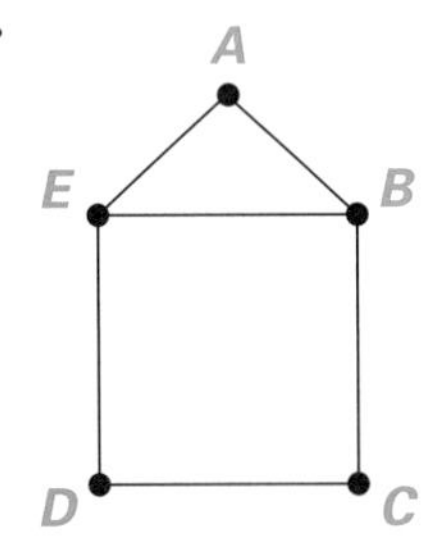

ii.

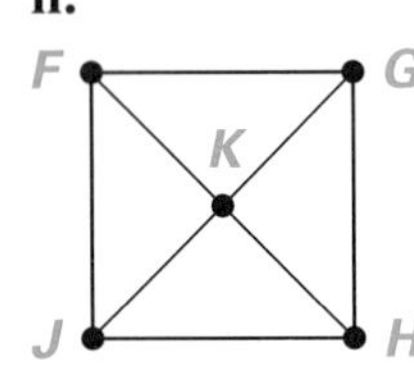

iii.

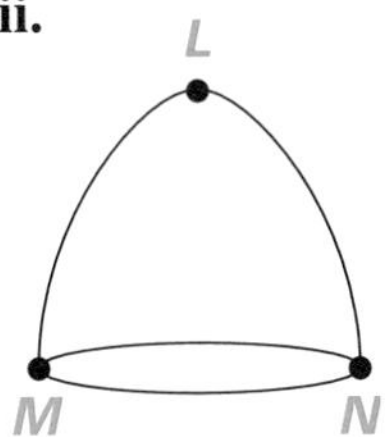

3. The sums of the numbers in each row of a matrix are called the **row sums** of the matrix.

a. Find the row sums of each of the matrices in Part c of Activity 2. What do these row sums represent in the graphs?

b. Is it possible to tell by looking at the matrix for a graph whether the graph has an Euler path or an Euler circuit? Explain your response.

Checkpoint

In this investigation, you saw how a matrix can be used to represent and help analyze a graph.

a A matrix corresponding to a graph that has five vertices, A, B, C, D, and E, in that order, has a 2 in the third row, fifth column. What does the 2 represent? What does a 1 in the first row, second column mean?

b Explain the differences between the row sums of matrices for graphs with and without Euler circuits. Explain the differences between the row sums for graphs with and without Euler paths.

Be prepared to share your thinking with the entire class.

Matrices like those you have been constructing are called **adjacency matrices**. Each entry in an adjacency matrix for a graph is the number of direct connections (edges) between the corresponding pair of vertices.

On Your Own

Check your understanding of adjacency matrices by completing the tasks at the top of the next page using the matrices below.

i.

$$\begin{array}{c} \\ A \\ B \\ C \end{array} \begin{array}{c} \begin{array}{ccc} A & B & C \end{array} \\ \begin{bmatrix} 0 & 2 & 0 \\ 2 & 0 & 1 \\ 0 & 1 & 0 \end{bmatrix} \end{array}$$

ii.

$$\begin{array}{c} \\ P \\ Q \\ R \\ S \end{array} \begin{array}{c} \begin{array}{cccc} P & Q & R & S \end{array} \\ \begin{bmatrix} 0 & 1 & 2 & 1 \\ 1 & 0 & 1 & 2 \\ 2 & 1 & 0 & 2 \\ 1 & 2 & 2 & 0 \end{bmatrix} \end{array}$$

a. Does each of the graphs whose adjacency matrix is given at the bottom of page 270 have an Euler circuit? An Euler path? How can you tell without drawing the graphs?

b. Draw and label a graph corresponding to each adjacency matrix in Part a. Find an Euler circuit or Euler path if there is one.

c. If a graph has an Euler path, can you tell from the adjacency matrix at what vertex the path begins or ends? Explain.

MORE Modeling • Organizing • Reflecting • Extending

Modeling

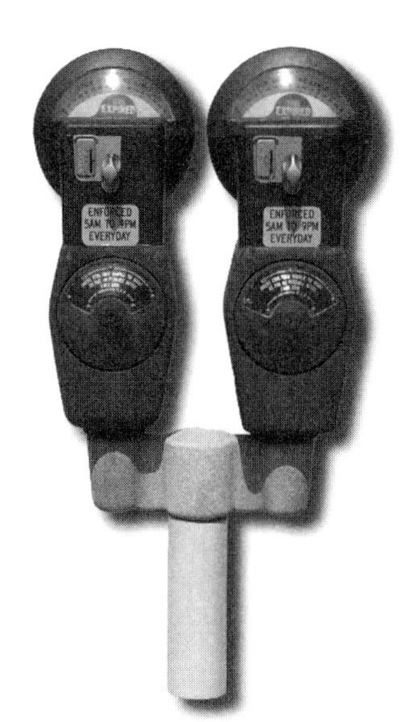

1. The diagram shown here represents the streets of downtown Springfield. The ovals represent parking meters. A city employee, Dianna, must regularly check these meters for expiration.

a. Model the situation of checking the meters with a graph. Use the vertices to represent where one block ends and another block begins. Use the edges to represent the presence of meters.

b. What is the most efficient route for Dianna to use when checking the meters?

c. Is this route an Euler path? Explain.

d. Eulerize the graph.

2. An auditorium floor is arranged for an art show as diagrammed in the floor plan at the right. Artwork will be exhibited on all four sides of the display boards (shown as the three rectangles). The art show organizers plan to mark off a route for visitors to follow so that they can view each piece of art exactly one time. Assume visitors can view the art displayed on only one side of an aisle on each walk down the aisle.

a. Model this situation with a graph. What do the vertices represent? What do the edges represent?

b. Suppose visitors hang up their coats outside the West door and pick them up there on their way out. How would you route visitors to allow for viewing each piece of art exactly one time?

c. Is it possible to route people from the West Door to the South Door so that they can view all of the artwork while not going past a row of exhibits more than once? How does the graph model show this?

d. As an exhibit coordinator, how would you add, delete, or rearrange the display boards so that a route exists that satisfies the conditions in Part c?

e. Is the route that you have designed an Euler path? Explain.

3. The map below shows the trails in Tongass State Park. The labeled dots represent rest areas scattered throughout the park.

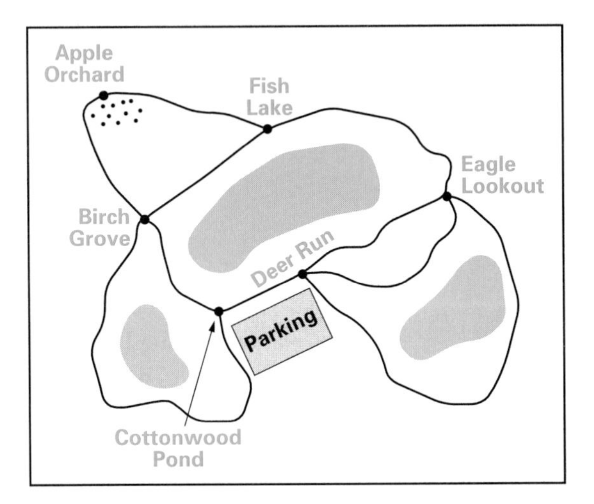

a. How would a graph model of this situation differ from the map? Is it necessary or useful to draw a graph model in this situation? Why or why not?

b. Construct an adjacency matrix related to the park map.

c. Is it possible to hike each of the trails in the park once and return to your car in the parking lot? Explain your answer by using the adjacency matrix from Part b and your knowledge of Euler paths and Euler circuits.

d. The Park Department has received money to build additional trails. Between which rest stops should they build a new trail (or trails) so that people can hike each trail once and return to their cars?

e. Does your solution to Part d Eulerize the graph? Why or why not?

f. Find two ways to Eulerize the graph. For each way, which paths would be repeated?

4. Certain towns in southern Alaska are on islands or isolated by mountain ranges. When traveling between these communities, you must take a boat or a plane. Listed below are the routes provided by a local airline.

Routes between:

Anchorage and Cordova
Anchorage and Juneau
Cordova and Yakutat
Juneau and Ketchikan
Juneau and Petersburg
Juneau and Sitka
Petersburg and Wrangell
Sitka and Ketchikan
Wrangell and Ketchikan
Yakutat and Juneau

a. Make a graph model of the airline routes.

b. In what ways is your graph model like the map? In what ways is it different?

c. An airline inspector wants to evaluate the airline's operations by flying each route. It is sufficient to fly each route one-way. Can the inspector start in Juneau, fly all the routes exactly once, and end in Juneau?

d. How would an adjacency matrix for the graph show whether or not there is a route as described in Part c?

Organizing

1. In this task, you will examine further the tradition of tracing continuous figures exhibited in cultures around the world.

a. The Malekula live on an island in the South Pacific chain of some eighty islands that comprise the Republic of Vanuatu. As with the Bushoong and Tshokwe in Africa, the Malekula also have figures that represent objects or symbols of the culture. For example, figure i represents a yam. Figure ii is called "the stone of Ambat."

- Can you trace each of these figures without lifting your pencil or repeating any edges?
- Describe any *symmetry* you see in each figure. Be as complete with your descriptions as possible.

i.

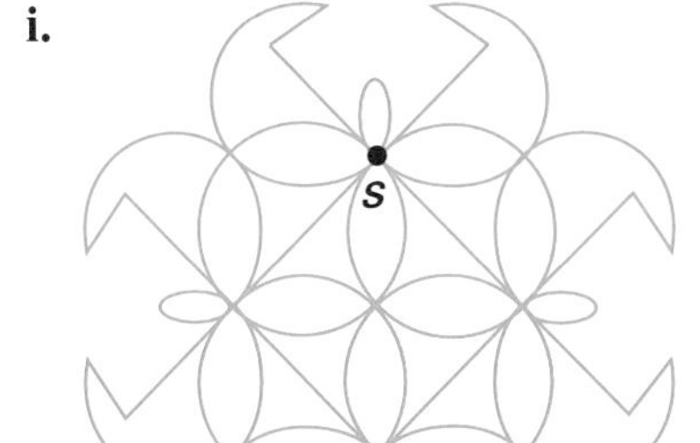

ii.

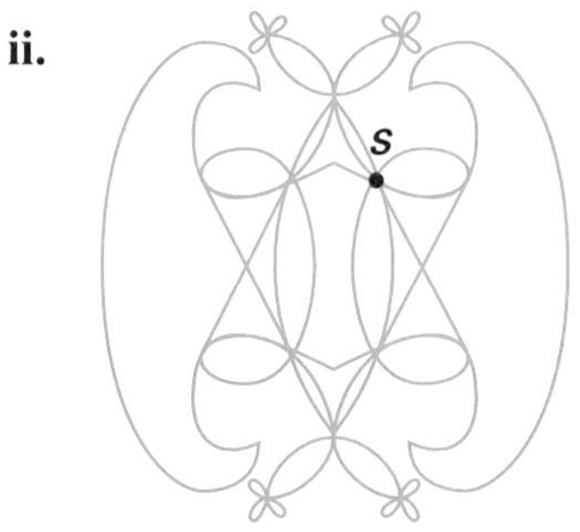

b. The ancient inhabitants of Pre-Inca Peru built ground-cover figures, some hundreds of yards in length. For each of the two figures below:

- discuss its traceability;
- discuss its symmetry;
- discuss how it seems to be related to similar figures you examined from other cultures.

i. **ii.**

2. Explain why the row sum of an adjacency matrix of a graph is the degree of the vertex corresponding to that row.

3. Consider the regular pentagon at the right as a graph.

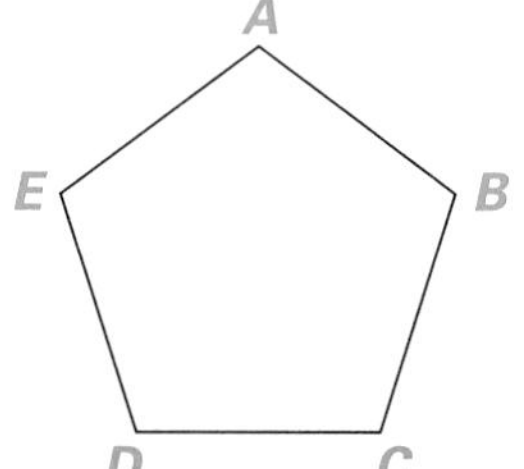

a. Write the adjacency matrix for this graph.

b. Modify a copy of this graph by adding all the *diagonals* (segments connecting pairs of vertices). Write the adjacency matrix for this modified graph.

c. Write a description of the adjacency matrix for a graph in the shape of a regular polygon with n sides. How would you modify the description of the adjacency matrix if the graph consisted of the polygon *and* its diagonals?

4. Eulerize the graph at the right using a minimum of repeated edges. Write an argument for why your solution involves the minimum number of repeated edge-tracings.

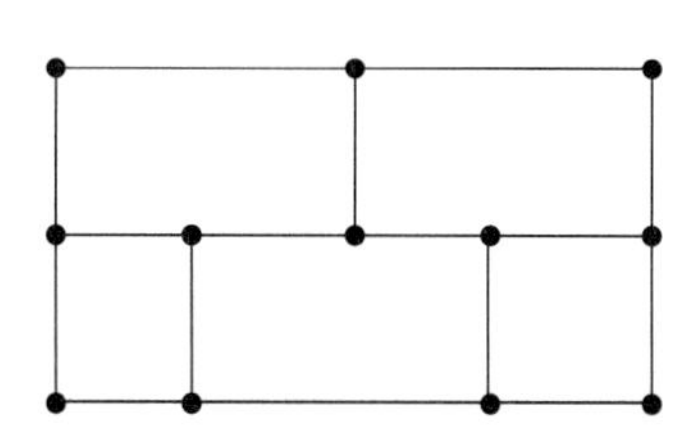

Reflecting

1. In a world atlas, find the countries mentioned earlier in this unit: The Democratic Republic of Congo in Africa, Peru in South America, and the Republic of Vanuatu in the South Pacific. Why do you think the concept of tracing a continuous line to create a pattern is evident in so many cultures?

2. Obtain a map of your town or city or of a nearby town or city (perhaps from a telephone book). Select one section of the town or city that is approximately 6 blocks by 6 blocks and design an efficient street-sweeping route for that area. Then design an efficient postal-carrier walking route for that area. Discuss some of the reasons why these two routes may be different.

3. Answer the following, based on your experiences in the investigations of this lesson.

   a. Is it possible for a vertex-edge graph to be a mathematical model for two different situations? Explain your reasoning.

   b. Is it possible for a matrix to be the adjacency matrix for two different vertex-edge graphs? Explain.

4. In what real-world situations would it be important to have an Euler circuit rather than a non-circuit Euler path?

Extending

1. Identify a real-world application of Euler paths or circuits, different from those in this lesson. Prepare a class presentation about how Euler paths or circuits are used in that application.

2. The stone of Ambat (figure ii in Organizing Task 1 Part a) can be traced in three stages, each beginning and ending at S. Two of the stages are given at the right. Find the third stage and then put them all together to produce a complete tracing.

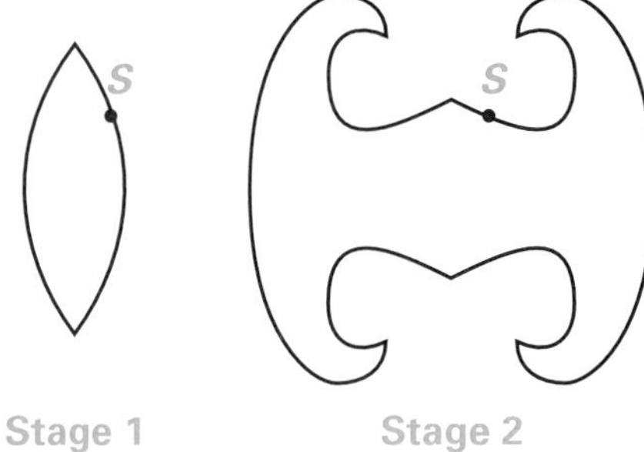

3. A **loop** is an edge connecting a vertex to itself. When constructing an adjacency matrix for a graph with loops, a 1 is placed in the position in the matrix that corresponds with an edge joining a vertex to itself. An example of such a graph and its adjacency matrix is shown at the right. (Extending Task 4 on page 264 presents a practical situation that can be modeled by a graph with a loop.)

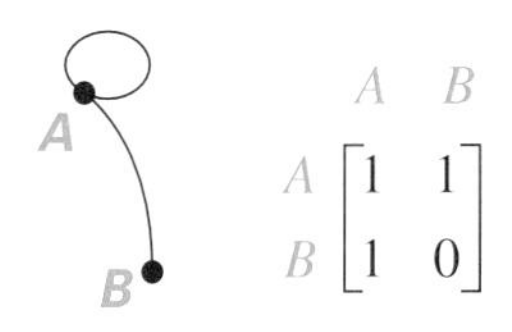

$$\begin{array}{cc} & \begin{array}{cc} A & B \end{array} \\ \begin{array}{c} A \\ B \end{array} & \begin{bmatrix} 1 & 1 \\ 1 & 0 \end{bmatrix} \end{array}$$

   a. You have learned in this unit that the degree of a vertex is the number of edges touching the vertex, except that a loop counts for two edge touchings. What is the degree of vertex A?

b. What is the row sum of the first row of the adjacency matrix? In Investigation 4 of this lesson, you found a connection between row sums of an adjacency matrix and the degree of the corresponding vertex. Does this connection still hold for graphs with loops like the one on the previous page?

c. Try to draw graphs with the following adjacency matrices.

i. $\begin{bmatrix} 2 & 3 \\ 3 & 0 \end{bmatrix}$ **ii.** $\begin{bmatrix} 0 & 1 & 2 \\ 1 & 1 & 1 \\ 2 & 1 & 0 \end{bmatrix}$ **iii.** $\begin{bmatrix} 0 & 2 & 1 \\ 2 & 0 & 2 \\ 1 & 1 & 0 \end{bmatrix}$

d. Some matrices cannot be adjacency matrices for graphs. Write a description of the characteristics of a matrix that *could* be the adjacency matrix for a graph.

4. Investigate the minimum number of edges needed to Eulerize graphs in the shape of polygons with all diagonals drawn from one vertex.

a. What is the minimum number of edges needed to Eulerize a graph in the shape of a pentagon with all the diagonals from one vertex as shown below? In the shape of a hexagon (6 sides) with all diagonals from one vertex? A heptagon (7 sides)? An octagon (8 sides)?

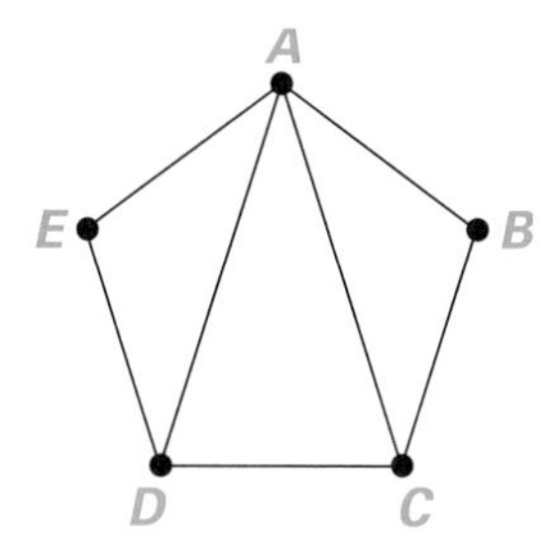

b. Organize your data from Part a in a table. Do you see any pattern relating the number of vertices and minimum number of edges needed to Eulerize these graphs? If so, write a rule. If not, collect additional data until you see a pattern and can write a rule.

c. Suppose you know the minimum number of edges needed to Eulerize a polygon with n vertices with all diagonals from one vertex. Is it possible to write an equation using *NOW* and *NEXT* to describe the minimum number of edges needed to Eulerize a polygonal graph of this sort with $n + 1$ vertices? Explain your response.

Lesson 2 Managing Conflicts

Have you ever noticed how many different radio channels there are? Each radio station has its own transmitter which broadcasts on a particular channel, or frequency.

The Federal Communications Commission (FCC) makes sure that the broadcast from one radio station does not interfere with the broadcast from any other radio station. This is done by assigning an appropriate frequency to each station. The FCC requires that stations within transmitting range of each other must use different frequencies. Otherwise, you might tune into "ROCK 101.7" and get Mozart instead!

Think About This Situation

Seven new radio stations are planning to start broadcasting in the same region of the country. The FCC wants to assign a frequency to each station so that no two stations interfere with each other. The FCC also wants to assign the fewest possible number of new frequencies.

a What factors need to be considered before the frequencies can be assigned?

b What method can the FCC use to assign the frequencies?

INVESTIGATION 1 Building a Model

Suppose that because of geographic conditions and the strength of each station's transmitter, the FCC determines that stations within 500 miles of each other must be assigned different frequencies. Otherwise their broadcasts will interfere with each other. The locations of the seven stations are shown on the grid at the right. A side of each small square on the grid represents 100 miles.

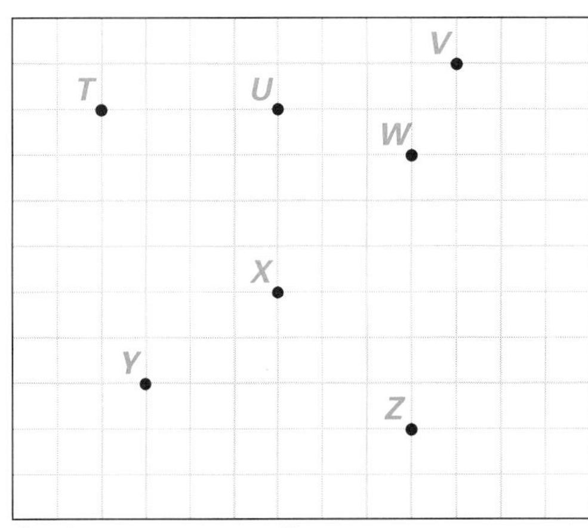

Scale: ⊢⊣ = 100 miles

1. Working on your own, figure out how many different frequencies are needed for the seven radio stations. Remember that stations 500 miles or *less* apart must have different frequencies. Stations more than 500 miles apart can use the same frequency. *Try to use as few frequencies as possible.*

2. Compare your answer with others in your group.

   a. Did everyone use the same number of frequencies? Reach agreement in your group about the minimum number of frequencies needed for the seven radio stations.

   b. Suppose one person assigns two stations the same frequency and another person assigns them different frequencies. Is it possible that both assignments are acceptable? Explain.

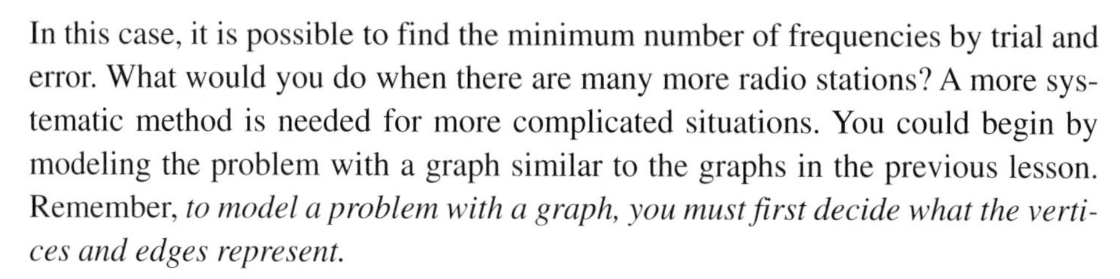

In this case, it is possible to find the minimum number of frequencies by trial and error. What would you do when there are many more radio stations? A more systematic method is needed for more complicated situations. You could begin by modeling the problem with a graph similar to the graphs in the previous lesson. Remember, *to model a problem with a graph, you must first decide what the vertices and edges represent.*

3. Working on your own, begin modeling this problem with a graph.

   a. What should the vertices represent?

   b. How will you decide whether or not to connect two vertices with an edge? Complete this statement:

   Two vertices are connected by an edge if. . . .

   c. Now that you have specified the vertices and edges, draw a graph for this problem.

4. Compare your graph with others in your group.

   a. Did everyone in your group define the vertices and edges in the same way? Discuss any differences.

   b. For a given situation, suppose two people define the vertices and edges in two different ways. Is it possible that both ways accurately represent the situation? Explain your reasoning.

   c. For a given situation, suppose two people define the vertices and edges in the same way. Is it possible that their graphs look different but both are correct? Explain your reasoning.

5. A common choice for the vertices is to let them represent the radio stations. Edges might be thought of in two ways as described in Parts a and b below.

a. You might connect two vertices by an edge whenever the stations they represent are 500 miles or *less* apart. Did anyone in your group do this? If not, draw a graph where two vertices are connected by an edge whenever the stations they represent are 500 miles or *less* apart.

b. You might connect two vertices by an edge whenever the stations they represent are *more* than 500 miles apart. Did anyone in your group do this? If not, draw a graph where two vertices are connected by an edge whenever the stations they represent are *more* than 500 miles apart.

c. Compare the graphs from Parts a and b.

- Are both graphs accurate ways of representing the situation?
- Which graph do you think will be more useful and easier to use as a mathematical model for this situation? Why?

6. For the rest of this investigation, you will use the graph where edges connect vertices that are 500 miles or less apart. Make sure you have a neat copy of this graph.

a. Are vertices (stations) X and W connected by an edge? Are they 500 miles or less apart? Will their broadcasts interfere with each other?

b. Are vertices (stations) Y and Z connected by an edge? Will their broadcasts interfere with each other?

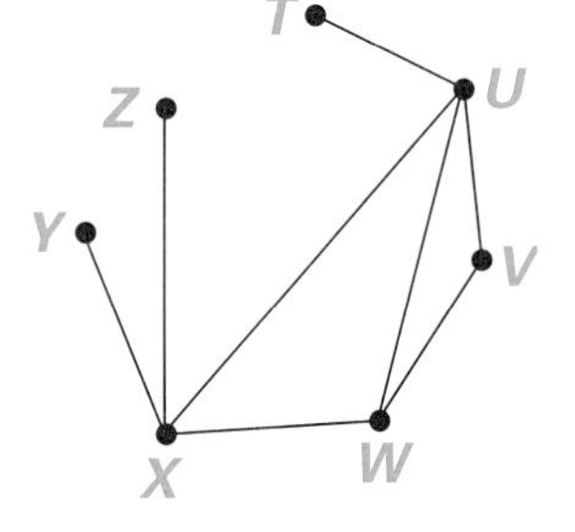

c. Compare your graph to the graph at the right.

- Does this graph also represent the radio-station problem?
- What criteria can you use to decide if two graphs both represent the same situation?

7. So far you have a model that shows all the radio stations and which stations are within 500 miles of each other. The goal is to assign frequencies so that there will be no interference between stations. You still need to build the frequencies into the model. So, as the last step in building the graph model, represent the frequencies as **colors**. To **color a graph** means to assign colors to the vertices so that two vertices connected by an edge have different colors.

You can now think about the problem in terms of *coloring the vertices of a graph.* The following table contains statements about stations and frequencies in the left-hand column. Corresponding statements about vertices and colors are in the right-hand column. Write statements to complete the right-hand column of the table.

Statements about stations and frequencies	Statements about vertices and colors
Two stations have different frequencies.	Two vertices have different colors.
Find a way to assign frequencies so that stations within 500 miles of each other get different frequencies.	
Use the fewest number of frequencies.	

8. Now use as few colors as possible to color your graph for the radio station problem. That is, assign a color to each vertex so that any two vertices that are connected by an edge have different colors. You can use colored pencils or just the names of some colors to do the coloring. Color or write a color name next to each vertex. Try to use the smallest number of colors possible.

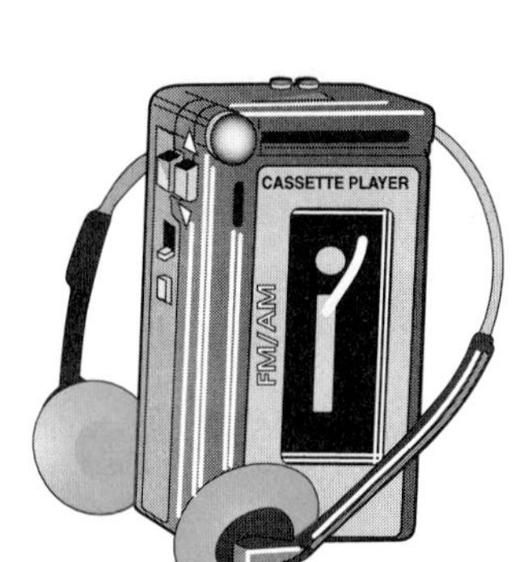

9. Compare your coloring with that of another group.

a. Do both colorings satisfy the condition that vertices connected by an edge must have different colors?

b. Do both colorings use the same number of colors to color the vertices of the graph?

c. Reach agreement about the minimum number of colors needed. Explain, in writing, why the graph cannot be colored with fewer colors.

d. Suppose one group assigns two vertices the same color and another group assigns them different colors. Is it possible that both assignments are acceptable? Why or why not?

e. What is the connection between graph coloring and assigning frequencies to radio stations? As you answer this question, compare the results of this activity to those in Activity 2.

10. Think about the strategy you used in Activity 8 to color the radio station graph with as few colors as possible.

a. Write down a step-by-step description of your coloring strategy. Write the description so that your strategy can be applied to graphs other than just the radio station graph.

b. Use the description of your strategy to color a copy of the graph at the right.

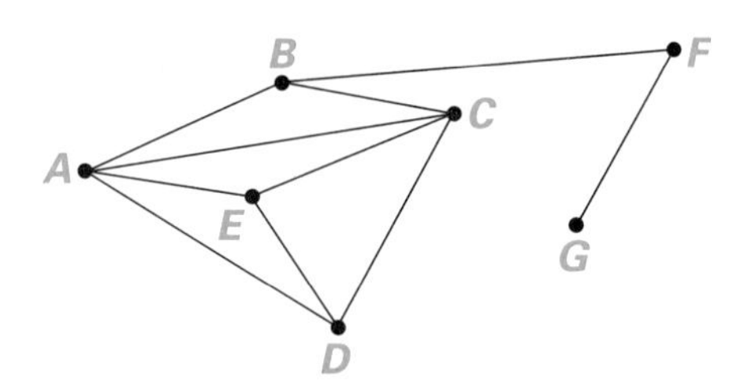

c. Refine the directions for your coloring strategy so that any one of your classmates could follow the directions.

11. Exchange your written coloring directions with another group. Then do the following:

a. Use the other group's directions to color a second copy of the graph in Part b of Activity 10. The other group will be doing the same thing with your directions.

b. Compare your colorings with the other group's colorings.

- Are they the same?
- Are they each legitimate colorings?
- Do they each use the least number of colors possible? Reach agreement with the other group about the minimum number of colors needed to color the graph.

c. Discuss any problems that came up with either group's coloring directions. If necessary, rewrite your directions so that they work better and are easier to follow.

As you saw in the previous lesson, a careful list of directions for carrying out a procedure is called an *algorithm*. Designing and applying algorithms is an important method for solving problems. There are many possible algorithms for coloring the vertices of any graph, including the ones you developed.

Checkpoint

Some problems can be solved by coloring the vertices of an appropriate graph model.

a What do the vertices, edges, and colors represent in the graph model that you have been using for the radio station problem?

b How does "coloring a graph" help solve the radio station problem?

c In what ways can two graph models differ and yet still both accurately represent a given situation?

d What are some strengths and weaknesses of the graph-coloring algorithm created by your group?

Be prepared to share your thinking and coloring algorithm with the entire class.

Graph-coloring algorithms continue to be an active area of mathematical research with many applications. It has proved quite difficult to find an algorithm that colors the vertices of any graph using as few colors as possible. You often can figure out how to do this for a given small graph, as you have done in this investigation. However, no one knows an efficient algorithm that will color *any* graph with the *fewest* number of colors. This is a famous unsolved problem in mathematics. At the time this book was written, at least, the problem was still unsolved. . . .

On Your Own

Copy the grid with the seven radio stations on page 277. (Your teacher may have a copy ready for you.) Add three more stations to the grid so that at least two of them are within 500 miles of one of the existing seven stations. Use graph coloring to assign frequencies optimally to all ten stations so that their broadcasts do not interfere with each other.

INVESTIGATION 2 Coloring, Map Making, and Scheduling

Now that you know how to color a graph, you can use graph coloring to solve many other types of problems.

For example, there are six clubs at King High School that all want to meet once a week for one hour, right after school lets out. The problem is that several students belong to more than one of the clubs, so not all the clubs can meet on the same day. Also, the school wants to schedule as few days per week for after-school club meetings as possible. Below is the list of the clubs and the club members who also belong to other clubs.

Clubs and Members

Club	Students belonging to more than one club
Varsity Club	Christina, Shanda, Carlos
Math Club	Christina, Carlos, Wendy
French Club	Shanda
Drama Club	Carlos, Vikas, Wendy
Computer Club	Vikas, Shanda
Art Club	Shanda

Your goal is to assign a meeting day (Monday–Friday) to each club in such a way that no two clubs that share a member meet on the same day. Also, you want to use as few days as possible.

1. Consider this problem as a graph-coloring problem.

a. Working on your own, decide what you think the vertices, edges, and colors should represent in the club-scheduling problem.

b. Compare representations with the other members of your group. Decide as a group which representations are best. Complete these three statements.

The vertices represent … .

Two vertices are connected by an edge if … .

The colors represent … .

c. Draw a graph that models the problem.

d. Color the club-scheduling graph using as few colors as possible.

e. Use your coloring to answer these questions.

- Is it possible for every club to meet once per week?
- What is the fewest number of days needed to schedule all the club meetings?
- On what day should each club meeting be scheduled?
- Explain how your coloring of the graph helps you answer each of these questions.

Another class of problems in which graph coloring is useful involves coloring maps. You may have noticed in your geography or social studies course that maps are always colored so that neighboring countries do not have the same color. This is done so that the countries are easily distinguished and don't blend into each other. In the following activities, you will explore the number of different colors necessary to color *any* map in such a way that no two countries that share a border have the same color. This is a question that mathematicians worked on for many years, resulting in a lot of new and useful mathematics.

2. Shown here is an uncolored map of a portion of southern Africa in 1980. Using a copy of this map, color the map so that no two countries that share a border have the same color.

a. How many colors did you use? Try to color the map with fewer colors.

b. Compare your map coloring with that of other classmates.

- Are the colorings different?
- Are the colorings legitimate; that is, do neighboring countries have different colors? If a coloring is not legitimate, fix it.

c. What was the fewest number of colors that were needed to color this map?

3. In Activity 2, you found the fewest number of colors needed to color the Africa map. Now think about the fewest number of colors needed to color *any* map.

a. Do you think you can color *any* map with, say, 5 different colors? Can the map of Africa be colored with 5 colors?

b. The map below has been colored with 5 colors. Is it possible to color the map with fewer than five colors? If so, copy it onto your own paper and color it with as few colors as possible.

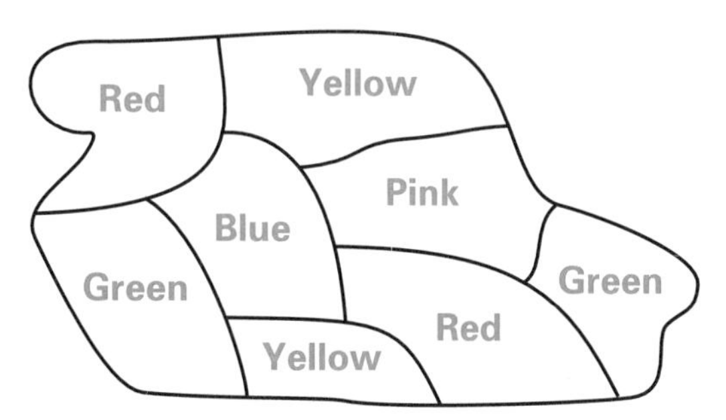

c. What do you think is the *fewest* number of colors needed to color *any* map? Make a conjecture now. Then, over the next few days, check your conjecture on other maps outside of class. Revise your conjecture as necessary. Compare it to the conjectures of your classmates. Conclude your outside-of-class investigation of map coloring by examining Extending Task 3 on page 293.

Maps can be colored by working directly with the maps, as you have been doing. But it is also possible to turn a map-coloring problem into a graph-coloring problem. This can be helpful since it allows you to use all the properties and techniques for graphs to help you understand and solve the map-coloring problems.

4. To build a graph model for a map-coloring problem, first think about what you did with the radio station and club-scheduling problems. In both of those problems, the edges were used to indicate some kind of *conflict* between the vertices. The vertices in conflict were connected by an edge and colored different colors. A crucial step in building a graph-coloring model is to decide what the conflict is. Once you know the conflict, you can figure out what the vertices, edges, and colors should represent.

a. What was the conflict in the club-scheduling problem? What was the conflict in the radio station problem?

b. Make and complete a table like the one at the top of the following page.

Modeling Conflicts

Problem	Conflict	Vertices	Connect with an Edge if:	Colors
Radio station problem	*Two radio stations are in conflict if* ________.	*radio stations*	*stations 500 miles apart or less*	*frequencies*
Club-scheduling problem	*Two clubs are in conflict if* ______.			
Map-coloring problem	*Two countries are in conflict if* ______.			

5. Consider the map on page 283.

a. Use the vertices and edges to represent aspects of the map as you determined in Activity 4.

b. Color the vertices of the graph. Remember that coloring always means that vertices connected by an edge must have different colors. Also, as usual, use as few colors as possible.

c. Compare your coloring with that of other classmates.

- Are all the colorings legitimate?
- Reach agreement on the fewest number of colors needed to color the graph.
- Is the minimum number of colors for this *graph*-coloring problem the same as the minimum number of colors for the *map*-coloring problem in Part c of Activity 2? Explain.

Checkpoint

In this lesson, you have seen three different problems that can be modeled by graph coloring:

- assigning frequencies to radio stations
- scheduling club meetings
- coloring maps

The title of this lesson is "Managing Conflicts." Explain how graph coloring allows you to "manage conflicts" in each of the three problems.

Be prepared to share your explanations with the entire class.

On Your Own

Hospitals must have comprehensive and up-to-date evacuation plans in case of an emergency. A combination of buses and ambulances can be used to evacuate most patients. Of particular concern are patients under quarantine in the contagious disease wards. These patients cannot ride in buses with non-quarantine patients. However, some quarantine patients can be transported together. The records of who can be bused together and who cannot are updated daily.

Suppose that on a given day there are six patients in the contagious disease wards. The patients are identified by letters. Here is the list of who cannot ride with whom:

A cannot ride with B, C, or D.

B cannot ride with A, C, or E.

C cannot ride with A, B, or D.

D cannot ride with A or C.

E cannot ride with F or B.

F cannot ride with E.

The problem is to determine how many vehicles are needed to evacuate these six patients. Use a graph-coloring model to solve this problem. Describe the conflict and state what the vertices, edges, and colors represent.

MORE

Modeling • Organizing • Reflecting • Extending

Modeling

1. A nursery and garden center plants a certain number of "mix-and-match" flower beds. Each bed contains several different varieties and colors. This allows customers to see possible arrangements of flowers that they might plant.

However, the beds are planted so that no bed contains two colors of the same variety. For example, no bed contains both red roses and gold roses. Also, no bed contains two varieties of the same color. For example, no bed contains both yellow tulips and yellow marigolds. This is done so that the customer can distinguish among and appreciate the different colors and varieties. A list of the varieties and colors that will be planted follows.

Flower Beds

Varieties	Colors
Roses	Red, Gold, White
Tulips	Yellow, Purple, Red
Marigolds	Yellow, Orange

The nursery wants to plant as few mix-and-match beds as possible. In this problem you will determine the minimum number of mix-and-match flower beds.

a. The varieties and colors listed above yield eight different types of flowers, such as red roses, red tulips, and yellow tulips. List all the other types of flowers that are possible.

b. It is the types of flowers from Part a that will be planted in the mix-and-match beds. The problem is to figure out the minimum number of beds needed to plant these types of flowers so that no bed contains flowers that are the same variety or the same color. First, you need to build a graph-coloring model.

- What should the vertices represent?
- What should the edges represent?
- What should the colors of the graph represent?

c. Draw the graph model and color it with as few colors as possible.

d. What is the minimum number of mix-and-match beds needed?

e. Using your graph coloring, recommend to the nursery which types of flowers should go in each of the mix-and-match beds.

f. When using a graph-coloring model, you connect vertices by an edge whenever there is some kind of conflict between the vertices. What was the conflict in this problem?

2. A local zoo wants to take visitors on animal feeding tours. They propose the following tours:

Tour 1 Visit lions, elephants, buffaloes.

Tour 2 Visit monkeys, hippos, deer.

Tour 3 Visit elephants, zebras, giraffes.

Tour 4 Visit hippos, reptiles, bears.

Tour 5 Visit kangaroos, monkeys, seals.

The animals should not be fed more than once a day. Also, there is only room for one tour group at a time at any one site. Can these tours be scheduled using only Monday, Wednesday, and Friday? Explain your answer in terms of graph coloring.

3. You often can color small maps directly from the map, without translating to a graph model. However, using a graph model is essential when the maps are more complicated. The map of South America shown here can be colored either directly or by using a graph-coloring model.

a. Color a copy of the map of South America directly. Use as few colors as possible and make sure that no two bordering countries have the same color.

b. Represent the map as a graph. Then color the vertices of the graph with as few colors as possible.

c. Did you use the same number of colors in Parts a and b?

4. The figure shown here is part of what is called a *Sierpinski Triangle*. (The complete figure is actually drawn by an infinite process described in Extending Task 2.)

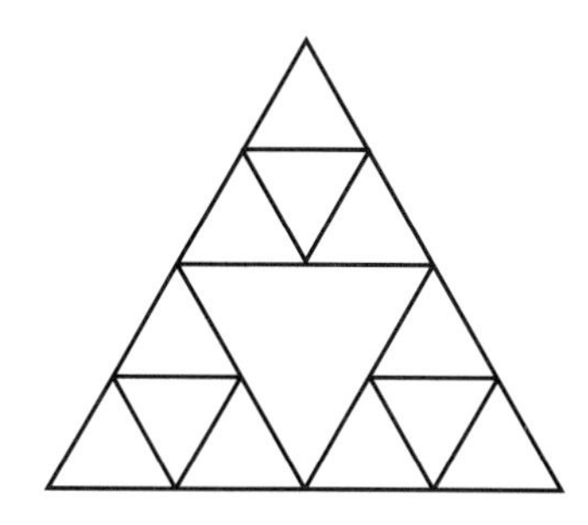

a. Think of this figure as a map in which each triangle not containing another triangle is a country. Make and color a copy of the map with as few colors as possible.

b. Construct a graph model for this map. Color the vertices of the graph with as few colors as possible. Compare the number of colors used with that in Part a.

c. Think of this figure as a map as Sierpinski did: the triangles with points upwards are countries and the triangles with points downwards are bodies of water that separate the countries. Using this interpretation of countries, color a copy of the map with as few colors as possible.

d. Construct a graph model for this second map. Color this graph with as few colors as possible. Did you use the same number of colors as in Part c?

Organizing

1. Shown here is Shanda's graph model for the radio-station problem from page 277.

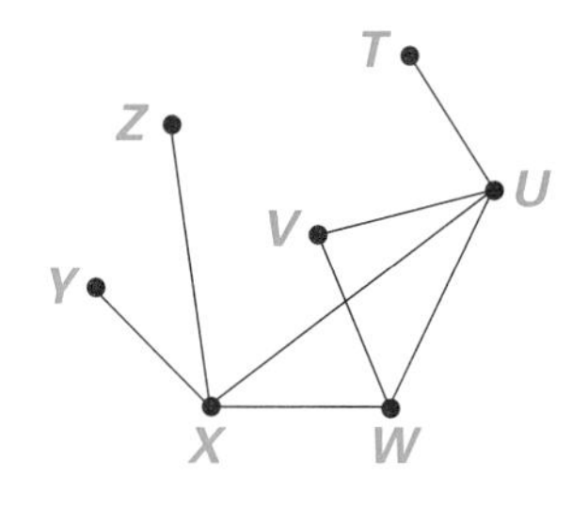

   a. Is this a legitimate model for the radio-station problem? Explain your reasoning.

   b. In Shanda's graph model, some edges cross at places that are not vertices. Can Shanda's graph be re-drawn without edge-crossings? Explain.

   c. Graphs that *can* be drawn in the plane with edges crossing only at the vertices are called **planar graphs**. Which of the graphs below are planar graphs?

   i.

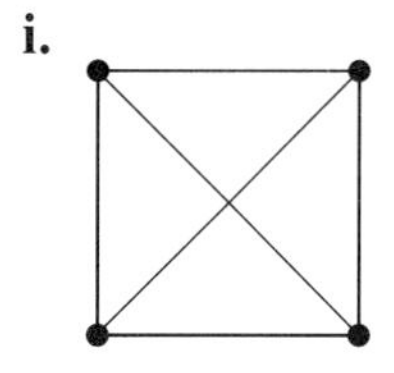

   ii.

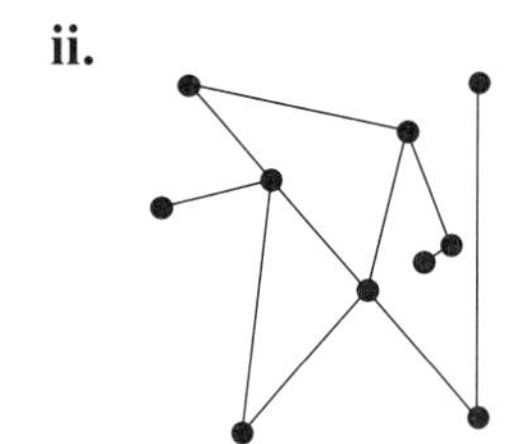

   iii.

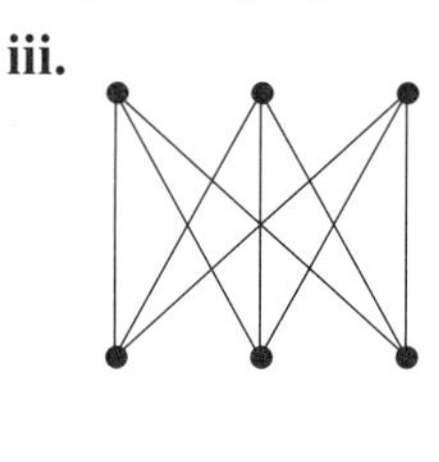

2. This task explores some properties of *complete graphs*. A **complete graph** is a graph that has an edge between every pair of vertices. Complete graphs with three and five vertices are shown below.

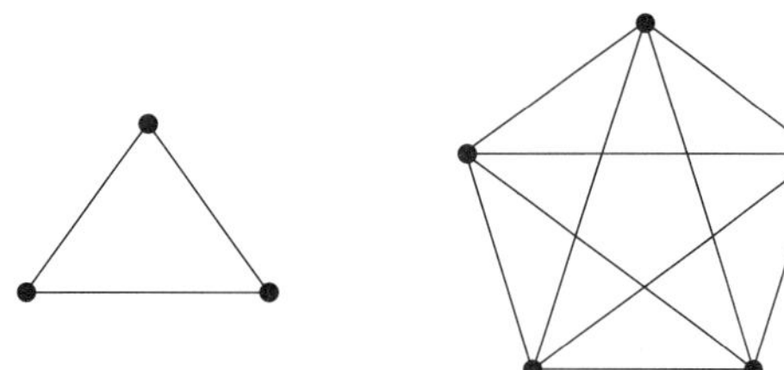

   a. Draw the complete graph with four vertices. Draw the complete graph with six vertices.

   b. Make a table that shows the number of edges for complete graphs with three, four, five, and six vertices.

   c. Look for a pattern in your table. How many edges does the complete graph with seven vertices have? The complete graph with n vertices?

   d. Recall the *NOW-NEXT* notation that you have used in previous units. Let *NOW* represent the number of edges for a given complete graph. Let *NEXT* represent the number of edges for a complete graph with one more vertex. Write an expression that shows how to calculate *NEXT* using *NOW*.

3. Refer to the definition of a complete graph given in Organizing Task 2.

a. What is the minimum number of colors needed to color the vertices of the complete graph with three vertices? The complete graph with four vertices? The complete graph with five vertices?

b. Make a table showing the number of vertices and the corresponding minimum number of colors needed to color a complete graph with that many vertices. Enter your answers from Part a into the table. Find several more entries for the table.

c. Describe any patterns you see in the table.

d. What is the minimum number of colors needed to color a complete graph with 100 vertices? With n vertices?

4. Besides coloring graphs, it is also possible to color polyhedra. Shown below are three of the five **regular polyhedra**.

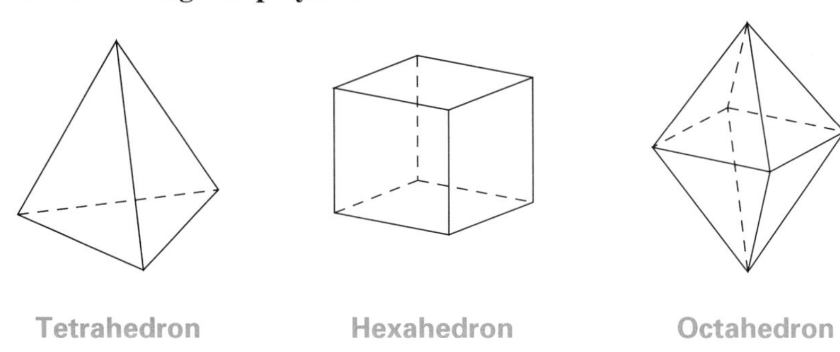

Tetrahedron   Hexahedron   Octahedron

Complete each coloring scheme below for each of the above polyhedra. Record your answers for each of these coloring schemes in a table like the one below.

a. Color the vertices. Use the minimum number of colors. (Vertices connected by an edge must have different colors.)

b. Color the edges. Use the minimum number of colors. (Edges that share a vertex must have different colors.)

c. Color the faces. Use the minimum number of colors. (Faces that are adjacent must have different colors.)

Coloring Polyhedra

	Minimum Number of Colors		
Regular Polyhedron	for Vertices	for Edges	for Faces
Tetrahedron			
Hexahedron			
Octahedron			

5. A **circuit** is a path that goes from vertex to vertex and ends where it started.

   a. Color the vertices of each of the circuits below using as few colors as possible.

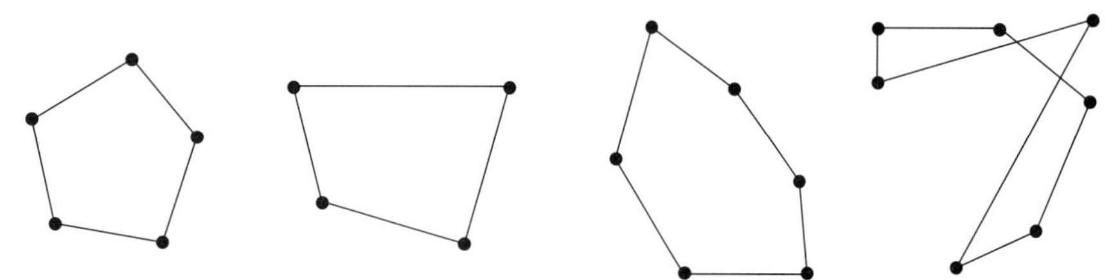

   b. Make a conjecture about the minimum number of colors needed to color circuits. Write an argument supporting your conjecture.

Reflecting

1. Throughout this course, and in this unit in particular, what you have been doing is called **mathematical modeling**. Below is a diagram that summarizes the process of mathematical modeling.

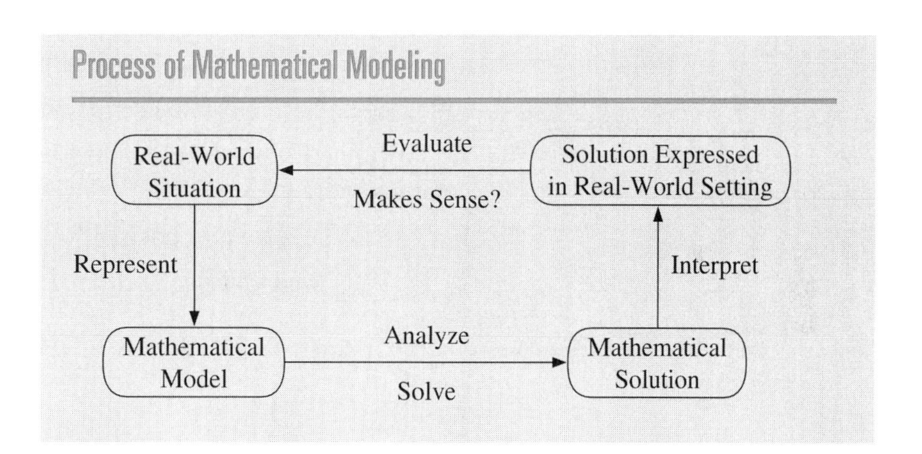

   Choose one example of mathematical modeling from this lesson. Use the example to illustrate each part of the diagram.

2. Think of a problem situation, different from those in this investigation, where you think graph coloring would be useful. Review the applications in this lesson to get started. Describe the problem situation, and then describe how you would solve it using graph coloring.

3. Think back over your work in Lessons 1 and 2. What did you find to be the most difficult part of modeling a problem with a graph? How did you resolve the difficulty?

4. Compare a current map of the southern part of Africa with the 1980 map on page 283.

a. Are there differences between the old map and the current one? If so, what are they?

b. Draw a graph model for the current map. Color the vertices of the graph with the fewest number of colors possible.

c. How many colors are used to color the map as it appears in a social studies book or atlas?

5. Research to find mathematicians who have worked on map coloring. Write a one-page report on one mathematician's contribution to the field.

Extending

1. Graph coloring is such an important application that several algorithms have been developed that are used on computers around the world. One commonly used algorithm is called the *Welsh and Powell algorithm*. Here's how it works:

i. Begin by making a list of all the vertices starting with the ones of highest degree and ending with those of lowest degree. (Recall that the *degree* of a vertex is the number of edges touching the vertex.)

ii. Color the highest uncolored vertex on your list with an unused color.

iii. Go down the list coloring as many uncolored vertices with the current color as you can.

iv. If all the vertices are now colored, you're done. If not, go back to step ii.

a. Follow the Welsh and Powell algorithm, step by step, to color the two graphs below.

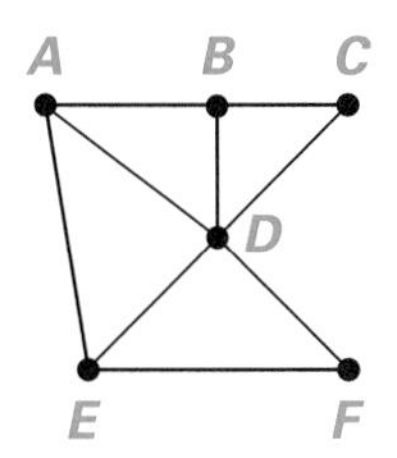

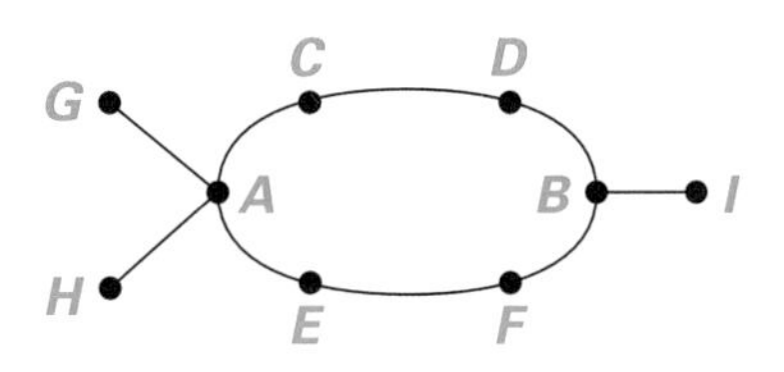

b. Does the Welsh and Powell algorithm always yield a coloring that uses the fewest number of colors possible? Explain your reasoning.

c. Use the Welsh and Powell algorithm to color the radio station graph and the club-scheduling graph from this lesson.

d. Compare the algorithm you wrote in this lesson to the Welsh and Powell algorithm. Are they similar? Which one do you like better? Why?

2. The Sierpinski Triangle is a very interesting geometric figure. If you try to draw it, you will never finish! That's because it is defined by a repetitive set of instructions. Here are the instructions:

i. Draw an equilateral triangle.

ii. Find the midpoint of each side.

iii. Connect the midpoints. This will subdivide the triangle into four smaller triangles.

iv. Remove the center triangle. (Don't actually cut it out, just think about it as being removed. If you wish, you can shade it with a pencil to remind yourself it has been "removed.") Now there are three smaller triangles left.

Sierpinski Triangle quilt made by Diana Venters.

v. Repeat steps ii–iv with each of the remaining triangles.

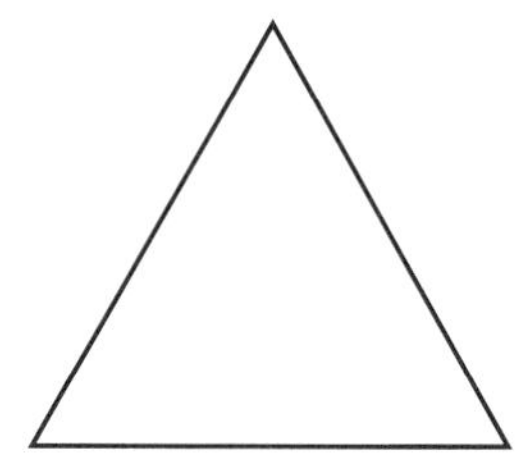
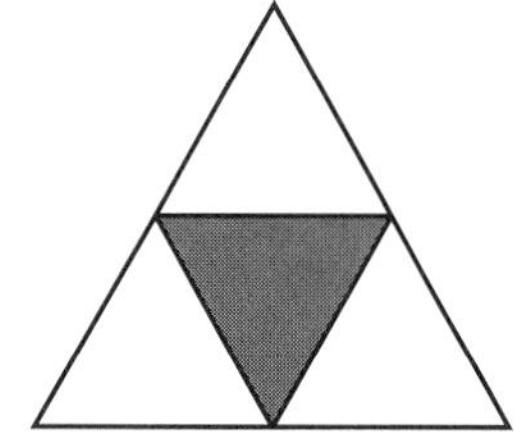
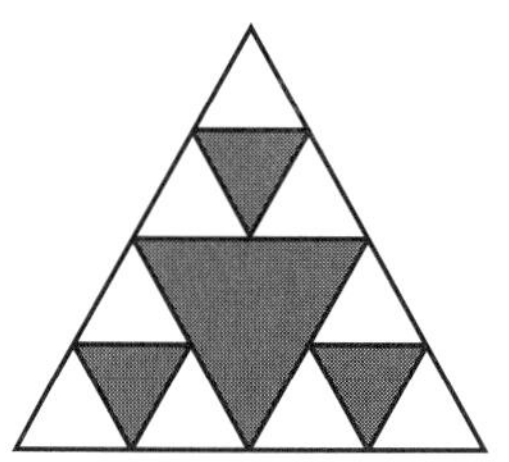

You never get finished with these instructions because there always will be smaller and smaller triangles to subdivide. The first few passes through the instructions are illustrated here.

a. On an enlarged copy of the third stage, draw the next few steps of the process.

b. If you think of the Sierpinski Triangle as a map (a very strange map with an infinite number of countries), what is the minimum number of colors needed to color the map?

3. In the 19th century, mathematicians made a conjecture about the minimum number of colors needed to color any map so that regions with a common boundary have different colors. This conjecture became one of the most famous unsolved problems in mathematics—until 1976 when the problem was solved. Based on your work in this investigation, how many colors do you think are needed to color any map? For this problem, you should only consider maps where the regions are connected. So, for example, do not consider a country that is split into two separate parts.

a. Try to draw a map that requires 3 colors and cannot be colored with fewer.

b. Try to draw a map that requires 4 colors and cannot be colored with fewer.

c. Try to draw a map that requires 5 colors and cannot be colored with fewer.

d. How many colors do you think are necessary to color any map? After you have worked on this problem for a while, search the Internet or a library for some recent information on graph theory. Find the answer and compare it to your answer. Write a brief report on your findings.

4. In this lesson, coloring a graph has always meant coloring the *vertices* of the graph. It also can be useful to think about **coloring the edges** of a graph. For example, suppose there are 6 teams in a basketball tournament and each team plays every other team exactly once. Games involving different pairs of teams can be played during the same round, that is, at the same time. The problem is to figure out the fewest number of rounds that must be played. One way to solve this problem is to represent it as a graph and then color the *edges.*

a. Represent the teams as vertices. Connect two vertices with an edge if the two teams will play each other in the tournament. Draw the graph model.

b. Color the edges of the graph so *edges that share a vertex have different colors.* Use as few colors as possible.

c. Think about what the colors mean in terms of the tournament and the number of rounds that must be played. Use the edge coloring to answer these questions.

- What is the fewest number of rounds needed for the tournament?
- Which teams play in which rounds?

d. Describe another problem situation that could be solved by edge coloring.

5. Described below is an interesting game involving a type of edge coloring which you can play with a friend.

- Place six points on a sheet of paper to mark the vertices of a regular hexagon, as shown here.
- Each player selects a color different from the other.
- Take turns connecting two vertices with an edge. Each player should use his or her color when adding an edge.
- The first player who is forced to form a triangle of his or her own color loses! (Only triangles whose vertices are among the six starting vertices count.)

a. Play this game several times and then answer the questions below.

- Is there always a winner? Explain.
- Which player has the better chance of winning? Explain.

b. Use the results of Part a to help you solve the following problem.

Of any six students who are in a room, must there be at least three mutual acquaintances or at least three mutual strangers?

Lesson 3 Scheduling Large Projects

Careful planning is important to ensure the success of any project. This is particularly true in the case of planning large projects such as the construction of a new high school, shopping mall, or apartment complex. Even smaller projects such as a party or a house remodeling job can profit from careful planning.

Think About This Situation

Suppose that you and some of your classmates are helping to plan a formal Spring Dance. You decide that a poster advertising the dance should be posted around the school a month before the dance.

a What are some tasks related to putting on a spring dance that should be completed before advertising posters are printed and posted? As a class, brainstorm as many tasks as possible.

b How can you make sure everything gets done on time?

INVESTIGATION 1 Building a Model

1. As you have seen before, as a first step in modeling a situation such as planning a dance, it is often helpful to make a diagram.

a. Working together in your group, draw a diagram that illustrates the schedule of tasks to be completed before posters for the spring dance can be displayed.

b. List two tasks that can be worked on at the same time by different teams.

c. Do some tasks need to be done before others? Tasks that need to be done before a particular task are called **prerequisites** for that task. Give one example of a task and a prerequisite for that task.

d. A prerequisite task might be done a long time before, or just before, a particular task. Construct a table that shows each task and the tasks that need to be done *just* before that task.

e. Does your diagram clearly show which tasks are prerequisites to others and which can be worked on at the same time? If not, make changes in your diagram to make it more accurate.

2. Exchange diagrams and tables with another group.

a. What are some similarities and differences between your group's table and diagram and the other group's table and diagram?

b. If necessary, modify your diagram and table so that they better show which tasks are prerequisites to other tasks.

Listed here are some of the tasks you may have found necessary in planning a spring dance. *These are the tasks that will be used for the rest of this investigation.* The order in which these tasks would need to be completed may vary from school to school.

Tasks
Book a Band or DJ (*B*)
Design the Poster (*D*)
Choose and Reserve the Location (*L*)
Post the Posters (*P*)
Choose a Theme (*T*)
Arrange for Decorations (*DC*)

At Marshall High School, the prerequisites for the various tasks are as follows:

- The tasks that need to be done just before booking the band are choosing and reserving the location *and* choosing a theme.
- The tasks that need to be done just before designing the poster are booking the band *and* arranging for decorations.
- There are *no* tasks that need to be done just before choosing and reserving the location.
- The task that needs to be done just before posting the posters is designing the posters.
- There are no tasks that need to be done just before choosing a theme.
- The task that needs to be done just before arranging for decorations is choosing a theme.

Tasks to be done *just* before a particular task are called **immediate prerequisites**.

3. Using the prerequisite information for Marshall High School, complete a table like the one below showing which tasks are immediate prerequisites for others. Such a table is called an **immediate prerequisite table**.

Dance Plans

Task	Immediate Prerequisites
Book a Band (B)	L, T
Design the Poster (D)	
Choose and Reserve the Location (L)	
Post the Posters (P)	
Choose a Theme (T)	
Arrange for Decorations (DC)	

4. Working on your own, complete a diagram like the one begun here showing how all the tasks are related to each other as follows.

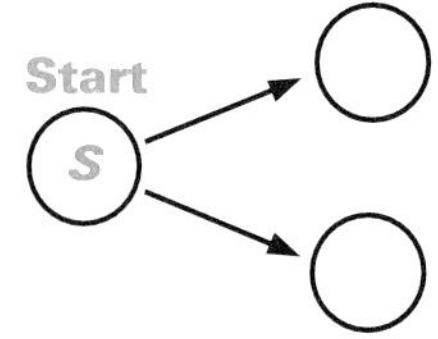

a. Copy the diagram onto your own paper. Place the circle labeled S at the far left of your paper. Put a new circle at the far right of your paper, where the diagram will end. Label it F for "Finish." The circles labeled S and F do not represent actual tasks. They simply indicate the start and finish of the project.

b. To the right of S are drawn two empty circles representing tasks that do not have any immediate prerequisites. Which tasks in the immediate prerequisite table from Activity 3 should be represented by the two empty circles? Label the two circles with those tasks.

c. Moving to the right again, draw a circle for each remaining task. Draw an arrow between two circles if one task (at the tail of the arrow) is an immediate prerequisite for the other task (at the tip of the arrow).

d. Finish the diagram by drawing connecting arrows from the final task or tasks to the circle marked F.

e. Your diagram may look a bit messy. Redraw your diagram so that it looks orderly.

5. Compare your diagram with the diagrams of other members of your group.

a. Do the diagrams look different? Explain the differences.

b. Does everyone's diagram accurately represent the information in the immediate prerequisite table?

c. As a group, decide on one organized, orderly diagram.

6. Use your diagram and the immediate prerequisite table to help you answer the following questions.

   a. Which of the following pairs of tasks can be worked on at the same time by different teams? Explain your reasoning.
      - Can tasks L and B be worked on at the same time by different teams?
      - Tasks L and T?
      - Tasks L and DC?
      - Tasks L and D?

   b. How do tasks that can be worked on at the same time appear in the diagram?

   c. Find one other pair of tasks that can be worked on at the same time.

   d. Explain, in terms of the school dance project and the individual tasks, why it is reasonable for the tasks you identified in Part c to be worked on at the same time.

7. How do tasks that are prerequisites appear in the diagram?

The diagram you have drawn is called a *directed graph*, or **digraph**. Digraphs are graphs that have *directed edges*. That is, the edges are arrows.

Checkpoint

The digraph showing how tasks involved in the dance project are related to each other is a mathematical model of the situation.

ⓐ What do the vertices of the project digraph represent?

ⓑ How are tasks that can be worked on at the same time represented in the project digraph?

ⓒ How are prerequisite tasks represented in the project digraph?

Be prepared to discuss your digraph and compare it to those of other groups.

On Your Own

"Turning around" a commercial airplane at an airport is a complex project that happens many times every day.

Suppose that the tasks involved are unloading arriving passengers, cleaning the cabin, unloading arriving luggage, boarding departing passengers, and loading departing luggage. The relationships among these tasks are as follows:

- Unloading the arriving passengers must be done just before cleaning the cabin.
- Cleaning the cabin must be done just before boarding the departing passengers.
- Unloading the arriving luggage must be done just before loading the departing luggage.
- All activities in the cabin of the airplane (unloading and boarding passengers and cleaning the cabin) can be done at the same time as loading and unloading luggage.

Construct the immediate prerequisite table and the project digraph for this situation.

INVESTIGATION 2 Finding the Earliest Finish Time

You have seen that a large project, like a school dance or "turning around" a commercial airplane, consists of many individual tasks that are related to each other. Some tasks must be done before others can be started. Other tasks can be worked on at the same time. A graph is a good way to show how all the tasks are related to each other.

The real concern in a large project is to get all the tasks done most efficiently. In particular, it is important to know the least amount of time required to complete the entire project. This minimum completion time is called the **earliest finish time (EFT)**.

1. There are many reasonable estimates that you and your classmates might make for how long it will take to complete each task of the school dance project. Experience at one school suggested the task times and prerequisites displayed in the following table. These task times will be used for the rest of this lesson.

Planning a Dance

Task	Task Time	Immediate Prerequisites
Choose & Reserve Location (*L*)	2 days	none
Choose a Theme (*T*)	3 days	none
Book the Band or DJ (*B*)	7 days	*L*, *T*
Arrange for Decorations (*DC*)	5 days	*T*
Design the Poster (*D*)	5 days	*B*, *DC*
Post the Poster (*P*)	2 days	*D*

Put these task times into the project digraph you constructed in the last investigation by entering the task times into the circles (vertices) of the digraph.

2. Now use the immediate prerequisite table and the project digraph to help you figure out how to complete the project most efficiently.

a. Using all the individual task times, what is the least amount of time required to complete the whole project (that is, what is the EFT for the project)? Each group member should write down a response *and* an explanation.

b. Compare responses and explanations with others in your group.

c. Is the earliest finish time for the whole project equal to the sum of all the individual task times? Explain.

d. How many paths are there through the project digraph, from *S* to *F*? List in order the vertices of all the different paths. For each path, compute the total time of all tasks on the path.

e. Which path through the graph corresponds to the earliest finish time for all the tasks? Write down your response *and* an explanation.

f. A path through the poster project graph that corresponds to the earliest finish time is called a **critical path**. Mark the edges of the critical path so that it is easily visible.

- What is the connection between the critical path and the EFT?
- What is the connection between the EFT and the path with the greatest total task time?
- What is the connection between the critical path and the path with the greatest total task time?

g. Compare your critical path from Part f to another group's critical path. If the paths are different, discuss the differences and decide on the correct critical path.

h. If all the posters are to be posted 30 days before the dance, how many days before the dance should work on the project begin?

3. Now explore what happens to the EFT and critical path if certain tasks take longer to complete than expected.

a. What happens to the earliest finish time if it takes 6 days, instead of 5 days, to design the poster (task D)?

b. What happens to the earliest finish time if it takes 9 days, instead of 7 days, to book the band (task B)?

c. What happens to the earliest finish time if one of the tasks *on the critical path* takes longer than expected to complete? A task on a critical path is called a **critical task**.

d. What happens to the earliest finish time if it takes 6 days, instead of 5 days, to arrange for the decorations (task DC)?

e. Suppose it takes 3 days, instead of 2 days, to choose and reserve a location (task L).

- What happens to the EFT?
- What happens to the critical path?

f. Suppose it takes 6 days, instead of 2 days, to choose and reserve a location (task L).

- What happens to the critical path?
- What happens to the EFT?

g. What happens to the earliest finish time and the critical path if one of the tasks that is *not on the critical path* takes longer than expected to complete?

Checkpoint

A critical path and the EFT for a project are two important related ideas.

a How can you find the EFT by examining a digraph for a project?

b Why is a critical path for a project "critical"?

Be prepared to share your thinking with the entire class.

On Your Own

Examine the digraph below.

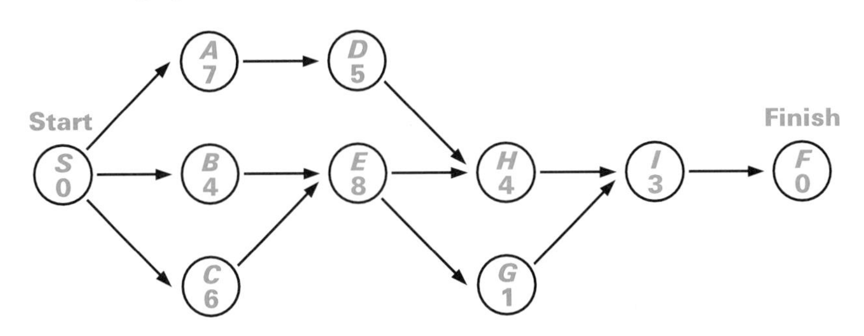

a. How many paths are there through this digraph, from S to F? List all the different paths, and compute the length of each path.

b. Find the critical path and the EFT. What are the critical tasks?

c. Are there any tasks that can have their task times increased by 3 units and yet not cause a change in the EFT for the whole project? If so, which tasks? If not, why not?

MORE

Modeling • Organizing • Reflecting • Extending

Modeling

1. Refer back to the beginning of the lesson and the entire list of tasks that your class came up with for the school dance project.

a. Assign reasonable task times to the various tasks.

b. Construct the immediate prerequisite table for the tasks.

c. Construct the project digraph.

d. Find a critical path and the EFT.

e. How do the critical tasks and EFT compare with those found in Activity 2 of this investigation?

2. Suppose that your school is planning to organize an Earth Day. You will have booths, speakers, and activities related to planet Earth and its environment. Such a project will require careful planning and coordination among many different teams that will be working on it.

Here are six tasks that will need to be done as part of the Earth Day project and estimates for the time to complete each task.

Planning Earth Day

Task	Task Time
Decide on Topics	6 days
Get Speakers	5 days
Choose Date and Location	3 days
Design Booths	2 weeks
Build Booths	1 week
Make Posters	6 days

a. Decide on immediate prerequisites for each of the tasks and construct the immediate prerequisite table.

b. Draw the project digraph.

c. Find the critical tasks and EFT.

3. Suppose that you and some friends are preparing a big dinner for 20 friends and family members.

a. List 4–8 tasks that must be done as part of this project.

b. Decide how long each task will reasonably take to complete.

c. Decide on the immediate prerequisites for each task and construct the immediate prerequisite table.

d. Draw the project digraph.

e. Find the critical tasks and EFT.

4. Shown below is the immediate prerequisite table for preparing a baseball field.

Preparing a Baseball Field

Task	Task Time	Immediate Prerequisites
Pick up Litter (L)	4 hours	none
Clean Dugouts (D)	2 hours	L
Drag the Infield (I)	2 hours	L
Mow the Grass (G)	3 hours	L
Paint the Foul Lines (P)	2 hours	I, G
Install the Bases (B)	1 hour	P

a. Find at least two tasks that can be worked on at the same time.

b. Draw the digraph for this project.

c. What is the EFT for the whole project?

d. Mark the critical path.

e. Do you think that the task times given in the table are reasonable? Change any times that you think are unreasonable. Use your new times to find the critical tasks and EFT.

5. Shown below is the immediate prerequisite table for building a house. Assume three specialists are working on each task.

Building a House

Task	Task Time	Immediate Prerequisites
Clear Land (*C*)	2 days	none
Build Foundation (*BF*)	3 days	*C*
Build Upper Structure (*U*)	15 days	*BF*
Electrical Work (*EL*)	9 days	*U*
Plumbing Work (*P*)	5 days	*U*
Complete Exterior Work (*EX*)	12 days	*U*
Complete Interior Work (*IN*)	10 days	*EL*, *P*
Landscaping (*L*)	6 days	*EX*

a. Find at least two tasks that can be worked on at the same time.

b. Draw the digraph for this project.

c. Mark the critical path.

d. What is the EFT for the whole project?

e. Suppose each specialist works 8 hours per day and is paid an average of $20 per hour. What will the total labor costs be?

f. Suppose some plumbing supplies will be late in arriving, so installing the plumbing will take 10 days. How does this affect the EFT and critical path?

Organizing

1. Reproduced below is the digraph from the "On Your Own" task on page 302. The critical path is shown by the dashed arrows. Verify that the EFT is 21.

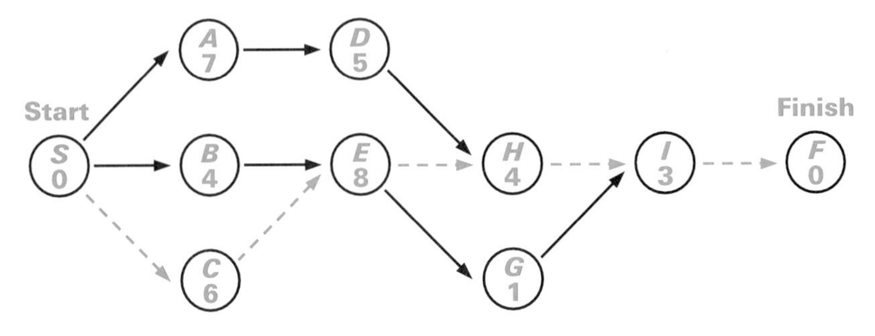

a. How do the critical path and EFT change if:

- the time for task *C* decreases by 5?
- the time for task *C* increases by 5?
- the time for task *D* increases by 2?
- the time for task *D* increases by 3?
- the time for task *D* decreases by 4?

b. Write a summary describing how changes in times for tasks on and off the critical path affect the EFT and the critical path.

c. Construct the immediate prerequisite table for the project digraph.

2. What are some similarities and differences between Euler paths and critical paths?

3. Recall the concept of an adjacency matrix from Lesson 1 (page 270).

a. How might you modify the concept of an adjacency matrix to make it useful with digraphs?

b. Make an adjacency matrix for the digraph at the right.

c. How are adjacency matrices for digraphs different from adjacency matrices for graphs that do not have directed edges?

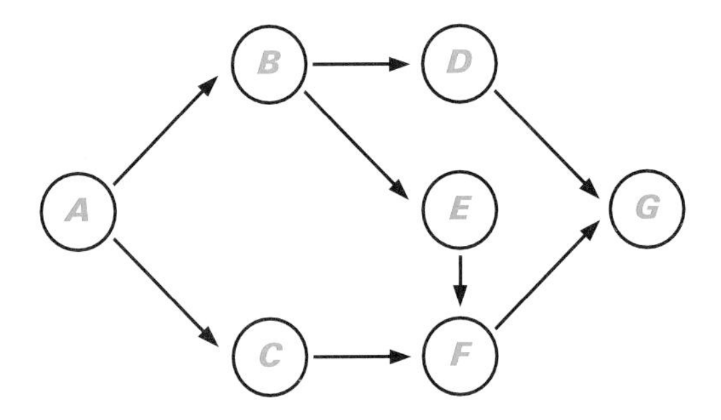

4. It is possible for a project digraph to have more than one critical path. That is, there can be more than one path through the project digraph that has maximum length.

a. Consider a modified version of the school dance poster project digraph, below.

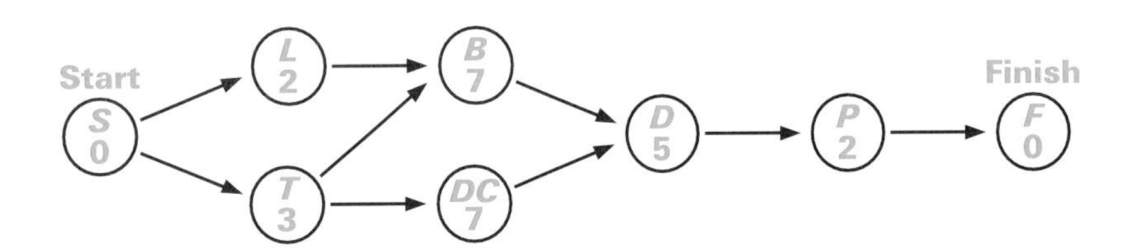

- Find the EFT for the project.
- How many critical paths are there? That is, how many paths are there with maximum length?
- List all the critical tasks.

b. Consider a modified version of the project digraph from the "On Your Own" task on page 302, shown below.

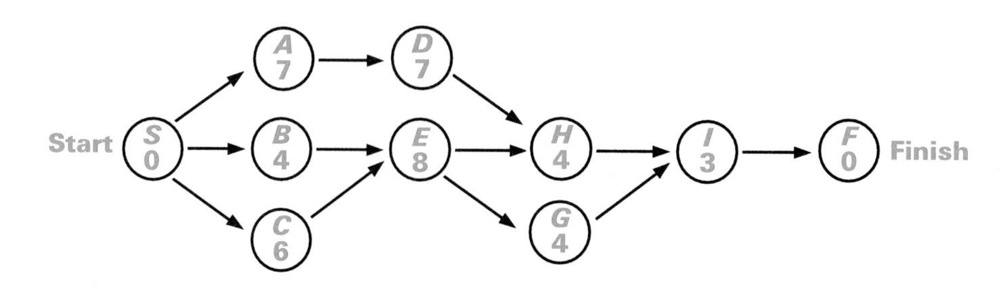

- Find the EFT for the project.
- How many critical paths are there? That is, how many paths are there with maximum length?
- List all the critical tasks.

Reflecting

1. Did you find anything particularly difficult in finding EFTs and critical paths for projects? If so, what did you do to overcome your difficulties?

2. Reproduced below is the project digraph from the "On Your Own" task on page 302. If you were a manager of the project, which tasks might require less of your attention or supervision? Why?

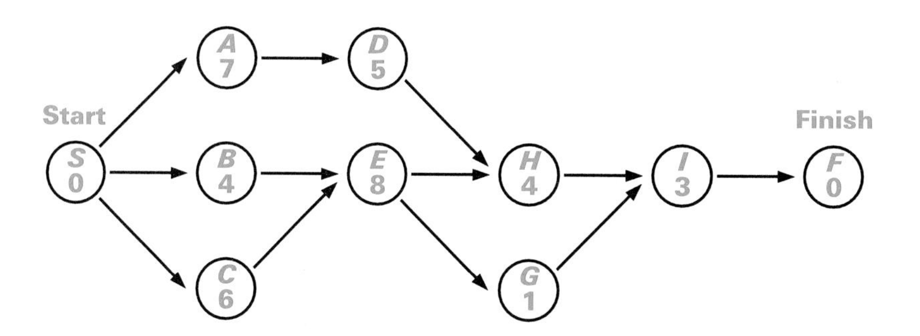

3. Do you think the method of immediate prerequisite tables and critical paths is useful? Why or why not?

4. Write a brief description of a project you or someone you know is working on, or of a project you read about in the paper or heard about on TV. Explain how this critical path method might be used to help organize and manage the project.

Extending

1. Write an algorithm for finding a critical path in a digraph.

2. Explain why the digraphs below could *not* be used to model a simple project.

a.

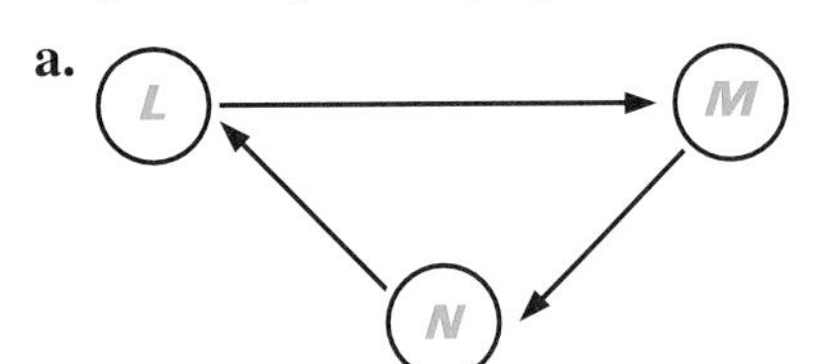

b.

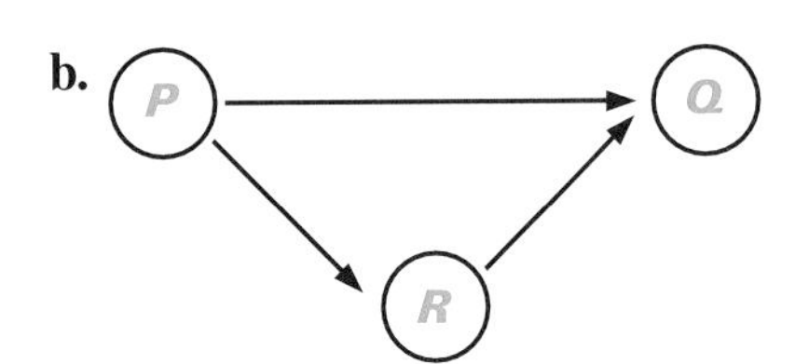

c. Are there any four-vertex configurations that can never occur in a project digraph? Explain and illustrate.

3. Refer to the immediate prerequisite table for the project of preparing a baseball field for play in Task 4 of the Modeling section.

a. What is the EFT for the project?

b. Suppose that there are only two people available for preparing the field. Does this change the EFT? Explain. If it does change the EFT, what is the new EFT?

c. Suppose that there are plenty of people to help prepare the field. However, mowing the grass in the time allotted (3 hours) can be done only if two people are mowing at the same time and, unfortunately, there is only one mower that is working. Propose a plan for how to deal with this problem. Will your plan change the EFT? Explain. If it does change the EFT, what is the new EFT?

4. Think of a project that could be modeled by this project digraph. Describe the project and each task.

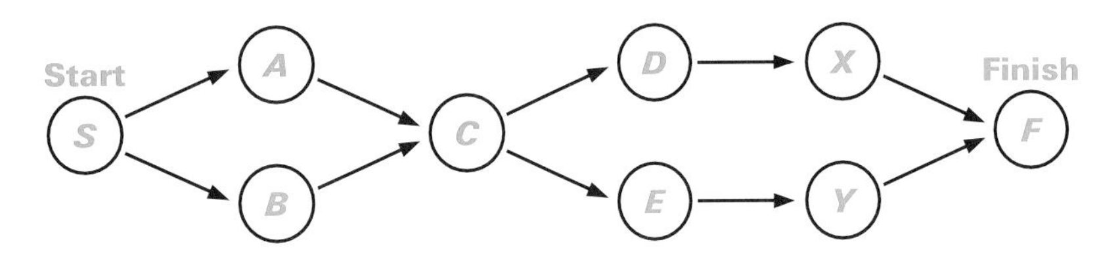

INVESTIGATION 3 Scheduling a Project

So far in this lesson, you have found a useful model for representing the school dance project—a digraph. Also, you know how to use the model to find the earliest finish time for the project—just find a critical (longest) path in the digraph. In this investigation, you will find out how to schedule the project.

Before working further on the project digraph, consider the simpler digraph below, which represents a simpler project. Tasks in the project are represented by letters, and the numbers represent the number of days needed to complete the tasks.

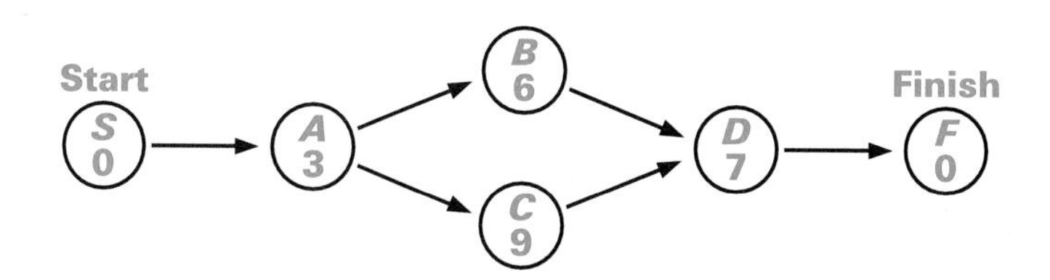

1. Working together as a group, find the EFT and record the critical path for this project.

2. Next, you will find the **earliest finish time (EFT)** for *each* task.

   a. The EFT for a given task is the least amount of time required to finish that task. Keep in mind that in order to finish the task, all of its prerequisites must be finished as well.

      - What is the least amount of time needed to finish task A?
      - What is the least amount of time needed to finish task C?
      - Find the EFT for the rest of the tasks in this project.

   b. Make a large copy of the project digraph on your own paper. Write the EFT for each task just above and to the right of the vertex representing the task in the digraph. Label it as "EFT".

   c. How does the EFT for the last task compare to the EFT for the whole project?

   d. Describe the method you used to figure out the EFTs.

3. Now you will figure out the **earliest start time (EST)** for each task.

   a. The EST for task B is 3.

      - Does this mean that the earliest start time for task B is "at the beginning of day 3" or "after 3 days"? Explain.
      - Find the EST for each of the remaining tasks.
      - Write the earliest start time for each task just above the EFT number on the digraph. Label it "EST".

   b. Write down the method you used to find the EST for each task.

   c. How can you use the EFT for a particular task to compute the EST for that task?

4. The **latest start time (LST)** for a task is the *latest* time that you can start work on that task and still stay on schedule to finish the whole project by the EFT for the entire project.

   a. Figure out the LST for each task *on* the critical path. Explain how you did it.

   b. Figure out the LST for each task *off* the critical path. Explain how you did it.

   c. Enter the LST numbers just to the left of the EST numbers on the digraph. Label them "LST".

5. Another important piece of information that is used to help manage a project is called **slack time**.

   a. The slack time for a given task is the difference between the latest start time (LST) and the earliest start time (EST). That is, slack time = LST – EST. Why do you think this is called "slack time"?

   b. Figure out the slack time for each task.

   c. What does it mean for a task to have a slack time of 3 days? 0 days?

   d. What can you say about the slack time for critical tasks?

   e. Are there other ways to determine the slack time, besides calculating LST – EST? Explain.

6. Now you are ready to schedule the project.

   a. When should you schedule work to begin on each task?

   b. Which of the numbers EFT, EST, or LST did you choose as the scheduled time to begin? Explain your reasoning.

   c. Put an asterisk (*) next to the number on the digraph that is the scheduled time to begin each task.

   d. Do some tasks have some flexibility concerning their scheduled time to begin? Which ones? Why?

   e. Which tasks have no flexibility with respect to their scheduled time to begin? Why?

7. Create a table like the one shown here. For each task, enter the EST, LST, EFT, and slack time. Also put "yes" or "no" for each task depending on whether or not it is a critical task.

Task Analysis

Task	EST	LST	EFT	Slack Time	Critical Task?
A	___	___	___	___	___
B	___	___	___	___	___
C	___	___	___	___	___
D	___	___	___	___	___

Checkpoint

In this investigation, you learned how to use digraphs to help schedule a large project.

a You determined the numbers EFT, EST, LST, slack time, and scheduled time to begin. Summarize what these numbers mean and how they are related to each other.

b How can you use the digraph for a given project and the numbers EFT, EST, LST, slack time, and scheduled time to begin, to make sure that the project gets done on time?

Be prepared to share your management ideas with the entire class.

The method you have been using to model and manage the poster project is called the Critical Path Method (**CPM**) or the Program Evaluation and Review Technique (**PERT**). It was developed in the late 1950s to aid in the development of defense systems. This method is used extensively in business and industry. In fact, it is one of the most frequently used mathematical management techniques. And now you know how to do it!

On Your Own

Consider the project digraph below.

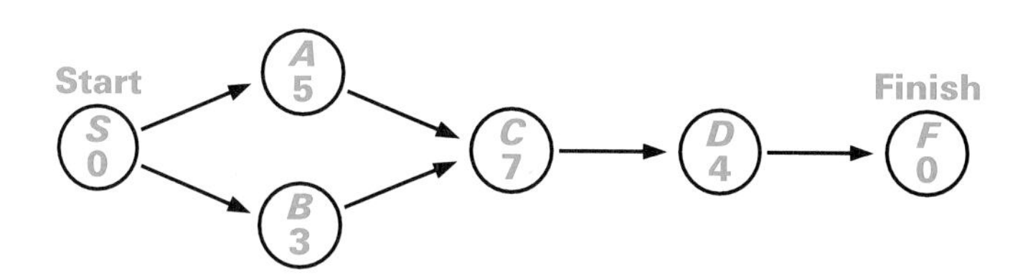

a. Determine the critical path and the earliest finish time for the whole project.

b. Find the EFT, EST, LST, and slack time for each task in the project.

c. Which tasks should a manager watch particularly closely? Explain your reasoning.

d. Are there some tasks that have some flexibility in terms of when they begin and end? Which ones? How much flexibility in scheduling the tasks would a manager have? Explain.

INVESTIGATION 4 Uncertain Task Times

In many applications of PERT, the projects are so complex that computer processing of the digraphs is required. Also, it is unlikely that you would know the exact time required to complete each of the many project tasks. In this investigation, you will modify the critical path technique you have learned so far to take into account uncertain task times. You will also use software to help reduce the time and effort required for your analysis.

Suppose a car manufacturer is considering a new fuel-efficient car. Since this new car will cost a lot of money to produce, the company wants to be sure that it is a good idea before they begin production. To help them decide if it is a good idea, they do a feasibility study. (In a feasibility study, you look at things like estimated cost and consumer demand to see if the idea is practical.) The feasibility study itself is a big project that involves many different tasks. As is often the case, the exact time needed to finish each of the tasks in the feasibility study is not known. However, for each task they have an estimate of the *best time* (the shortest amount of time to finish the task), the *worst time* (the longest amount of time needed to finish the task), and the *most likely time* needed to finish the task. All of these estimates, and the prerequisite information, are summarized in the following table.

Tasks, Times, and Prerequisites for the Feasibility Study

Task	Best Task Time	Most Likely Task Time	Worst Task Time	Immediate Prerequisites
Design the Car (*A*)	5 weeks	7 weeks	10 weeks	none
Plan Market Research (*B*)	1 week	2 weeks	4 weeks	none
Build Prototype Car (*C*)	8 weeks	10 weeks	16 weeks	*A*
Prepare Advertising (*D*)	2.5 weeks	4 weeks	7 weeks	*A*
Prepare Initial Cost Estimates (*E*)	1.5 weeks	2 weeks	4 weeks	*C*
Test the Car (*F*)	3.5 weeks	5 weeks	8 weeks	*C*
Finish the Market Research (*G*)	2 weeks	3.5 weeks	6 weeks	*B*, *D*
Prepare Final Cost Estimates (*H*)	1 week	2.5 weeks	5 weeks	*E*
Prepare Final Report (*I*)	0.5 week	1 week	2 weeks	*F*, *G*, *H*

1. Think about how to analyze this information to arrive at a schedule for the feasibility study project.

   a. Draw a digraph for this project. Just show the tasks, not the task times.

   b. How can you use all the information about best, worst, and most likely task times to figure out the critical tasks and schedule for this project? Write down as many methods and ideas as you can.

   c. Which method do you think is the best? Why?

For the remainder of this lesson you should use critical path software to find critical tasks and EFTs.

2. This activity illustrates how the PERT software designed for TI-82/83/92 calculators can be used. Your software may work differently.

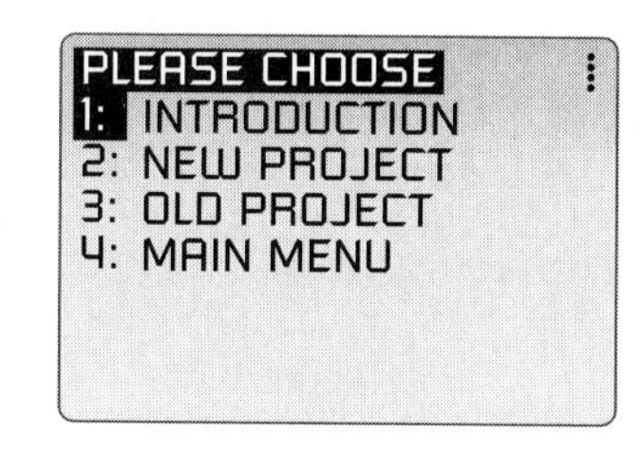

   Load or access the software on your calculator. Continue pressing ENTER until you see a screen similar to the one displayed here.

   a. Press 2 in order to enter the data for the project. Enter the data for *best task times* into your calculator. Follow the screen prompts. When the screen prompt asks for the number of prerequisites, it is asking for the *immediate* prerequisites. Check your data entries by choosing SHOW DATA from the *main* menu.

   b. Find the critical tasks and the EFT for the whole project, based on the best task times, by choosing SCHEDULE from the main menu. Mark the critical path on your digraph.

   c. Using EDIT from the main menu, enter the project a second time using the *most likely task times.* Then find the EFT for the whole project and the critical tasks. Using a different color pen or pencil, mark the corresponding critical path on your digraph.

   d. Using the *worst task times*, enter the data into your calculator. Then find the EFT and critical tasks. Draw the critical path on your digraph using a different color pen or pencil.

3. Compare the project EFTs and critical paths that you found in Activity 2.

   a. Do you get a different critical path and EFT depending on the times you use?

   b. Is there some way to combine the three EFTs you calculated in Activity 2? Explain.

   c. If just one EFT must be reported to the president of the company, what should be reported? Explain your reasoning.

4. Now consider one way to combine the best, worst, and most likely task times.

a. For each task, describe some way to compute an "average" time for completing the task.

b. In a later unit, you will learn more about how to analyze mathematically "most likely" and "average" time. For now, one way to estimate an "average" time is to multiply the most likely time by 4, add the best time, add the worst time, and divide by 6.

- Explain why this is a reasonable computation of average time.
- Sometimes an average like this is called a *weighted average.* Why do you think it is called a *weighted* average?
- Compute the average time for each task in the feasibility study project using the formula:

$$\text{Average time} = \frac{\text{best time} + 4(\text{most likely time}) + \text{worst time}}{6}$$

c. Use the average times from Part b and the PERT software to compute the EFT for the whole project and the critical tasks.

d. How do the EFT and critical tasks compare with your answers to Activity 2?

Checkpoint

Task times are uncertain in some large projects.

a What are some possible methods for planning a project when the exact times for finishing certain tasks are not known?

b Which method do you think is the best? Why?

Be prepared to share your thinking with the entire class.

On Your Own

Suppose all the worst task times doubled in the new car feasibility study project.

a. Find the EFT and critical tasks, using the average task time method used in Activity 4.

b. Draw the project digraph and highlight the critical tasks.

MORE

Modeling • Organizing • Reflecting • Extending

Modeling

1. Refer back to Task 2 of the Modeling section on page 302. In that task, you constructed an immediate prerequisite table and drew a project digraph for the Earth Day project. Now schedule that project. Include EFT, EST, LST, and slack time for each task.

2. On a large construction project, there usually is a general contractor (the company that coordinates and supervises the whole project) and smaller contractors (the companies that carry out specific parts of the project, like plumbing or framing).

   Suppose that the company responsible for putting in the foundation for a building estimates the times shown in the following prerequisite table. The general contractor wants the foundation done in 13 days. Can the foundation crew meet this schedule? If so, explain. If not, propose a plan for what they should do in order to shorten task times and finish on schedule.

Putting in a Foundation

Task	Task Time	Immediate Prerequisites
Measure the Foundation (*A*)	1 day	none
Dig Foundation (*B*)	4 days	*A*
Erect Forms (*C*)	6 days	*B*
Obtain Reinforcing Steel (*D*)	2 days	*A*
Assemble Steel (*E*)	3 days	*D*
Place Steel in Forms (*F*)	2 days	*C*, *E*
Order Concrete (*G*)	1 day	*A*
Pour Concrete (*H*)	3 days	*F*, *G*

3. A task-times-prerequisite table for putting on a school play is shown below.

Planning a School Play

Task	Best Task Time (days)	Most Likely Task Time (days)	Worst Task Time (days)	Immediate Prerequisites
Choose a Play (*A*)	7	9	14	none
Tryouts (*B*)	5	8	12	*A*
Select Cast (*C*)	3	5	10	*B*
Rehearsals (*D*)	25	35	40	*C*
Build Sets and Props (*E*)	20	22	25	*A*
Create Advertising (*F*)	4	5	6	*C*
Sell Tickets (*G*)	10	12	15	*F*
Make/Get Costumes (*H*)	20	25	30	*C*
Lighting (*I*)	7	10	14	*E*, *H*
Sound and Music (*J*)	9	10	12	*E*
Dress Rehearsals (*K*)	5	6	9	*D*, *I*, *J*
Opening Night (*L*)	1	1	1	*G*, *K*

a. You must report to the principal how long it will take before the play is ready to open. Based on EFTs, what will you report? Explain your method and reasoning. (Remember that you may use critical path software, if it is helpful.)

For Parts b through d, consider only the best task times.

b. Suppose that because of a conflict with another special event, you find out that you must complete the project in 6 days fewer than the "best task time" EFT. In order to meet this new timetable, you recruit some more helpers and put them to work. This will allow you to cut the time on some of the tasks. For which task or tasks should you cut time in order to meet the new deadline? Show which tasks you will shorten and how this will result in a new EFT that is 6 days shorter than the "best task time" EFT. (Shortening task times like this is sometimes called **crashing** the task times.)

c. Another way to attempt to shorten the EFT is to figure out a way to change some of the prerequisites. Suppose you decide to change the prerequisites for setting up the lighting by doing that task whether or not the set and props are built. Thus, "*E*" is no longer an immediate prerequisite for "*I*." How much time will this save for the "best task time" EFT?

d. Describe at least one other reasonable rearrangement of prerequisites. By how many days does your rearrangement increase or decrease the "best task time" EFT?

4. The music department of City High School is doing a production of *A Christmas Carol*. One scene has many villagers singing carols. The costume committee needs to schedule sewing dresses for this scene. From previous productions, times for various tasks in making a dress have been recorded. The times are listed below.

Dress Production

Task	Best Task Time (minutes)	Worst Task Time (minutes)	Immediate Prerequisites
Interface Collar (*A*)	5	10	none
Put Gathers in the Sleeves (*B*)	30	45	none
Interface the Cuffs (*C*)	5	10	none
Put Gathers in the Skirt (*D*)	30	45	none
Sew the Bodice (*E*)	30	60	none
Sew the Collar (*F*)	15	30	*A*
Attach the Cuffs to the Sleeves (*G*)	15	30	*B*, *C*
Sew Buttons/Buttonholes (*H*)	30	60	*G*
Attach the Skirt and Bodice (*I*)	30	45	*D*, *E*
Insert the Zipper (*J*)	30	60	*I*
Attach the Sleeves to the Dress (*K*)	30	60	*H*, *J*
Attach the Collar to the Dress (*L*)	30	60	*F*, *K*
Hem the Dress (*M*)	30	60	*J*

a. Using the best task time, find the EFT and critical tasks for this project.

b. Using the worst task time, find the EFT and critical tasks for this project.

c. In the investigation, you used "average" task times to compute an EFT for the project. Use a similar idea here to find a better estimate for this project's EFT.

5. Recall the school dance project from Investigation 1 (page 295). The immediate prerequisite table and project digraph for that project are reproduced below.

Planning a Dance

Task	Task Time	Immediate Prerequisites
Choose & Reserve Location (*L*)	2 days	none
Choose a Theme (*T*)	3 days	none
Book the Band or DJ (*B*)	7 days	*L*, *T*
Arrange for Decorations (*DC*)	5 days	*T*
Design the Poster (*D*)	5 days	*B*, *DC*
Post the Poster (*P*)	2 days	*D*

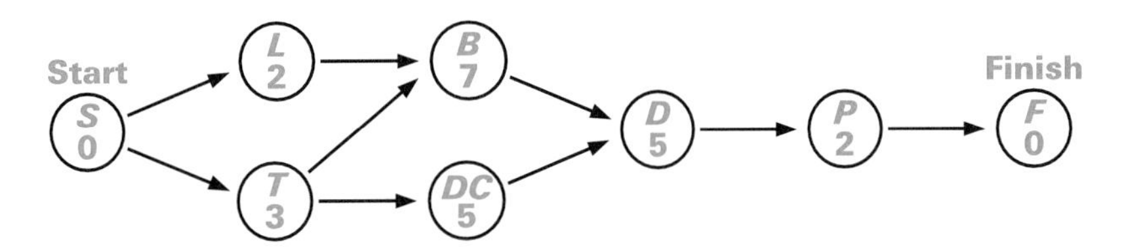

Schedule this project by finding the EFT, EST, LST, slack time, and scheduled time to begin for each task. (Note: If you use the PERT calculator software, you will first need to relabel the tasks alphabetically.)

Organizing

1. The adjacency matrix for a graph can be used to enter the graph into a computer. You also can get information about the graph and the project just by looking at the adjacency matrix.

a. Write the adjacency matrix for this digraph, ignoring *S* and *F*.

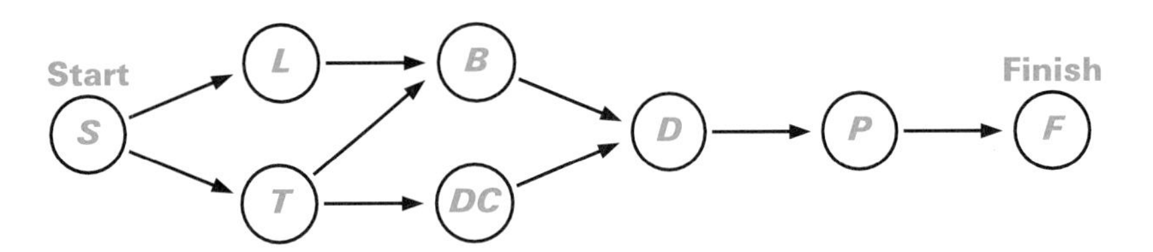

- Add up all the numbers in row *B*. What does this sum mean in terms of *B* as an immediate prerequisite?
- Add up all the numbers in column *B*. What does this sum mean in terms of how many immediate prerequisites *B* has?
- What does a row of all zeroes mean in terms of prerequisites?
- What does a column of all zeroes mean in terms of prerequisites?

b. The **indegree** of a vertex is the number of arrows coming into it. The **outdegree** is the number of arrows going out of it.

- What are the indegree and outdegree of *B*?
- What is the outdegree of *T*?
- Ignoring *S* and *F*, what is the indegree of *T*? The outdegree of *P*?
- What do indegree and outdegree mean in terms of prerequisites?
- How can you compute indegree and outdegree by looking at the rows and columns of the adjacency matrix for a digraph?

2. Describe one of the methods you suggested for determining the critical tasks and EFT for the feasibility study project. (See Activity 1 of Investigation 4 on page 312.) Then, use that method to actually find the critical tasks and EFT.

3. Write an algorithm for finding the LST for a task.

Reflecting

1. Interview some adults in business who use ideas from PERT to schedule projects, or do some library or Internet research. Find out how EFTs and critical paths are used. Write a brief report on what you discover.

2. Consider the project digraph for some project.

   a. If a task has many immediate prerequisites, will it have many arrows coming into it or many going out?

   b. If a task is an immediate prerequisite to many other tasks, will it have many arrows coming into it or many going out?

3. Suppose you are using critical path software like PERT for scheduling a project. What additional advantages would you gain if you use a digraph as well?

4. In this unit, you have seen how vertex-edge graphs can be used to represent and analyze relationships in many different contexts. Some examples are prerequisite relationships among tasks in a large project, conflicts between club meetings or radio stations, or connections between locations in a street network. You even can use graphs to represent and analyze relationships among the new concepts that you are learning in this course. This is done using a type of graph called a **concept map**. In a concept map, the vertices represent ideas or concepts and edges join concepts which are connected. The start of a concept map for this unit is shown here.

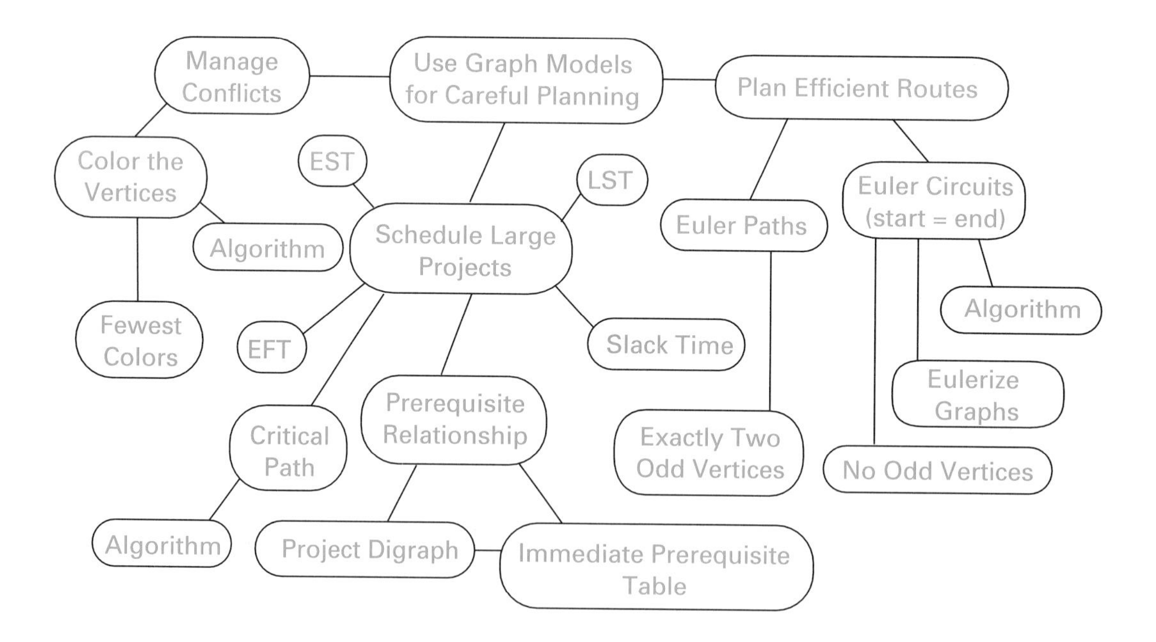

a. Explain why there is an edge between the vertices "Color the Vertices" and "Manage Conflicts."

b. Explain why there are edges connecting the vertices "Prerequisite Relationship," "Immediate Prerequisite Table," and "Project Digraph."

c. Obtain a copy of this concept map. Add other concepts and draw other edges that show concepts which are related.

d. Study the completed concept map. Should any of the edges be directed? Explain your reasoning.

Extending

1. Another interesting number for a project task is the latest finish time (LFT) for the task.

a. Find the LFT for each task in this project digraph.

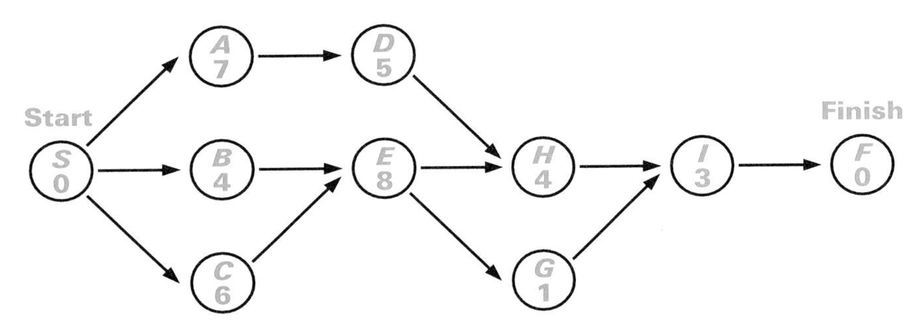

b. In terms of managing a project, why do you think it would be useful to know the LFT?

2. Write a software program that will determine and report each of the following for each task in a project.

a. EFT

b. EST

c. LST

d. slack time

3. In an immediate prerequisite table, the task times given are the times needed to complete each task using the available resources. For example, the table on the following page shows the task times for preparing the school baseball field for a game. Several people are available to work on the project. However, each task time assumes that only one person is working on that particular task.

Preparing a Baseball Field

Task	Task Time	Immediate Prerequisites
Pick up Litter (*L*)	4 hours	none
Clean Dugouts (*D*)	2 hours	*L*
Drag the Infield (*I*)	2 hours	*L*
Mow the Grass (*G*)	3 hours	*L*
Paint the Foul Lines (*P*)	2 hours	*I*, *G*
Install the Bases (*B*)	1 hour	*P*

a. What is the EFT for preparing the field?

b. In order to finish by this EFT, what is the fewest number of people needed to work on the project? Explain.

c. Suppose you have to complete the project by yourself. That is, you must do every task. How long will it take you to complete the project?

d. Suppose that you and two friends are hired to prepare the field. You and your friends decide that you might work together or separately to complete any task. If you work together on a task, then you can finish that task in less time than is shown in the prerequisite table above. The only task that must be done by someone working alone is mowing the grass, since there is only one mower. How would you assign duties in order to complete the project in the least amount of time?

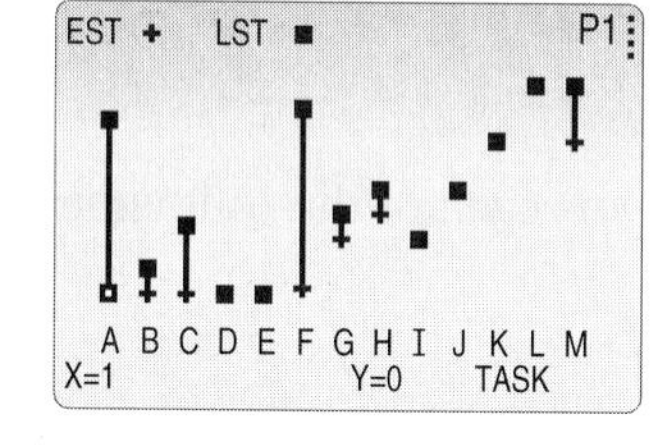

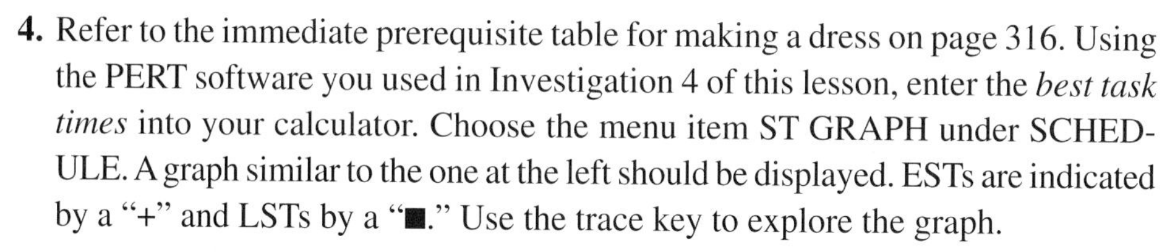

4. Refer to the immediate prerequisite table for making a dress on page 316. Using the PERT software you used in Investigation 4 of this lesson, enter the *best task times* into your calculator. Choose the menu item ST GRAPH under SCHEDULE. A graph similar to the one at the left should be displayed. ESTs are indicated by a "+" and LSTs by a "■." Use the trace key to explore the graph.

a. What do the vertical lines represent?

b. Tasks *A* and *F* have the largest slack times. What are the slack times for *A* and *F*? Does this mean that each of tasks *A* and *F* can be delayed by an amount of time equal to their slack times? Explain your reasoning.

5. So far in this lesson, you have scheduled a project by finding the EFT, EST, LST, slack time, and scheduled time to begin. For a specific project you would go even further and assign days to each task. Reconsider the school dance poster project. Assume that the first day of work on this project will be Tuesday, March 15. Make a timeline chart showing which days will be allocated for each of the tasks. Such a timeline chart is called a **Gantt chart**.

Lesson 4

Looking Back

In this unit, you have used graph models to solve problems related to careful planning. The problems you have explored include finding efficient routes, managing conflicts, and scheduling projects and events. The models you have used include Euler circuits and paths, graph coloring, and critical paths. In this final lesson of the unit, you will put it all together to solve problems that might involve any of the graph models.

For each of the problems below, decide which graph model will be the most useful representation. Use that model to solve the problem. Be prepared to explain your solution.

1. One city's Department of Sanitation organizes garbage collection by setting up precise garbage truck routes. Each route takes one day. Some sites that need garbage collection more often are on more than one route. However, if a site is on more than one route, the routes should not visit that site on the same day. Here is a list of routes and the sites that are on other routes in addition to that route.

Route 1: Site A, Site C

Route 2: Site D, Site A, Site F

Route 3: Site C, Site D, Site G

Route 4: Site G

Route 5: Site B, Site F

Route 6: Site D

Route 7: Site C, Site F

a. Can all seven routes be scheduled in one work week (Monday–Friday)?

b. Set up a schedule for the garbage truck routes, showing which routes run on which days of the week.

2. Suppose that you are the editor for a school newspaper. Study the following background information about the publishing process.

It takes 10 days for the reporters to research all the news stories. It takes 12 days for other students, working at the same time as the reporters, to arrange for the advertising. The photographers need 8 days to get all the photos. However, they can't start working until the research and the advertising arrangements are complete. The reporters need 15 days to write the stories after they have done the research. They can write while the photographers are getting photos. It takes 5 days to edit everything after the stories and the photos are done. Then it takes another 4 days to lay out the newspaper and 2 more days to get it back from the printer.

Write a report to your teacher-advisor explaining how long it will take to turn out the next edition of the paper. State which steps of the publishing process will need to be monitored most closely. Include diagrams and complete explanations in your report.

3. The security guard for an office building must check the building several times throughout the night. The figures below are the floor plans for office complexes on two floors of the building. An outer corridor surrounds each office complex. In order to check the electronic security system completely, the guard must pass through each door at least once.

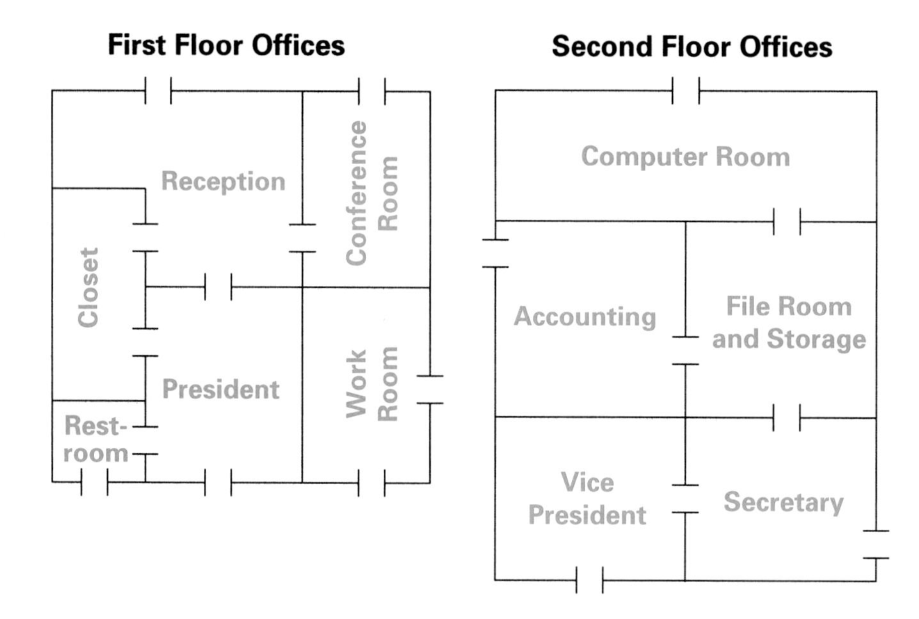

a. For each office complex, can the guard walk through each door exactly once, starting and ending in the outer corridor? If so, show a route the guard could take. If not, explain why not.

b. If it is not possible to walk through each door exactly once starting and ending in the outer corridor, what is the fewest number of doors that need to be passed through more than once? Show a route the guard could take. Indicate the doors that are passed through more than once.

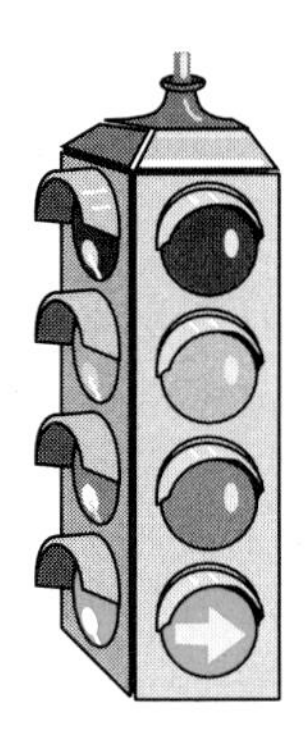

4. Traffic lights are essential for controlling the flow of traffic on city streets, but nobody wants to wait at a light any longer than necessary. Consider the intersection diagrammed below. The arrows show the streams of traffic. There is a set of traffic lights in the center of the intersection.

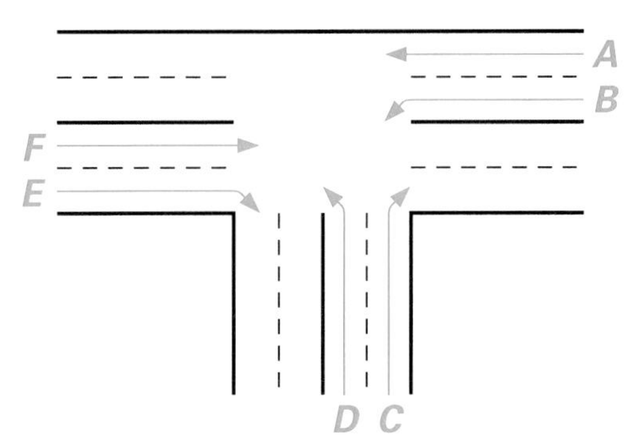

a. Can traffic streams *B* and *D* have a green light at the same time? How about *B* and *C*? List all the traffic streams that conflict with *B*.

b. Streams of traffic that have a green light at the same time are said to be on the same green-light cycle. What is the fewest number of green-light cycles necessary to safely accommodate all six streams of traffic?

c. For each of the green-light cycles you found in Part b, list the streams of traffic that can be on that cycle.

Checkpoint

In this unit, you have used graph models to study problems involving efficient routes, managing conflicts, and scheduling projects.

a For each of the problems in this lesson:

- Which graph model did you use?
- What did the vertices and edges represent?
- Explain why you chose the graph model you used.

b For each of the graph models you have studied—Euler circuits and paths, graph coloring, and critical paths—describe the types of problems that can be solved using the model.

Be prepared to share your descriptions and explanations with the entire class.

On Your Own

Write, in outline form, a summary of the important mathematical concepts and methods developed in this unit. Organize your summary so that it can be used as a quick reference in future units and courses.

Index of Mathematical Topics

Page numbers in italic can be found in Course 1, Part B.

U

V

W

X

Y

Index of Contexts

Page numbers in italic can be found in Course 1, Part B.

Photo Credits

We would like to thank the following for providing photographs of Core-Plus students in their schools. Many of these photographs appear throughout the text.

Janice Lee, Midland Valley High School, Langley, SC
Steve Matheos, Firestone High School, Akron, OH
Ann Post, Traverse City West Junior High School, Traverse City, MI
Alex Rachita, Ellet High School, Akron, OH
Judy Slezak, Prairie High School, Cedar Rapids, IA
The Core-Plus Mathematics Project

Cover, Photodisc; 1, Susan Van Etten; 13, Earl Gustie/Chicago Tribune; 14, Daimler Chrysler; 27 (right), Chuck Savage/CORBIS Stock Market; 27 (left), Robert Holmes/CORBIS; 29, Jack Demuth; 30, David Young-Wolff/PhotoEdit; 35, Oscar (c) A.M.P.A.S. Reprinted by permission of Academy of Motion Picture Arts and Sciences; 37, M. Timothy O'Keefe/West Stock; 46, Mary Kate Denny/PhotoEdit; 49, Kevin Tanaka/Chicago Tribune; 51, John Evans; 53 (left), Volvo cars of North America, LLC; 53 (right), John Evans; 56, 64, Aaron Haupt; 69, North Wind Picture Archives; 71, Aaron Haupt; 72, Ron Kimball/Ron Kimball Stock; 76, Daimler Chrysler; 78 (top), James Blank/ West Stock; (bottom left), CORBIS; (bottom, right), Terry Donnelly/West Stock; 82, 83, Ford Motor Company; 84, Walter Kale/ Chicago Tribune; 87, Laura Sifferlin; 91, Feature Photo Service; 94, PhotoDisc; 95, Geoff Butler; 97, The Harold E. Edgerton 1992 Trust, courtesy of Palm Press, Inc.; 118, Aaron Haupt; 118, Texas Instruments Incorporated, Dallas, Texas; 125, Walter Neal/Chicago Tribune; 126, ETHS Yearbook Staff; 131, Matt Meadows; 132, Pictures/West Stock; 140 (top), Chuck Berman/Chicago Tribune; 142, Stan Polich/ Chicago Tribune; 148, AP/Wide World Photos; 155, Walter Kale/Chicago Tribune; 156, file photo; 157, Andrew Child/West Stock; 168, Aaron Haupt; 169, Matt Marton; 174, UPI/Corbis-Bettmann; 175, TempSport/CORBIS; 179, Jack Demuth; 187, Laura Sifferlin; 193, B.F. Peterson/West Stock; 197, John Evans; 202, Michael Newman/PhotoEdit; 203, Photodisc; 206, Walter Kale/Chicago Tribune; 208, Corbis-Bettmann; 210, Jim Prisching/Chicago Tribune; 212, SuperStock; 215, Aaron Haupt; 216, Jim Brown; 219, Carl Wagner/ Chicago Tribune; 229, Photodisc; 230, Earl Gustle/Chicago Tribune; 233, Photo courtesy of Panasonic Consumer Electronics Co. (800) 222-4213; 239, Matt Meadows; 243, Nuccio DiNuzzo/Chicago Tribune; 244, Photodisc; 246, Robert Fila/Chicago Tribune; 249, AP/Wide World Photos; 252, Ingersoll-Rand's Charge Air; 261, FOTOPIC/West Stock; 282, Melanie Carr/West Stock; 286, Don Normark/ West Stock; 293, Diana Venters; 295, Jerry Tomaselli/Chicago Tribune; 299, Chicago Tribune; 303 (bottom), David H. Wells/Corbis; 303 (top), Ovie Carter/Chicago Tribune; 311, Corbis-Bettmann; 314, John Greim/West Stock; 321 Hung T. Vu/Chicago Tribune.